P9-EDR-817

Reagan and Public Discourse in America

STUDIES IN RHETORIC AND COMMUNICATION
General Editors:
E. Culpepper Clark
Raymie E. McKerrow
David Zarefsky

"Hear O Israel":
The History of American Jewish Preaching, 1654–1970
Robert V. Friedenberg

A Theory of Argumentation
Charles Arthur Willard

Elite Oral History Discourse:
A Study of Cooperation and Coherence
Eva M. McMahan

Computer-Mediated Communication:
Human Relationships in a Computerized World
James W. Chesebro and Donald G. Bonsall

Popular Trials:
Rhetoric, Mass Media, and the Law
Edited by Robert Hariman

Presidents and Protesters:
Political Rhetoric in the 1960s
Theodore Otto Windt, Jr.

Argumentation Theory and the Rhetoric of Assent
Edited by David Cratis Williams and Michael David Hazen

Rhetorical Dimensions of Popular Culture
Barry Brummett

A Voice of Their Own:
The Woman Suffrage Press, 1840–1910
Edited by Martha M. Solomon

Reagan and Public Discourse in America
Edited by Michael Weiler and W. Barnett Pearce

Edited by Michael Weiler
and W. Barnett Pearce

Reagan and Public Discourse in America

The University of Alabama Press Tuscaloosa and London

The paper on which this book is printed meets the minimum requirements of American National Standard for Information Science-Permanence of Paper for Printed Library Materials, ANSI Z39.48-1984.

Library of Congress Cataloging-in-Publication Data

Reagan and public discourse in America / edited by Michael Weiler and W. Barnett Pearce.
p. cm.—(Studies in rhetoric and communication)
Includes bibliographical references and index.
ISBN 0-8173-0585-8
1. Reagan, Ronald—Language. 2. Reagan, Ronald—Oratory. 3. Rhetoric—Political aspects—United States—History—20th century. 4. Communication in politics—United States—History—20th century. 5. United States—Politics and government—1981–1989. I. Weiler, Michael, 1949– . II. Pearce, W. Barnett. III. Series.
E877.2.R425 1992
973.927'092—dc20 92-1349

British Library Cataloguing-in-Publication Data available

In memory of Robin Carter (1941–1992)

Contents

Acknowledgments

Several of the essays in this volume arose from a colloquium on the rhetoric of the Reagan presidency held at the University of Massachusetts in April 1988, under the auspices of the Department of Communication, the U Mass Debating Union, and the Faculty of Social and Behavioral Sciences.

Earlier versions of some segments of "Acting like a President" (chapter 4) were presented in public lectures at Louisiana State University, 4 March 1984; at Wabash College, 30 January 1986; and at Idaho State University, 7 September 1988.

Various portions of "The Transformation of Actor to Scene" (chapter 6) were presented at conventions of the Speech Communication Association and the Eastern Communication Association. The authors wish to thank Debra Madigan, who prepared those manuscripts, and the students in political communication classes who lived through the several incarnations of the essay.

Chapter 7, "A Rhetorical Ambush at Reykjavik," profited greatly from the keen eye and mind of Stephen Littlejohn, to whom the authors express their gratitude.

The author of "Civil Religion and Public Argument" (chapter 12) would like to thank Karlyn Kohrs Campbell, Celeste Condit, and Randall Lake for their comments, and G. Thomas Goodnight for organizing the Abortion Colloquium at Northwestern University, where the essay was presented and its claims tested. Work on the essay was begun while the author, under Northwestern's CIC Program, was a traveling scholar at the University of Minnesota.

The editors would like to thank Laurie Garvin for her help in compiling the index and proofreading the final copy.

Reagan and Public Discourse in America

Introduction

The Rhetorical Analysis of the Reagan Administration

W. Barnett Pearce
Michael Weiler

Although he no longer occupies the White House, Ronald Reagan's presidency continues. Scholars and politicians alike inscribe its history, and this social process of remembering continually adds to its form and substance.

In principle, no act or statement is ever completed nor its meaning fully determined. All acts and utterances are "poetic" in that they give or lend form to states of affairs that have as yet uncertain identities and are themselves formed by subsequent events and statements.[1] An event so extended and complex as a presidency is particularly susceptible to multiple interpretations. These interpretations are best understood as a *part* of that presidency (for example, giving it form by summarizing, abstracting, and refocusing) rather than as stories *about* it, which describe without affecting it.

For better or worse, the way Reagan's presidency is inscribed is as much a part of its political history as the events and policy decisions that occurred during it. Definitions of the Reagan "legacy" will be contested for years as interpretations of its meaning are formed, sifted, and reevaluated. As with all historical explanation, subsequent events will have much to do with how the Reagan years are appraised and to what uses those appraisals are put.

Critical interpretations began well before the last official "photo opportunity," when the president and first lady boarded the plane that would return them to private life. Long before George and

Barbara Bush and a cast of thousands celebrated (on nationwide television, of course) the inauguration of the forty-first president of the United States, an army of self-appointed scholars and pundits already had begun the process of inscribing the successes and failures of the Reagan administration.[2]

Most critics have focused on the effects of Reagan's policies on the economy, international relations, and domestic tranquility. Those who have praised the Reagan presidency generally have echoed the official Oval Office line: Reagan invigorated the economy, restored U.S. military strength, and reduced unnecessary and burdensome regulation of business. Just as important to these observers, Reagan's insistence on framing political issues within a background of "traditional" values and a personalistic morality inspired the American people with renewed optimism about their country's future and a patriotic commitment to its God-given destiny among nations of the world.[3]

Most early contributors to the orgy of postmortemizing, however, have assumed a more critical posture. Three distinct themes have emerged already. The first emphasizes domestic problems largely ignored and, in some cases, aggravated by the Reagan administration's policies. Leading examples are educational underachievement, a rapidly decaying physical infrastructure, environmental pollution, the savings and loan debacle, and, as a less significant but nonetheless depressing footnote, the Housing and Urban Development scandal. These last two are attributed often to the Reaganites' deregulatory zeal,[4] the others to their "malignant neglect" of the needs of American citizens, especially the poor.[5]

A second theme concerns the Reagan administration's foreign policy. As the Soviet hegemony in eastern Europe has crumbled, and as economic stagnation and the movements for national independence have forced the Soviet government to turn urgently inward, the budget-bursting levels of military spending and the anti-Communist fervor of the Reagan years seem at best quaint relics of a past era and at worst wasteful and wrongheaded.[6] The rhetoric that construed the USSR as the "Evil Empire,"[7] though abandoned by Reagan himself in his second term, has seemed to leave Americans ill prepared to grasp the significance of Mikhail Gorbachev's *perestroika* and the movement toward democratization in eastern Europe or to seize the opportunities that these dramatic developments afford.[8] In addition, Reagan's preoccupation with threats from Libya and Nicaragua, in both instances personalized by his diatribes against the heads of those states, are seen as examples of distorted priorities if not as outright policy failures.[9] Finally, the still-continuing saga of the Iran-Contra affair, complete with alleged deals between the Central Intelligence Agency and Panamanian dictators and the criminal convictions of Reagan's chief of staff, John Poindexter, and his "Agent 007," Oliver North, have raised questions about whether the National Security Council, the State Department, the intelligence agencies, and the congressional committees charged with overseeing foreign policy shared a common agenda—or even knew what the others were doing—during Reagan's tenure.[10] These questions, in turn, have led to increased scrutiny of the U.S. role in other parts of the world, from Grenada and El Salvador to Cambodia and the Persian Gulf.

Third, there is the matter of what sort of leader Reagan was. For him especially, this has seemed to be a question largely separate from issues of policy. His personal popularity remained a formidable strength even during periods when many of his policies were unpopular.[11] Yet here, too, the former president seems increasingly vulnerable. Stripped of the impeccable stage management of Michael Deaver, Reagan appears more and more to be what his opponents always said he was: a not particularly intelligent, compassionate, or serious man, in over his head in discussions of policy, used (with or without his knowledge) by powerful groups of self-interested backers, but able for a remarkably long time to survive and even flourish through a combination of sloganeering, public relations, and luck. A year and a half after his departure from Washington, Reagan's testimony in the trial of Admiral Poindexter found him confused and shaken, incapable of remembering events or recognizing photographs of people with whom he had conducted the nation's business. His jaunty wink to his former chief of staff as he entered the courtroom merely underlined the irony that in order to avoid personal responsibility for the mistakes and crimes of Irangate, the former president was willing to sacrifice his most trusted subordinates.[12] By conceding, apparently without shame, that he did not recall or had not known numerous important details of what had occurred on his watch, Reagan in effect allowed Poindexter and North to bear alone the consequences of the Iran-Contra affair. Perhaps the cruelest blows to Reagan's image, however, have come from a series of "kiss and tell" books written by White House insiders. Even the most loyal zealots of the Reagan "revolution" have provided ammunition for those who would paint the former president in hues a good deal less bright than those in which his image was portrayed throughout most of the 1980s.[13]

Critics write from a particular perspective, expressing in their choice of what to analyze and how to proceed their own interests, training, and biases. It cannot be otherwise, so long as criticism is a human enterprise. Honest criticism, however, acknowledges and discloses those perspectives.

The authors of this book are concerned with the quality of public discourse in the United States. We are committed to a vision of a democratic state in which important tasks are done discursively and decisions that affect the public are made in discourse in which the public participates. We do not have a naive notion of "direct" democracy and understand that the public must often participate through representatives. However, we believe that the process is imperiled if the quality of the discourse is such that the public does not identify with it.

This book takes the form of an assessment of the Reagan administration's impact on public discourse. To some extent, the selection of the Reagan administration is a convenient choice. It is the most recent completed presidency, it was a full two-term presidency (the first since Eisenhower), and it was widely believed to have reversed certain trends and cast us on a new course. In another sense, it is an obvious choice. More so than any president since John Kennedy or Franklin Roosevelt, Reagan's impact depended on and was constituted by his rhetorical practices. As many of the contributors to the present volume show, these practices were matters of primary concern to the strategists in the Oval Office.

Any assessment of the Reagan administration must include an analysis of its rhetoric. All presidencies since Kennedy's have been "rhetorical,"[14] and all presidents use many genres of discourse in which to accomplish their "deeds . . . in words."[15]

Conventional wisdom during the 1980s attributed much of Reagan's success to his ability as an orator. This was altogether a very successful theory: those who believed that Reagan embodied the rhetorical conventions of his era cite several data points, use the theory to explain otherwise puzzling phenomena, and find in it new support for an old critique of rhetoric.

His poised delivery in front of television cameras led many to refer to him as the "Great Communicator." Kathleen Hall Jamieson has noted that the new medium of choice for American politics, television, and Reagan's oratorical style were ideally suited to each other.[16] Even his previous career fitted the new fusion of popular culture and civic life.[17]

There was an unprecedented gap between Reagan's personal popularity, which remained consistently high, and public approval of his policies, which fluctuated considerably. This "paradox of the Reagan presidency"[18] was explained by his ability to appeal directly to the voters. As a result, other political leaders and institutions were reluctant to challenge him, and in the minds of the electorate he remained somehow above the fray.

Ever since Plato scolded the Sophists for putting the power of rhetoric into unworthy hands, the specter of the too-powerful rhetor—with images of Elmer Gantry, Adolph Hitler, Senator Joseph McCarthy—has haunted the champions of a free marketplace of ideas. Those who disagreed with Reagan's policies could fault him for being too skilled a rhetor; like Plato they accused him of making the worse reason appear to be the better by using the tricks of rhetoric.

We presume a more substantive notion of rhetoric. Our assessment of Reagan's administration is rhetorical because we focus on the way the administration adapted its actions to the situations it confronted in the context of discursive formations not fully of its own choosing, and on the way its actions affected those situations and contexts. We are not particularly interested in "grading" Reagan as a public speaker; our interest in detailed descriptions of his style and strategies derives from our more substantive interest in the effect they had on the quality of public discourse.

Our emphasis on the quality of public discourse is justified for two reasons. First, we believe that this is where the Reagan administration will prove to have had its most lasting, if not its greatest, impact. Some have suggested that the Reagan legacy, especially as appraised in terms of the administration's legislative agenda, will be brief. Citing the unprecedented gap between Reagan's personal popularity and public support for his policies, some prophets employ familiar metaphors of cycle and pendulum to forecast a liberal revival.[19]

A very different conclusion derives if the Reagan legacy is appraised in terms of its impact on the way Americans talk about public issues and what issues are discussed. The administration forged a discursive hegemony that excluded many voices, disempowered others, and forced the discussion of a wide range of issues into structures congenial to its own agenda. Even as Reagan's personal popularity and the consequences of his specific policy initiatives recede, his impact on the quality of public discourse will remain, the stubborn residue of a presidency that otherwise might come to appear increasingly aberrational. The habits of discourse that were in part initiated but in larger part reinforced during the 1980s will continue to exert substantial influence on the content of the political agenda and on how the policy problems it poses are or are not resolved.

Second, we believe that the Reagan administration's impact on the quality of public discourse will prove to be pervasive and profound. Not only were Reagan and his associates effective in setting the agenda for what would and would not be contested in the public sphere, but, at a more fundamental level, they were able in large part to structure the relation between assertion and evidence in public debate and to limit the range of viable positions. Unhappily, the discourses in which the Reagan administration sought and exercised power provide resources inadequate for the public deliberation of many of the most important issues of our time. The global ecological crisis, the declining role of the United States in the world economy, a restructured East-West relationship and its impact on North-South structures, the continuing misery of political refugees around the world and the homeless at home: these topics do not play as well in Reaganesque discourse as do constitutional amendments to criminalize flag burning, a "war" on drugs framed in military/police terms, and use of the U.S. military as a way of policing the world.

Of course, the discursive dynamic of the Reagan legacy cannot be appraised apart from the policies he advocated and the leadership he exerted to generate public support for them. But neither can it be analyzed exclusively in these terms. For the habits of discourse inculcated and reinforced during the Reagan era will continue to affect for years to come the questions of who in the United States will wield power, how effectively, and to what ends.

The savings and loan fiasco, for example, is part of the policy legacy of the administration (and a compliant Congress). Insofar as failed savings and loan associations will continue to cost American taxpayers tens and even hundreds of billions of dollars for the repayment of insured depositors, the policies that led to these failures are of obvious historical significance. But we are as interested in the shape of the public discourse in which the crisis has been handled as we are in its fiscal consequences. The quality of public discussion of the issue, the process through which some actors have been identified as culprits while others have not, the ways in which a tax-funded bailout has been justified to the public: these elements are as much a part of the Reagan legacy as the economic facts themselves.[20] Without understanding these discursive processes, one cannot understand how the Reagan administration seized and held political power.

It is important to remind ourselves that Reagan did not single-handedly create a new public discourse for the 1980s. Just as his personal political consciousness was a product of a particular ideological environment and history, so too was his discourse the product of the particular discursive resources available to him when he ran for and won the presidency in 1980. In a sense, Reagan plunged into public discourse more than he invented it. But the waves from that startlingly dramatic entry reverberate still, pushed outward by eight years of purposeful splashing in America's political waters. Reagan's unique articulation of these discursive elements with the historical circumstances of the 1980s defines his impact on public discourse during that decade, and after. The habits of discourse that his presidency in part adopted, in part reinforced, and in part created, will be difficult to break, as indeed such habits always are. Describing and explaining the dynamic process through which the Reagan administration influenced public discourse in the United States is essential to any thorough appraisal of the Reagan legacy.

Six of the chapters in this book emerged from a three-day conference at the University of Massachusetts at Amherst; the other six were selected from those subsequently contributed by authors who learned of the project.

Part 1 explores the invention of Reagan's rhetorical practice. Long considered one of the most useful canons of rhetorical criticism, *invention* refers to the process by which arguments are created. Traditionally, it describes a systematic process in which a rhetor identifies materials and shapes them into the rhetorical forms best suited to particular occasions. As used here, *invention* acknowledges that the Reagan administration did not emerge *ex nihilo*, nor did it fabricate its rhetorical practices from thin air. The discursive structures that existed in 1980 prefigured what the administration could do. Weiler and Pearce describe three discourses that Reagan's rhetorical practices wove into a hegemony and warn that the result was to move "deliberation" out of the public arena. Goodnight describes how the Reagan administration adapted to the rhetorics of war and peace. Carter takes the perspective of an "outsider," showing how Reagan's rhetoric played in London.

The style of Reagan's discourse is the focus of part 2. Traditionally, style refers to elaborate taxonomies of figures of thought and speech from which a rhetor would choose; as used here, it consists of a broader array of perspectives from which to assess the form of rhetorical practices. Auer focuses on aspects of thespian form imported from Reagan's first career into his performance as president. Jasinski explores two figures of speech, antithesis and oxymoron, that illuminate the form of Reagan's discursive structures. Blankenship and Muir employ Kenneth Burke's dramatistic model to characterize the way Reagan himself became the central object in the administration's rhetoric.

Parts 3 and 4 use the discursive structure of particular issues as the means to assess the effect of the Reagan administration. Pearce, Johnson, and Branham use the "ambush" sprung by Mikhail Gorbachev at the Reykjavik meeting as a way of illuminating the structure of Reagan's foreign policy discourse and of the "conversation" between the United States and the Soviet Union. Bass analyzes Reagan's discourse about Nicaragua, finding in it a "paranoid style." Young explores the discourse in which the administration explained the destruction of a civilian Iranian airliner by American forces, comparing it to that used by the Soviet Union when its forces shot down a Korean airliner.

Domestic policies are examined in part 4. Weiler assesses the Reagan administration's discursive structures about social welfare as an "attack." Burnier and Descutner trace the way in which cities were constructed as marketplaces in Reagan's discourse. Palczewski, in her analysis of the discursive structures of civil religion, characterizes Reagan as the "public priest" of the antiabortion movement.

As a whole, the book is a sustained jeremiad about the declining quality of public discourse in the discursive hegemony fashioned by the Reagan administration. Each of the chapters presents an intensive study of a particular facet of that discursive hegemony; taken together, they constitute a compelling argument for three conclusions. First, the Reagan administration had a major impact on public discourse. Second, the most significant aspect of that impact was on the discursive structures in which issues are framed and in which argumentation occurs. This perspective is crucial. It envisions the sphere of the rhetorical as encompassing the institutions, media, policies, and participants in the public arena, not as limited to the messages that are fashioned by and transmitted among those participants. That is to say, the rhetorical criticism in this volume is not focused on speeches or campaign advertisements; to the extent that specific rhetorical practices are analyzed, it is in service to an understanding of the larger context. Within this frame, it is possible to draw the third conclusion: the impact of the Reagan administration was detrimental to the quality of public discourse.

These important claims are not made lightly. If true, they have far-reaching implications. We offer this book as a contribution to public deliberation about the current state of public discourse, an analysis of how we reached this point, and a basis for deciding where and how to go from here.

Part I

The Invention of Reagan's Discourse

1

Ceremonial Discourse

The Rhetorical Ecology of the Reagan Administration

Michael Weiler
W. Barnett Pearce

No single rhetor or group of rhetors creates public discourse. Rhetors constitute themselves as such within a preexisting rhetorical environment. At the same time, however, the addition to that environment of a new rhetorical element alters it irretrievably and sometimes fundamentally. We understand the administration of Ronald Reagan as such an element. We will examine Reagan's rhetoric in its interactions with the system of public discourse that gave it birth and yet, in important ways, was transformed by it. We advance a view of public discourse designed to account both for the discursive environment that Reagan faced as the 1980s began and the ways in which his own rhetoric exploited, regenerated, and reshaped the resources of that environment during his presidency. We refer to this discursive process as the rhetorical ecology of the Reagan administration. Our summary term for the direction in which Reagan's rhetorical interventions helped to move public discourse in the 1980s is *ceremonialization.*

We argue that Reagan's public discourse was fashioned chiefly from three subsystems of discourse: populism, civil religion, and national security. His appropriations from these subdiscourses were rendered in a style and by artistic means suited to the ceremonial trend in public discourse, a trend already well under way but much strengthened, nonetheless, during the Reagan years.

Reagan sustained his personal popularity by selecting from these subsystems elements that appealed at times to the meanest and most fearful impulses of his audiences and at times to their most optimistic, patriotic, and nostalgic natures, but always at the price of impoverishing public discussion. He used populist discourse to ally himself with "the people" against the government, thus blocking rather than facilitating a serious debate about government's appropriate role in a capitalist economy. He appealed to the doctrines of America's "civil religion" to invest his policy agenda with divine approval, thus demonizing his opponents rather than confronting their objections. He invoked the national security rationale for his military budget proposals and foreign adventures, thus placing these topics outside the realm of partisan political discussion.

In the sense in which we speak of it, Reagan's impoverishment of the public sphere did not render that sphere empty or void but substituted for the possibility of rational political discussion within it an intellectually unnourishing form of talk. We will use the three discursive subsystems we have named to describe the inventional content of Reagan's rhetoric, but these alone cannot account for its impact. The claims and appeals he selected from them were particularly appropriate to a ceremonial mode of discourse, but that mode comprises noninventional characteristics as well. We will point to several, including Reagan's extensive use of moral references, his reliance on visual images both as a part of his discourse and as scenic accompaniments to it, and his employment of pathetic rather than logical appeals.[1]

That these characteristics fit particularly well the rhetorical exigencies of the times was no accident: the Reagan administration used a sophisticated array of market-research techniques to guide the work of sculpting a presidential image that accorded with the ideological predilections and emotional instincts of voters. The net effect of this potent combination of discursive resources and "effeminate" style[2] was to ceremonialize the president's discourse during his eight years in office. Even when forced to address a particular issue, his was a rhetoric that *announced decisions and called on the faithful for support*. Reagan, however, was far more in his element when his topic could be addressed not in concrete policy terms but by way of a generalized sense of "feeling good about America." Here he could rely on visual images, employ easy-to-grasp slogans and endearing anecdotes, and articulate a moral vision more comforting in its uncomplicated fervor than comprehensive in its grasp of an array of viewpoints and relevant evidence.

By transmuting the rhetoric of the presidency into ceremonial discourse, sustained by an infrastructure of Hollywood entertainment values and Madison Avenue market-research techniques, Reagan enacted what other politicians had only attempted: a vision of the public sphere belonging more to Steven Spielberg than Thomas Jefferson, a long-running movie designed to make audiences laugh and cry, rather than a political space in which they might actively participate. No one has described this phenomenon more accurately than Ronald Reagan's public relations wizard Michael Deaver, a man, appropriately enough, who in the early days of the administration was said to have equal policy-making influence with Reagan's top advisers, Edwin Meese and James Baker. When interviewed by Bill Moyers for his 1989 documentary series "The Public Mind," Deaver identified himself as a "Hollywood producer" in his management of Reagan's 1980 presidential campaign and boasted of his ability to get the television news media to present the candidate as Deaver had designed him. When Moyers asked, "Should we be proud of what we've done in this regard?" Deaver replied: "Well, I couldn't change that. If I had tried to do what I thought was the right way to go about it, we would have lost the campaign. People would have been bored to tears. In a democracy that's interested in where their leaders are going to stand, in what they are going to do on the issues, that [a more substantive approach] would have been the right thing to do, but this country isn't interested in it. They want 'feel good' and 'fuzz' and to not be upset about all of this. They want to just sit in their living rooms and be entertained. No, I don't feel good about that at all."[3]

The ceremonialized version of public discourse that Deaver describes produces two major effects. First, it diminishes the discursive terrain for dealing with complex public issues. It contracts and impoverishes the public sphere. It tends to exclude moments of moral doubt and strategic indecision. In such an environment a president must respond to even the most intractable problems with solutions at once simple and obvious. There can be no room for qualification, no allowance for unforeseen contingency. Since real social problems cannot be solved this way, ceremonial public discourse must become increasingly detached from political reality. In Murray Edelman's trenchant phrase, politics becomes a case of "words that succeed and policies that fail."[4]

It should not surprise us that during the Reagan years the role of factual documentation in presidential rhetoric diminished. Indeed, his frequent misstatements of fact were treated by his handlers as unimportant. "The President misspoke himself" was considered an adequate excuse for even the most outrageous perversions of fact.[5]

Those who did not agree with the president's views on issues were excluded from his conversations. Neither dialogue nor forensic disputation played an important role in the administration's public discourse. Scripted speeches were favored over press conferences.

Not surprisingly, Reagan's finest hour was in leading the nation in mourning the deaths of the *Challenger* astronauts (Reagan read his script—written by his chief speechwriter, Peggy Noonan—over a background of striking pictures of the *Challenger* crew and the explosion of the shuttle); his most egregiously poor showing was his attempt to answer specific questions about the Iran-Contra affair.[6]

A second effect of the ceremonialization of public discourse, and in our view a corollary of the first, is that it creates what it portrays itself as responding to: an apathetic electorate, uninterested in the campaign, uninformed about the issues, and increasingly alienated from the practice of national political power. A recent two-year study by the Markle Commission on media and the electorate found that an "astonishing number of Americans are indifferent to national elections. In a nationwide poll conducted in September 1988, 49 percent could not identify Lloyd Bentsen as the Democratic vice-presidential candidate, and 37 percent could not identify Dan Quayle as the Republican candidate.[7]

Though Michael Deaver may consider himself responding to *the way the electorate is*, we believe that passive publics are made, not born. Voters may come to accept their status as inactive sponges waiting to soak up the received Word, not out of preference but from an overwhelming sense of helpless resignation.[8] Insofar as political leaders make discursive choices designed to foreclose rather than to facilitate dialogue and debate, voter apathy is bound to continue and even increase.

Though the issues of a shrunken public sphere and its nonparticipating public are hardly unique to the 1980s,[9] we believe that through the ceremonialization of its discourse, the Reagan administration aggravated them significantly. What follows is our explanation of how this happened and what it might mean for American politics in the future.

Rhetorical Ecology

We think that "public discourse" as a system can be imagined most usefully as a kind of ecosystem in which various individual discursive subsystems interact in relations of conflict and mutual dependence. Rhetors are forced to act within the confines of the ecosystem, and their discourses must reflect the web of relationships among its species and their surroundings. But as the rhetorical ecosystem evolves, as any living thing must, so too do its discursive possibilities, and within the system there is ample room for authorial creativity and cleverness. The rhetorical options available are thus constrained but not determined by the intertexuality of or "spaces" in the array of discourses that confront rhetors. Context both fits rhetorical action and is reconstructed by it.

Our view of public discourse sees it as static and dynamic, conservative and subversive. On the static/conservative side, we believe that the meaning of a text is neither determined by nor even necessarily consistent with the intentions of the person who produced it. Discourses generate meaning at different levels, and authorial intention is merely one of them. As Michel Foucault has argued, discourses may be defined also by patterns of inclusion and exclusion of which authors (and audiences) are unconscious.[10] These patterns may be so pervasive and longstanding as to constitute "a historical, modifiable, and institutionally constraining system" of discourse, "reinforced and renewed by a whole strata of practices."[11]

But just as language is, in an ideological sense, rigid and inflexible, so too does it possess inherently the potential for change. As the Italian Marxist Antonio Gramsci put the matter: "The whole of language is a continuous process of metaphor, and the history of semantics is an aspect of the history of culture; language is at the same time a living thing and a museum of fossils of life and civilizations."[12] It is not necessary, therefore, to deny the recalcitrance of discursive practices in order to emphasize the dynamic, recursive character of public discourse. Habits of talk are constantly being reinforced and reshaped, articulated to changing historical circumstances and personal histories of individual rhetors, and sometimes subverted. But they are still habits. Rhetors do not create discourse but are themselves reshaped by it. One thinks of the effect of human civilization on the biological environment. What may appear to be human control of the natural sphere is, in another sense, human accommodation to it and, in the extreme case (for example, the greenhouse effect), determination by it. To theorize the public sphere and its discourse is to suggest a kind of *rhetorical ecology* in which the intentional, strategic activities of many rhetors are in inescapable tension with, yet accommodative to, multiple patterns of intertextuality.

Similarly, we consider the relationship of language and power more complicated than a grossly materialistic (not necessarily Marxist) view might suggest. Though the vocabularies we use to name objects, motives, and relationships may reflect power relations in a given society, it is in the nature of language to escape the control of ruling groups.[13] Claims, by their nature as such, invite counterclaims. Even in the most Orwellian world, the creative potential of language thwarts the efforts of rulers to circumscribe the range of social meanings. So long as politics is talked about at all, oppositional possibilities persist.

We are most concerned, therefore, with discursive practices that place topics outside the realm of public discussion. Jurgen Habermas has examined the process by which the range of public discourse in Western liberal societies has been constricted. He has noted how the "scientization of the public sphere" has effectively disqualified members of the public as participants in political dialogue in all policy categories.[14] At the level of political symbolization, Murray Edelman has described a similar process, noting how the names we attach to various problems can remove them from public political, as opposed to private professional, consideration. "The most fundamental and long-lasting influences upon political beliefs," he argues, "flow . . . from language that is not perceived as political at all, but nonetheless structures perceptions of status, authority, merit, deviance, and the causes of social problems."[15]

Which and how many topics are part of public discourse is at least as important as what is said about them. In our analysis of the Reagan administration's discourse, we will attend to what it helped take out (or keep out) of the public sphere as well as what it helped put in. We will categorize (as three sybsystems of discourse) the inventional resources Reagan chiefly relied on, and we will describe the stylistic practices that characterized his use of them. We will suggest that the administration at once was shaped by and helped to shape public discourse during the 1980s, and that analyzing this process is crucial to an understanding of the nature and possibilities of public discourse in the decades to come.

Reagan's Subsystems of Discourse

The rhetorical ecology of the Reagan administration's discourse could be characterized in many ways, but we have chosen to emphasize three discursive subsystems: populism, civil religion, and national security. These categories allow us to describe what we think are the most significant contributions of the administration to public political discussion during the 1980s. Though they by no means exhaust the full range of those contributions, they do allow us to emphasize inventional elements we consider essential to the Reagan ethos.

In discussions of discourse the issue of intentionality arises. In this section, however, we are not concerned with the deliberative process through which the rhetorical agent chooses particular ways of producing certain audience effects. We will not attempt to trace the path between Reagan's intentions and the illocutionary status or perlocutionary effects of his discourse.[16] The three subdiscourses we describe constitute the rhetorical ecology in which Reagan's utterances occurred. As we have suggested, no rhetor, not even the president, can enter into discourses in which what is said does not acquire meanings that extend far beyond the literal (in Austin's term, locutionary) utterance. It is irrelevant to our analysis whether Reagan deliberately used phrases and images that invoke populism, civil religion, or national security. Evidence of deliberateness indeed may exist, but even if Reagan and his handlers had not fully intended to use these discourses, the results would have been the same.

Populism

In any nation with a representative form of government, political leaders must devise strategies to identify themselves with at least a majority of those who vote in elections. Ruling officials must present themselves not only as willing to represent the interests and values of the public, but as actually sharing them. To share the values and interests of "the people" is to be, in an important sense, one of them and thus to be identified with them.

But this identification is not complete. However close an ideological correspondence between leader and voter there may be (or appear to be), leaders are still leaders and voters are not. And not only the possession of political power but social and economic class may separate leaders from voters. Leaders are likely to be richer and better educated than most voters, and their social circle will likely magnify such differences. In other words, political leaders are, by any reasonable standard, an economic and social, as well as political, elite.

Populist appeals, as distinguished from other subsystems of political discourse, are directed at denying and obscuring this elitism: denying it by pointing to social and economic (and even political) similarities between the leader and most voters; obscuring it by focusing voters' attention on the power of alternatively defined elites against which the leader can claim to be fighting. The politician's statement "I believe that the Constitution of the United States is divinely inspired," though it may express the views of most voters, is not a populist appeal. By contrast, the statement "The liberals in Washington believe our cherished Constitution is nothing more than a front for the selfish economic interests of our Founding

Fathers" does have obvious populist overtones. It identifies an elite (liberals) whose views offend the popular consensus, and its collective pronouns (*our*) place the speaker on the side of "the people" against the view of the arrogant debunkers of that consensus.

This formula of identification—*with* the people *against* the elites—is the basis of all populist appeals. Its positive and negative terms can be employed separately depending on what is appropriate to a specific rhetorical situation. However, they are not really separate, but two sides of the same coin; they must be combined for full effect. References to the wisdom of the common people and homey anecdotes of a politician's humble past are helpful as far as they go, but without an elite toward which the majority can direct their irritations, their frustrations, and even their hatred, populist discourse cannot achieve more than a fraction of its potential.

Jimmy Carter understood this during his successful campaign for the presidency in 1976 but seemed to forget it once he had won the office. After the election, his talk of the "honesty and decency of the American people" continued, but the Georgia peanut farmer's heroic struggle against Washington insiders (so many of them *lawyers*) gradually faded from view. Of course, once a candidate becomes an elected official, it is hard to maintain the role of the untainted outsider. Yet Ronald Reagan, from the day he entered the Oval Office until the last ride to Santa Barbara, never stopped attacking the bureaucratic, congressional, and journalistic elites "inside the Beltway." His was a presidency that relied, perhaps more than any in this century, on populist discourse—and with stunning success. Despite the depressing effect of the Iran-Contra scandal, Reagan left office with his personal popularity largely intact, a testament in part to how much Americans still considered this millionaire and former Hollywood star "one of them." How did he pull it off? A part of the answer rests on his skilled employment of positive populism, that set of appeals that emphasize the quiet strengths of common people and, in the same attractive sense, the "commonness" of the leader himself.

Reagan's first inaugural address set the pattern for eight years of such appeals. The speech's theme was the capacity of ordinary people to do extraordinary things. In the best Dale Carnegie tradition, Reagan traded on the notion, inherent to populist discourse, that when you compliment people a lot, they can't help liking you for it. "Those who say that we're in a time when there are no heroes," he argued, " . . . just don't know where to look. You can see heroes every day going in and out of factory gates. . . . You meet heroes across a counter—and they're on both sides of that counter"[17] and so on. He returned to this theme repeatedly. "We hear much of special interest groups," Reagan observed. And then, describing a slightly modified version of Richard Nixon's famous "silent majority," he continued, "Well, our concern must be for a special interest group that has been too long neglected. It knows no sectional boundaries, or ethnic and racial divisions and it crosses political party lines. It is made up of men and women who raise our food, patrol our streets, man our mines and factories, teach our children, keep our homes and heal us when we're sick. . . . They are, in short, 'We the people.'"[18]

The address concluded with a reading from the diary of Martin Treptow, a U.S. soldier killed in World War I. In simple, straightforward language, Treptow expressed his patriotic commitment to his country. Here was an illustration of the common heroism of which *every* American citizen is capable. Lest the point be missed, Reagan had observed earlier that "I have used the words 'they' and 'their' in speaking of these heroes. I could say 'you' and 'your' because I'm addressing the heroes of whom I speak—you, the citizens of this blessed land."[19]

Examples from Reagan's discourse of this kind of flattery could be multiplied almost endlessly. Of course, such appeals are common in political rhetoric. But the extent to which Reagan used them is not. One might expect perhaps two or three references of this type in a typical inaugural address, but in Reagan's case, virtually the entire speech used populist appeals in one form or another. Though few Reagan speeches approached this degree of emphasis, his reliance on such appeals was always extraordinary.

Positive populism consists not just in flattering "the people" but in making yourself one of them. In the case of a president leading a president's life, this is obviously difficult to do. Nor do voters expect or even find it desirable for the president of the United States actually to lead an ordinary sort of life (recall Jimmy Carter's ill-considered attempts to exude the common touch by carrying his own bags and wearing cardigan sweaters). The potential for a president to identify himself with the lives of common people lies instead in the success with which he constructs a past history consistent with what most people would consider admirable ordinariness.

If this history is relatively recent then so much the better. Richard Nixon's famous "Checkers" speech, with its references to eighty-dollar-a-month apartments and Republican cloth coats, is an example.[20] Reagan, however, was forced by his years as a Hollywood actor and then as governor of California to travel back in time several decades. The radio sportscasting career of "Dutch" Reagan became a favorite theme.[21] He was fond of telling the story of how he and his partner would create for their local audience fictitious play-by-play accounts of baseball games already concluded. Though he did not attempt (as Nixon had done) to construct systematically an early life of poverty and struggle, Reagan could strike a humble note when the opportunity arose. When asked by Walter Cronkite in 1981 whether life as the president was a difficult adjustment, he replied, "I'm not surprised about the confinement of living in the White House. I lived above the store when I was a kid, and it's much like that."[22] When an Iowa news reporter asked him whether he had been inside the Iowa state capitol before, he answered, "No, I don't recall ever being in here before. I do recall sleeping out on the lawn. I'm old enough to remember when air conditioning was only found in motion picture theatres, and on some of those hot Iowa nights, this Capitol lawn out here would literally be covered with whole families that would bring a blanket and come over to the lawn and sleep all night."[23]

Reagan was simply a master of the populist anecdote. His ability to express the essence of ordinary life in endearing and reassuring images, and at the same time to associate himself with them via his personal life history, contributed significantly to his avuncular ethos. These stories, together with his incessant praise of the wisdom and heroism of ordinary people, make Reagan one of American political history's most successful positive populists. But this success could not have been nearly as great had he not coupled such appeals with a virulently negative populism. To get people to love him, he had to get them to hate someone else.

For members of the Republican party and especially for conservative Republicans, negative populist discourse since World War II (and indeed for sometime before) has been rooted in opposition to Franklin Roosevelt and the New Deal. Roosevelt's long tenure in office, an aberration in a century of national leadership dominated by Republican presidents, has provided Republicans with the inventional material for a whole range of negative populist appeals. First of all, Roosevelt himself was something of a patrician, well educated, well off, and exuding a northeastern sophistication and confidence. Second, the New Deal involved an unprecedented degree of federal government intervention in areas regarded previously as private preserves, especially in the economic sphere. New Deal programs required the authorization of new cadres of Ivy league–educated government experts and the creation of new bureaucracies, all designed to plan and implement initiatives *from* Washington *to* the rest of the country.

From its earliest days, the New Deal was condemned by its opponents as socialistic and un-American. In sympathetic accounts, this was mostly a case of corporate business interests fighting a redistribution of profits to the lower and middle economic classes. And, indeed, big business did lead the fight against the New Deal. It should be emphasized, however, that the victories the corporate business sector achieved during and especially after the Roosevelt presidency were dependent in large part on mobilizing public sentiment against New Deal programs. Populist appeals were a crucial element in this strategy. And this same anti–New Deal populism has set the pattern for conservative rhetoric ever since. Although, during the 1960s and after, the War on Poverty became the primary target, the arguments used against it remained basically the same.[24]

Indeed, one could begin an analysis of antigovernment populist discourse much earlier in American, and for that matter in European, history. Distrust of central authority is a major element of liberal political thought. Insofar as government activity is seen to reflect a concentration of political power, the traditional liberal would see it as dangerous regardless of how that power might be used in specific instances. And such fears are expressed not only by those whose own concentrated power (business) might be checked by a stronger central government. In the age of bureaucracy, critiques of government authority come as often from the left and center as from the right,[25] and almost always, the interests of the common people are at their core.

The wellsprings of political discourse always run deeper than any single analysis is prepared to go. The roots of Reagan's negative populism can be found both close to the surface of the history of the 1980s, and far below. In his case, however, there is good reason for emphasizing the New Deal. His historical narratives tend always to follow the same sequence. America's story from 1776 to 1932, hazily sketched, is one of uninterrupted political freedom and economic progress. Then the New Deal came along and introduced welfare statism, which, thanks mostly to a Democrat-dominated Congress and despite the efforts of right-minded Republican presidents, continues to grow. The expansion was particularly rapid during the late 1960s and all through the 1970s as Johnson's Great Society programs made their mark. Since 1981, however, the trend toward collectivism has been reversed, but there is still much to do, especially given a revanchist Congress determined to thwart the Reagan revolution.[26] The New Deal, then, was America's watershed event; in effect, it dispossessed a chosen people of their historic birthright. For forty years we have wandered in the socialist wilderness; now it is time to emerge and claim that which is truly ours, a free and prosperous future.[27]

Negative populism enters the story as a three-headed monster called the government. Sometimes "the government" is the federal executive bureaucracy; sometimes it is the U.S. Congress; sometimes it is "Washington" in general. Always, the government acts against the interests of the people. Consider this account of the federal bureaucracy:

> The unprecedented growth of Federal spending of the past two decades did more than precipitate economic stagnation. The expanding Federal monolith undermined the system of checks and balances and the division of power that long protected the freedom of our people. The people, who, after all, were meant to be in charge, are now so far removed from the decisionmakers that they have little say in policies that dramatically affect their lives. And, all too often, those policies don't work. Thomas Jefferson said, "Were we directed from Washington when to sow and when to reap, we should soon want for bread." More and more people realize that things don't seem to work anymore, because too much power flows to too many unaccountable people in Washington D.C.[28]

Reagan makes his case against the bureaucracy on grounds of both freedom and economic efficiency. The former need not be traded for the latter as New Dealers might have thought; the free course is also the prosperous course.

At many points during his presidency, Reagan used populist characterizations of the federal bureaucracy to justify his policies. Deep reductions in federal block grants to states and cities were presented as attacks on centralized intervention in local affairs. Deregulation of business was a way of freeing the economy from federal control. Restrictions on eligibility for welfare programs spared former benefit recipients from enervating government paternalism. In each of these cases, as in his domestic rhetoric generally, Reagan used the theme of the people's liberation from elite control as moral grounding for conservative economic and social policies.

Any president faced with hostile majorities in one or both houses of Congress, as Reagan was during his eight years, will likely seek to blame the legislative branch for the country's problems. No president ever gets his complete program enacted, and so whatever goes wrong can be attributed to congressional inaction or misdirection. Harry Truman's highly successful campaign against the "Do Nothing Congress" of 1948 is a famous example. In Truman's case, however, the attacks came in the last year of his first term as he geared up for reelection. With Reagan, the salvos began almost immediately and quickly became a central rhetorical strategy, even during the first two years of his first term, a time when the Congress was busy passing a great deal of what he asked for. Six months after taking office, Reagan could be heard issuing warnings. "I'm not here to criticize the Congress," he remarked disingenuously. "I'm here to say that you and I and the rest of the Nation will support them if they act responsibly and courageously. We'll help them shoulder the burden of taking tough but necessary action. . . . But if they don't finish the job, America will have merely delayed the day of reckoning, a day which will cause us to slip once again into the terrible quicksand of built-in inflation, high interest rates, and government out of control."[29]

Here, as in all his populist rhetoric, Reagan is the Washington outsider, who, together with the public, stands apart from the government, fearing it, condemning it, and attempting to tame it. "The professional politicians," he noted in an election-eve broadcast in 1984, "are set in their ways. . . . They don't care how the American people had voted. They ridiculed our new ideas. House Speaker Tip O'Neill even warned us that things might not move as fast as we think they should because, 'You're in the big leagues now.'"[30]

Perhaps it surprised O'Neill, the down-to-earth Boston Irishman, that he could come to embody the worst sort of elitist arrogance. But here he was, ridiculing the ideas of the people's champion and, by extension, the people themselves. The people were not to be mocked. Empowered by the Reagan revolutionary vanguard, they struck back. "We did something that shocked the old guard here in Washington," Reagan concluded, "we took our case to you, the people. And you gave us your support. . . . You got their attention. Together we took command of a rudderless ship, adrift in a sea of confusion."[31]

One of the strategic strengths of populist discourse is its way of cutting through political "confusion" and complexity. Negative populism's simple us-*versus*-them formula allows members of the public to project their frustrations, disappointments, and fears onto a single scapegoat. In doing so, they escape the causal complexities their problems embody, and particularly they avoid having to admit that the fault may lie as much in themselves as in their stars. It does not matter that the scapegoated group's membership, motives, or modes of operation are not always clearly defined. Indeed, so much the better. The more specific one is about these matters, the more difficult it is to lay a broad range of sins at the scapegoat's door. It is enough that the members of the scapegoat group hold power, are out for themselves, and are against "us." In discussions of particular policies, some or all of these details may be filled in. The point is, however, that as the image of an arrogant, inaccessible, and even conspiratorial elite is reinforced repeatedly, the need to provide prima facie grounds for blaming it in particular cases is reduced.

One way of dodging the need to specify the "who," "what," and "how" relative to the government's program of thwarting the popular will is to define it by location. The word *Washington* becomes at once a concrete reference to where the government is to be found and a way of obscuring just who the government consists of, what "they" want, and how they pursue their nefarious objectives. "The greatness of America doesn't begin in Washington," Reagan observed in 1984, "it begins with each of you—in the mighty spirit of free people under God, in the bedrock values you live by each day in your families, neighborhoods and workplaces."[32] "If we do nothing else in this administration," he said in an earlier speech, "we're going to convince that city that the power, the money, and the responsibility of this country begin and end with the people and not in some puzzle palace on the Potomac."[33]

Whether or not "that city" was thus convinced during his presidency, Ronald Reagan mined populist discourse with great profit. His electoral support was notably strong not just among traditionally Republican groups but among blue-collar workers, farmers, and white-collar workers at the low end of the pay scale. His continuously high approval ratings also attested to the breadth of his popularity. And this popularity coincided with his successful efforts to break labor unions and cut back unemployment benefits, reduce farm subsidies, and enact tax cuts heavily loaded toward the wealthiest taxpayers.

On such evidence, there is at least a preliminary case to be made for the importance of populist discourse in a program to foster the impression across economic and social classes that a politician (or ruling group) is acting in the general interest when in fact he is not. Populism, in its always mythical creation of "the people," is, in our age of democratic promises (if not performance), an essential feature of ideology in the negative sense, of a false political consciousness based on the obscuring of class differences.[34] In making people feel as one, with a leader and against a common enemy, populist discourse responds to the deepest desires of human beings for community, for certainty, and for simplicity. Even in highly repressive political systems, it can contribute to the power of the state. It is all the more important in liberal democracies where the consent of the governed must substitute for physical coercion.

Civil Religion

The term *civil religion* has a long history and multiple meanings. Particularly in French political and social thought, it refers to a national religion where creed and social practice legitimate the mores and institutions of the secular state. In 1967 Robert Bellah popularized an alternative usage as a description of important aspects of how Americans understood their history. In his sense, civil religion refers to a transcendental universalist religion of a nation, "an understanding of America's role in history and each person's role as an American citizen."[35]

In the United States, civil-religious discourse appropriates freely from the Judeo-Christian concept of a God who works on earth through a chosen people, and it unabashedly identifies "America" (as opposed to the less emotionally freighted "United States," and certainly as opposed to any particular administration)[36] as the Chosen Nation through which other nations will be redeemed. Although not specifically Christian, American civil religion's particular form is impossible to understand except in the context of this strand of the Judeo-Christian heritage.[37]

America's civil religion is based on a familiar story. First came the Puritan settlers of New England escaping religious persecution. Then a second wave of European immigration washed up on Ellis Island, a symbol of the safe haven from less prosperous societies (and hence, less righteous ones) that America, by the end of the nineteenth century, had become. During the twentieth century America gradually assumed its God-appointed role of military and economic dominance, rejecting its traditional isolationism but still maintaining a strong sense of its moral apartness, a superiority stemming from its unique history and the special characteristics of its ethnically diverse but politically united people.

In this story the flag, Fourth of July, Veterans Day, and the Statue of Liberty are not merely symbols of national identity but sacred dwelling places of a covenant by virtue of which America has special privileges and special responsibilities. Viewed in this way, it (almost) makes sense to talk of the "desecration" of the American flag, and, accordingly, to pass a constitutional amendment prohibiting profane treatment of it.

The "power phrases" of Reagan's oratory and his carefully stage-managed photo opportunities, by invoking this discourse, communicated more than they said. The "Miss Liberty" celebrations were not merely markers of national longevity; they were calls for renewed commitment to the national purpose, one in which the United States was seen as having a redemptive role in international relations. "I don't believe," asserted Reagan, "that our destiny is to watch this unique experiment in government slip from disrepair into decay. But if we remember that freedom rests, and always will, on the individual, on individual effort, on individual courage, and an individual faith in God—then we will have met the challenge of our generation and brought our nation safely through our turning point in history."[38]

Here, as in his references to America as "a shining city set on a hill," Reagan suggested that United States is not just a nation among nations, but an instrument of some higher force. Of course, he was not the only one to use such appeals. Probably every politician who has appeared before the American Legion or at a Fourth of July celebration has done so. However, as with his use of populist discourse, Reagan was distinctive, at least in recent American political history, in the success with which he brought a civil religion reserved for ceremonial occasions into persuasive and deliberative contexts. His much-practiced bromides would have sounded banal from the mouths of less skilled actor-politicians, but his ability to sound sincere and spontaneous each time he repeated them allowed him to frame public policy in the discourse of civil religion. His first inaugural speech set the pattern. "The crisis we are facing today," he warned, "[requires] our best effort, and our willingness to believe in ourselves and to believe in our capacity to perform great deeds; to believe that together with God's help we can and will resolve the problems which now confront us. And after all, why shouldn't we believe that?" he asked in conclusion. "We are Americans."[39]

The use of civil religion discourse had at least three consequences for public discourse in general. First, because it is a ceremonial discourse, it enhances a tendency to appeal to passion rather than reason; rational elements such as evidence, facts, and logic are subordinated. "Our alliances, the strength of a democratic system, the resolve of free people," he observed near the end of his first term, "all are beginning to hold sway in the world. We've helped nourish an enthusiasm that grows every day, a burning spirit that will not be denied: Mankind was born to be free. The tide of the future is the freedom tide."[40]

Whether or not these lofty claims are actually true, they are presented in a style designed not to engage reason but to arouse emotion. Over time, a steady diet of such discourse may gradually erode an audience's critical capacities to the point where factual truth no longer constrains rhetoric, much less founds it. "It's no accident," observed Reagan in 1982, "that we who are the freest people on earth have an educational system unrivaled in the history of civilization. We know that knowledge and freedom are inseparable, and we also acknowledge the right of every individual to both. They cannot be arbitrarily apportioned according to race, station or class."[41]

Any casual observer of public education in the United States today cannot but wonder at Reagan's approbatory use of the adjective *unrivaled* here. And while he may be correct that the educational system does not "arbitrarily" apportion its benefits according to race, station, or class, there can be little doubt that they are so apportioned. Yet, as part of the "America as Number One" theme of our civil religion, and composed in a passionately patriotic and upbeat style, these and similar claims have come to be, if not widely believed, then at least, and perhaps more significantly, seldom publicly refuted.

The irrelevance of factual evidence, though obvious throughout his discourse, received its quintessential statement in Reagan's assessment (on 4 March 1987) of whether there had been an "arms for hostages" deal with Iran. "A few months ago," he noted, "I told the American people I did not trade arms for hostages. My heart and my best intentions still tell me that is true, but the facts and the evidence tell me it is not."[42] Here is an astoundingly candid comment on what constitutes acceptable truth in public discourse. For Reagan, so long as his heart and intentions were engaged, almost anything, however palpably false, could be spoken (or "misspoken"). Of course, a less beloved politician scarcely could have hoped to get away with such a remark, and indeed some commentors chastised him for it. But Reagan's willingness to substitute heart for head in this case suggests that increasingly in our society feelings, not facts, authorize action; or perhaps more accurately, that feelings are a kind of fact more significant than the consequences of action.[43]

Second, Reagan's civil religion discourse, because of its explicit emphasis on the religious foundations of American political ideology, allowed him to align with a large, newly politicized constituency: the New Christian Right. In 1979 Reagan, then a candidate for president, appeared before the Religious Roundtable, a meeting of evangelical and fundamentalist Christians worried that what they perceived as the decline of the military and economic strength of the United States would undermine their base for worldwide evangelism. In his speech to the group, Reagan identified himself with their cause not only by using the appropriate code words but by cleverly displaying his sensitivity to his audience's difficulty in reconciling their overtly political activities with the constitutional principle of church-state separation. "I know that you cannot endorse me" he said. "What I want you to understand is that I endorse you." This allowed them to have their cake and eat it too; they had an official candidate without officially supporting one. Jerry Falwell, leader of the Moral Majority, gleefully advised his followers to "vote for the Reagan of their choice."

In Reagan's discourse explicit references to religious doctrine and practice turned up much more frequently than in the rhetoric of most 1980s politicians. "The Pledge of Allegiance," he noted, "now missing from too many of our classrooms, concludes with the affirmation that we are 'one nation under God . . . with liberty and justice for all.' America embraces these principles by design and would abandon them at her peril."[44] In one of his weekly radio broadcasts to the nation, in 1982, Reagan claimed that all of the various peoples who had immigrated to America since the seventeenth century had done so "with prayers on their lips and faith in their hearts. . . . It's said that prayer can move mountains," he recalled. "Well it's certainly moved the hearts and minds of Americans in their times of trial and helped them to achieve a society that, for all its imperfections, is still the envy of the world and the last, best hope of mankind."[45]

Reagan's civil religion was not just a political ideology serving a religious function, but a discourse infused to an extraordinary degree with explicitly religious references. These occurred not just in speeches on obviously religious issues such as prayer in the schools or in addresses to religious organizations, although in these cases the references were more pervasive. Throughout his discourse Reagan granted a voice to those often left out of public political discussion: that part of American society, typified by Christian fundamentalism, that wishes to be not just deeply religious but publicly so. No wonder they loved him for it.

Third, like any discourse, civil religion facilitates some discursive formations and impedes others. In particular, civil religion discourse blocks the formation of a counterdiscourse. A philosophy that possesses the rhetorical status of a religion is not something to question or debate, it is something for the faithful to embrace and the faithless to reject. Thus, opposition to an accepted civil religion reflects not the strength of claims against it but the spiritual weakness of its opponents.

Because our sacred story strongly differentiates between "us" and "them," "their" roles in it are secondary, the reactionary complements of our own. As a result, things that happen (for example, the wave of democratization in the Warsaw Pact countries) are normally interpreted as the results of "our" actions rather than the consequence of forces to which we are peripheral. "This then, is our historic task," Reagan observed in 1984, "—it always has been—to present to the world an America that is not just strong and secure, but an America that has a cause and a vision of a future where all peoples can experience the warmth and hope of individual liberty."[46]

Civil religion discourse treats morality and history as integrally related. Economic and military success become signs of moral rectitude. Among other things, this aspect of the discourse makes it difficult to pose certain questions or think certain thoughts. It has no vocabulary within which those who use it can even question the moral propriety of a successful military operation (for example, the invasion of Grenada, the bombing of Libya, the invasion of Panama) since success is itself a sign of propriety. In his nationally televised report on U.S. military operations in Grenada, Reagan said:

> In these last few days, I've been more sure than I've ever been that we Americans of today will keep freedom and maintain peace. I've been made to feel that by the magnificent spirit of our young men and women in uniform and by something here in our Nation's Capitol. In this city, where political strife is so much a part of our lives, I've seen Democratic leaders in the Congress join their Republican colleagues, send a message to the world that we're all Americans before we're anything else, and when our country is threatened, we stand shoulder to shoulder in support of our men and women in the Armed Forces.[47]

Before Reagan managed to merge policy and ceremony, embracing the trappings of civil religion was considered safe but not particularly powerful. A fulsome endorsement of God, motherhood, and apple pie might not help a candidate very much, but there was no available discourse in which these issues could be soundly attacked. Reagan, however, managed to turn civil religion discourse into a formidable political weapon. In using it he rendered mute those who would oppose him.

Thus Reagan's use of civil religion discourse, in effect, shut down the conversation about major policy issues in the public realm. By the transmutation of the forensic and deliberative arenas into civil-religious ceremonial grounds, the "essentially contested terms" of politics[48] were placed in a discursive realm in which they were uncontested and incontestable.

National Security

U.S. foreign policy, as its exercise has evolved historically, is chiefly within the domain of the executive branch. Its formulation is characterized by vertical, hierarchical patterns of communication. Policy is devised by the administration and then transmitted to the people, who are informed of decisions reached but who are seldom involved in the decision-making process. Because of the preemptive need for national security, and because of the secrecy with which its implementers argue it must be formulated and often conducted, what has come to be known as "national security discourse"[49] imposes limits on public discussion greater even than those we have noted under the headings of populism and civil religion.

The first amendment to the Constitution, guaranteeing individuals the rights to speech, press, and assembly, may be understood as providing ordinary citizens *access* to the public discourse about public policy. "Security talk, however," always "represents structures of authority and control."[50] The kind of discursive environment that the "free access" safeguards of the First Amendment imply is inherently inimical to the needs of national security, and an unresolved tension exists between the "privilege" of the executive to keep its own counsel in such matters and its responsibility to include the electorate in the discourse.

One of the "lessons" of World War II was that the United States should not again be caught unprepared to defend itself against attack, and those entrusted with the national defense could not help noticing, after war's end, the increased ability of would-be attackers to deliver enormously destructive weapons onto the American homeland. And not just the tools of war have changed in the nuclear era. Embedded in what has been officially called the "National Security Doctrine" since the late 1940s, a subtle shift occurred in the concept of "security" itself.[51]

Before World War II security was construed as the ability to respond to unpredictable events; since then it has come to be understood as the ability to anticipate and forestall every conceivable threat. The new concept of security entails the complete elimination of surprise (and hence a tremendous investment in "intelligence") and the preparation of an overwhelming response to all potential threats (and hence a tremendous investment in military capability). In a 1986 nationally televised address, Ronald Reagan offered a characteristically homey version of this postwar doctrine. "To those who think strength provokes conflict," he answered, "Will Rogers had his own answer. He said of the world heavyweight champion of his day: 'I've never seen anyone insult Jack Dempsey.'"[52]

A second, less noticed shift occurred in the concept of security. In the late 1940s, threats to the United States were perceived primarily in terms of armed attacks by other nations. But the concept has now expanded to include not only foreign relations but matters of trade, education, and research. Sales of high-tech equipment, law enforcement, choices in military procurement and the deployment of troops and equipment, and so forth are encompassed by national security discourse.

With the experience of the OPEC embargo and a better understanding of the interpenetration of the world's economies, the range of topics listed under "threats to national security" continues to broaden. It now includes the unfavorable balance of payments with Japan and the potential for a blockage of the oil-shipping lanes in the Persian Gulf. Even a campaign against drug abuse can be expressed in national security terms, as Reagan proved in a joint appearance with his wife, Nancy, on national television in 1986. "My generation will remember," he recalled, "how Americans swung into action when we were attacked in World War II. The war was not just fought by the fellows flying the planes or driving the tanks. It was fought at home by a mobilized nation—men and women alike—building planes and ships, clothing sailors and soldiers, feeding marines and airmen; and it was fought by children planting victory gardens and collecting cans. Well, we're in another war for our freedom, and it's time for all of us to pull together again."[53]

According to one critic, national security discourse originated as the alternative to offering the American people an "honest choice" between imperialism and republicanism after the Second World War.[54] Unwilling to admit that U.S. policy had become frankly imperialistic, imperialism was justified as the only responsible response to external "threats." The reality of those threats was "sold" to the American people through extensive government propaganda, including civil defense drills during the 1950s, premised on the danger of a Soviet nuclear attack. The marketing of this discourse was successful: cold war discourse has occupied the American political scene for over forty years.

Ronald Reagan did not invent national security discourse, nor did he initiate its expansion to cover so many areas of public life. However, he spoke this discourse as a first language, having learned it in Hollywood both as an actor and as an anti-Communist labor leader in the Screen Actors Guild. His descriptions of policies toward Nicaragua, Lebanon, Grenada, and the Soviet Union are classic restatements of its central themes, and its structure permeates his treatment of a wide range of topics whose relations to legitimate U.S. foreign policy goals often require tortured explanation. In his 1983 speech to the nation on Lebanon and Grenada, Reagan noted that recent events in the two countries, "though oceans apart, are closely related. Not only has Moscow assisted and encouraged the violence in both countries, but it provides direct support through a network of surrogates and terrorists."

"You know," he continued, "there was a time when our national security was based on a standing army here within our own borders. . . . The world has changed. Today, our national security can be threatened in faraway places. It's up to all of us to be aware of the strategic importance of such places and to be able to identify them."[55]

Once our national security interests come to be defined in vague terms such as "strategic importance," and once it becomes the business of the United States to forestall rather than react to any and all threats to such interests, virtually any foreign adventure can be "identified" as vital to the national security. Thus, by extension, a Marxist revolutionary regime in an impoverished Central American country can become a direct threat to the territorial integrity of the United States.

Reagan's televised speech in 1986, asking support for the Contras, is a particularly explicit example. Seeking $100 million in aid, he stressed that this is not "new money" but a "switch" of "a small part of our present defense budget—to the defense of our own southern frontier." The rationale by which U.S. opposition to the Sandinista government in Nicaragua is linked to the security of the United States involved both a resuscitation of the domino theory and a blatantly racist appeal to nativist fears:

> My Fellow Americans, I must speak to you tonight about a mounting danger in Central America that threatens the security of the United States. This danger will not go away; it will grow worse, much worse, if we fail to take action now. I am speaking of Nicaragua, a Soviet ally on the American mainland only two hours' flying time from our own borders. With over a billion dollars in Soviet-bloc aid, the Communist Government of Nicaragua has launched a campaign to subvert and topple its democratic neighbors. Using Nicaragua as a base, the Soviets and Cubans can become the dominant power in the crucial corridor between North and South America. Established there, they will be in a position to threaten the Panama Canal, interdict our vital Caribbean sea lanes and, ultimately, move against Mexico. Should that happen, desperate Latin peoples by the millions would begin fleeing north into the cities of the southern United States, or to wherever some hope of freedom remained.[56]

Like any discourse, "national security" contains patterns of equivalencies and oppositions, discursive formations and spaces which make some things easy to think and do, and others hard. The necessity for certainty and efficiency in countering perceived threats leads to a willful concentration of authority in the hands of a relatively few people in the executive branch of government. This elite group of foreign policy managers finds secrecy, deniability, and disinformation not only useful but legitimate.[57]

In national security discourse, the "agents" who think, act, threaten, conspire, agree, and so on are "states" or "ideologies," not the citizens of those states. As a result, the head of state (or its official delegate) is taken as the appropriate spokesperson. This discourse promotes a trained incapacity for hearing other voices either from other states or from disparate factions within our own. Indeed, its function is, in large part, to obscure the existence of such vocal discord. Accordingly, in Reagan's appropriation of this discourse, foreign states were often personified as their leaders, and their leaders as evil and alien adversaries of the United States. Not Libyans as a people but the eccentricities of Colonel Qaddafi were emphasized, and not Nicaraguans but Daniel Ortega's penchant for designer sunglasses.

As national security discourse blurs distinctions between foreign policy and national security policy, and between foreign and domestic policy, political power is more easily concentrated in the hands of those elite few who know how best to defend "us" from "them." Both the need for secrecy and the apparent complexity of national security issues in the nuclear age restrict the authority to determine security interests to a relatively small group of "expert" political leaders and bureaucrats. As with the discourse of civil religion, domestic dissent becomes tantamount to heresy,[58] and opponents are treated as enemies, but for an additional reason. Not only does it become unpatriotic to doubt the moral superiority of U.S. actions in the world, but because political leaders are the only people sufficiently informed and schooled in national security affairs, doubters can be branded as ignorant as well. The "if you knew what we know" argument has been a staple of government bureaucracies for a long time. As a part of national security discourse, however, it becomes particularly potent.

Overlapping Discourses

Subsystems of discourse can be distinguished for analytical purposes, but they do not actually exist as distinct loci of inventional resources. Discursive subsystems overlap each other in constantly shifting formations. These overlaps, as much as the three subsystems separately, are integral to our description of Reagan's public discourse; they are vital to understanding "what kind of 'medicine' this medicine man . . . concocted."[59] In this section, we discuss a few of these cases of overlap. For the most part, we will find that the three subsystems of discourse we have identified—populism, civil religion, and national security—are quite easily reconciled and combined.

Overlapping discourses are those that combine features of separate discursive subsystems with relatively little ingenuity; separate elements seem to fit together naturally without need of casuistic reconciliation. Civil religion and national security themes offer this potential. For example, in a 1981 interview Reagan told Walter Cronkite: "We're naive if we don't recognize in their [the Soviets'] performance of that, that they also have said that the only morality—remember their ideology is without God, without our idea of morality in the religious sense—their statement about morality is that nothing is immoral if it furthers their cause, which means they can resort to lying or stealing or cheating or even murder if it furthers their cause, and that is not immoral."[60] This is not the most articulate appraisal, perhaps, but it makes its point. The Soviets, because they have rejected the Western brand of religiously based morality, represent a doubly disturbing threat. They confront us not just with a rival empire but with an evil one. In renouncing God, they endanger both our national security (because they ignore international rules of good conduct) and our civil religion (because they call into question its moral and religious foundations).

Reagan's aggressive foreign policies often forced him to justify an interventionist stance. Civil-religious narratives were helpful here. In a 1984 speech at his alma mater, Eureka College, Reagan listed important historical events of twentieth-century American history and then mentioned "an additional development worth noting. That is the emergence of America's international role—our sudden designation as the champion of peace and human freedom in the struggle against totalitarianism. We didn't seek this leadership," he continued. "It was thrust upon us. In the dark days after World War II when much of the civilized world lay in ruins, Pope Pius XII said, 'The American people have a genius for splendid and unselfish action, and into the hands of America, God has placed the destinies of afflicted humanity.'"[61] U.S. interventions in the affairs of foreign nations are thus doubly justified. They serve our national security interest in opposing antidemocratic forces, and they are part of a pact with God to fulfill the role of humanity's last, best hope.

This combination of national security and civil-religious claims can be extended easily to include populist discourse as well. The connections are straightforward enough. America has been "chosen" to serve as the bulwark against totalitarianism. That means its people have been chosen as well. And the people have always been there in the crunch. World War II, Reagan noted in his speech on drug abuse, "was not just fought by the fellows flying the planes or driving the tanks. It was fought at home by a mobilized nation—men and women alike."[62] All Americans share equally in their nation's God-ordained mission. Theirs is both the duty and the glory of its fulfillment.

Populist and civil religion make a nice rhetorical blend in non–national security contexts as well. These usually take the form of stories about past exploits of the people. In a 1981 speech in support of his tax bill, Reagan recalled that "we survived a great depression that toppled governments. We came back from Pearl Harbor to win the greatest military victory in world history. Today's living Americans have fought harder, paid a higher price for freedom, and done more to advance the dignity of man than any people who ever lived. We have in my lifetime gone from the horse and buggy to putting men on the Moon and bringing them safely home. Don't tell me that we can't be trusted with an increased share of our earnings."[63] The important rhetorical move here is to attribute all of these accomplishments to the people, to "us," rather than to America or the nation.

In a later passage from the same speech, Reagan shifted from the past to the present, still lacing his references to America's civil religion with a heavy dollop of populism. "I happen to believe," he said, "that our free country was not put on this earth simply to make a government bigger. Families shouldn't have to work only to achieve survival. America was put here to extend freedom and to create richer and fuller lives for its people, and this is why our administration wants more than anything else to give the economy back to you, the American people."[64] Here is an endorsement of material progress and wealth, an attack on big government, and the implication of divine mission all rolled into one. And the concluding sentence, though essentially nonsensical, is the perfect cymbal crash to signal the end of Reagan's ideological crescendo.

So easily do elements of populist, civil-religious, and national security discourse combine that it is often difficult to separate them analytically. This is because, although it is possible to specify the ideological content of discursive subsystems, any particular rhetorical construction, embodying as it does the reductive capacities of language, may comprise two or more discourses at once within a single statement, phrase, or term. Michael McGee's distinction between ideology and ideographs captures this potential.[65] A single ideographic term, for instance, the word *freedom*, may evoke, depending on its context, a whole cluster of associations that themselves can be expressed as ideological claims of various kinds.

To chart the overlap of discourses, we must look not only for combinations of sentences, or even explicit combinations of terms, but also for the overlap implicit in a single word or phrase. Such analysis is beyond our scope here. We raise the issue, however, to suggest something of the range of possibilities inherent in the notion of discourse overlap. Just as a single animal species can serve as a linchpin for the overlap of several related ecosystems, so too can a single ideograph be the key to understanding the ecology of public discourse.

Reagan's Art

A goal of much theatrical art is to appear (amid carefully calculated artifice) spontaneous, sincere, and natural. The contemporary theater of politics, like productions on stage, pursues these same effects. Careful attention to scenery, lighting, and makeup, and, of course, painstaking rehearsal, all are necessary to political effectiveness. In this sense, Reagan's performance as a rhetor was an artistic success. Insofar as he shaped public discourse—and the label "the Great Communicator" attests to the widespread belief that he embodied the standards of his era—he did so by drawing on artistic resources to give the appearance of artless candor.

Ralph Whitehead, with many others, notes that Reagan was more sophisticated in his use of television than most of his opponents: "Reagan's great capacity was that he could look at this object, this black eye, and see it as if it were a person . . . sitting there. It's not! It's a black machine! It's like staring into your VCR or the hood of your car. . . . Reagan could look into a machine and treat it like a person and people thought that he was 'warm!'"[66] Commenting on Reagan's ability to present a speech written by a ghostwriter and displayed on a TelePrompTer, David Gergen, who worked for Reagan in 1982, said, "He could make the phone book sound good, providing you give him a page with Irish names."[67]

Reagan's success as a rhetor involved more, however, than just the fact that he found cameras "user friendly." He was part of a high-tech, state-of-the-art, sophisticated rhetorical enterprise that capitalized on the technological and social opportunities of the 1980s. In the context of traditional politics, there is something slippery and frustrating about Reagan's rhetoric. Superbly adapted to visual communication media, particularly television, it presents its claims in ways that do not satisfy even conventional tests of logic, coherence, and evidence.

The Effeminate Style

Kathleen Jamieson, in *Eloquence in an Electronic Age*, argues that "television invites a personal, self-disclosing style that draws public discourse out of a private self and comfortably reduces the complex world to dramatic narratives."[68] As an actor, Ronald Reagan had learned the techniques of self-presentation and storytelling; he understood "the power of television to argue visually in ways that defy traditional evidence."[69] In short, Reagan was the perfect politician for the electronic age.

Jamieson summarizes these elements of Reagan's rhetorical style under the label *effeminate*. Her point is historical. "Because it was presumably driven by emotion," she suggests, "womanly speech was thought to be personal, excessive, disorganized, and unduly ornamental. . . . Where womanly speech sowed disorder, manly speech planted order. Womanly speech corrupted an audience by inviting it to judge the case on spurious grounds; manly speech invited judicious judgement."[70]

Of course, political (or any kind of) rhetoric has never been and could never be entirely free of emotional, personal appeals or, for that matter, ornament. The question is, as it has been in much of our analysis of Reagan's discourse, one of degree. We find him pursuing the effeminate style to the exclusion of a balancing interest in more "manly" concerns of fact, analysis, organization, and ratiocination. Better suited to the combative oratory of the old-fashioned political stump, manly style does not play well on television. Thus it does not play well for Reagan either.

We have in Reagan a double irony. Despite his masculine, avuncular image, his rhetoric is "self-disclosive, narrative, personal, 'womanly.'"[71] And despite his fondness for nostalgic reminiscence, his style is supremely appropriate to the modern electronic age: "The intimate medium of television requires that those who speak comfortably through it project a sense of private self, unselfconsciously self-disclose, and engage the audience in completing messages that exist as mere dots and lines on television's screen. . . . Once condemned as a liability, the ability to comfortably express feelings is an asset on television."[72]

Visual Imagery

In 1989 Neil Postman observed, "The environment created by language and the printed word has now been moved to the periphery of culture, especially the printed word. And at its center, the image is taking over. Mostly the television image. But not only the television image. This is a culture that is inundated with visual imagery."[73] Ronald Reagan was not the first to exploit this cultural trend, but he did it more successfully than any national politician ever has—and not by accident. His chief of public relations, Michael

Deaver, describes how deliberately and systematically the Reagan public relations campaign was orchestrated throughout his administration: "We absolutely thought of ourselves when we got into national campaigns as producers. We tried to create the most entertaining, visually attractive scene to fill that box, so that the cameras from the networks would have to use it. It would be so good that they'd say, 'Boy, this is going to make our show tonight.'"[74]

This was not an occasional strategy, one that aimed mostly at important events. Rather, it was standard procedure designed to dominate television news coverage on a daily basis. The domination sought was largely unrelated to a specific legislative agenda. The administration's objective was to present Reagan in visual images that reinforced a generally favorable impression of his leadership. What he *said* as a character in these scenes was not so important as that he *acted* in them. Indeed, Reagan often seemed most effective when he did not speak at all. It was enough that he be seen striding toward an airplane or helicopter as if on the way to doing something presidential.

Sometimes Deaver's dramatizations did touch on specific policy areas. In one case, Reagan, during a trip to Fort Worth, Texas, led the press through a partially built house with construction workers all around busily applying the finishing touches. The administration had released figures that day of an increase in housing starts, and the president's visit was meant to underscore the improvement. That evening, the major networks dutifully played the tape of Reagan's visit as they reported the housing statistics.[75] Significantly, Reagan made no statement relating the administration's housing policy to the figures, or arguing for new initiatives. The point was simply to establish visually what could not be argued as easily or as validly in words: a relationship between the president's leadership and this particular piece of good economic news.

Language creates vivid images via metaphor. Certainly, Reagan's discourse was not bereft of imagery in this traditional sense. Yet, one finds in his speeches relatively few truly inventive, strikingly vivid constructions. There is not much poetic resource represented in the term *evil empire*, for example. Nor does the theme of his first inaugural speech, "A New Beginning," give much promise of vivid metaphors to come. But Reagan didn't seem to need them. In an age where the possibilities of language appear to have been subordinated to the visual image, Deaverism is the coin of the realm. Reagan understood this, and his particular talents placed him perfectly to take advantage of the new dynamics of political discourse in a televised world.

Audience Analysis

The use of polls is nothing new for presidents, but the Reagan entourage possessed an unprecedented sophistication in the technology of discerning what the public wanted and then giving it to them. In 1980 Reagan's top pollster, Richard Wirthlin, was selected by his peers as "Ad Man of the Year" for his work as director of consumer research for the Reagan campaign. The titles both of his award and his job are suggestive; in the Reagan years, the systematic application of principles of consumer marketing helped to achieve and maintain this president's astonishing record of personal popularity. Significantly, Wirthlin was a permanent fixture of the White House staff; his presence was not confined to election time.

For some years conventional political wisdom has been that "values" rather than stances on issues are the most important indices of voter preferences. Knowing this, however, does not tell political strategists how best to appeal to the values the electorate holds most intensely. Wirthlin's genius was to sense which issues could be linked to important values most memorably, and, just as important, how attitudes about a candidate could be shaped to fit into a kind of perceptual triangle with values, issues, and candidate forming the three sides. His hypotheses about these linkages would be based on extensive attitude-survey research. Then, once political ads were composed in keeping with the results, "focus groups" would be assembled to give an indication of likely voter reactions. Group members might be asked to respond not just to the ad as a whole, but even to individual words and phrases in it. The same techniques were used to pretest presidential speeches. The goal, said Wirthlin, was "to fine-tune your communication messages. That is, it's to develop what we call power phrases that capture the essence of what you want to communicate in the most compelling and positive way."[76]

One of the forms of data generated by the Reagan public opinion enterprise was "thermometer readings" of virtually all public figures. This was an index of the extent to which people felt "warm and positive" or "cold and negative" toward the person rated, and it guided the choice of television newspersons to whom interviews about various subjects were granted. "Hot buttons" were the topics that sparked an immediate, visceral response among the electorate. Here again, focus groups were interviewed to determine what these issues were, and on this basis advertising campaigns were developed to incite and mold public opinion.[77]

Wirthlin was not the first to use such techniques. Network television planners have employed them for years to market-test new shows; businesses use them to pretest new products. But his application of state-of-the-art marketing technology was admirably thorough and workmanlike. When problems in Reagan's image as a leader were identified, a sustained campaign could be mounted to change the public's perception of the president. This was often not a matter of changing policies or political views. The trick was to achieve perception change entirely at the level of public relations.

Reagan's political art, combining, as did his best movies, an actor's studied naturalness and spontaneity with the resources of a powerfully visual medium, contributed significantly to the potency of his discourse. The effeminate style, in its subordination of sound argument and evidence to emotion and personality, fit nicely with inventive strategies designed to foreclose rather than promote critical deliberation about public issues. Michael Deaver's carefully orchestrated visual dramas were as disengaging to the reason as they were engaging to the eye. And Richard Wirthlin's exhaustive search for viscerally appealing, hot-button issues helped create a discourse that shed more heat than light on the American political scene. In his use of noninventional, as well as inventional, resources, Reagan's public discourse established the same pattern. It emptied the public sphere of substance as it filled it with glitz and glitter. In a word, it made of public discussion a ceremony, and a ceremony only.

Epilogue

With the Reagan presidency past, and Reagan himself largely out of the public eye, does it make sense to talk of Ronald Reagan's "legacy" as something fundamental and lasting?[78] If only specific policies are considered, perhaps not. But we believe that the ceremonial form of public discourse that Reagan helped to create and reinforce during his decade on the national political scene will continue to exert its power against the realization of a genuinely democratic public sphere.

We reiterate that the trend toward ceremonialization of public discourse did not originate with Reagan, though we think we have shown that he contributed significantly to its power. It might be argued, however, that this trend will not long survive him. As economic problems in the United States deepen in the coming years, might we not expect that the "happy talk" of the Reagan years will give way to a more realistic form of discourse, one with which we are better able to confront the nature of our problems and to search for solutions?

Certainly, we hope so. But habits of discourse die hard. And the mere existence of increasing challenges does not guarantee that they will be met, discursively or legislatively. The 1988 presidential campaign, though coming too soon in Reagan's wake to provide an adequate test of the discursive possibilities of the post-Reagan era, nonetheless illustrates what we mean.

Though most electoral campaigns are never as intellectually nourishing as one might wish, 1988 seemed to reach a new low. Inflammatory rhetoric about flag burning and prison furloughs, rather than proposals of how to deal with the budget crisis or the savings and loan fiasco, occupied our television screens. As a *Washington Post* reporter, Paul Taylor, told the story:

> The pollster in the campaign is supposed to find the issues, frame the issues, and serve the issues back out to the voter. If you thought in 1988—in the summer or spring of 1988—what are going to be the issues in the Presidential campaign, who would have thought that in the campaign we'd talk about furloughs and flags. We did, and that's the power of the pollster. The George Bush campaign went out and discovered that these were "hot button" issues. They conducted focus groups in New Jersey and the focus group leader, with the pollster whispering in his ear, said, "Well, let me tell you about Michael Dukakis," and they clipped off three or four things, flags, furloughs, Boston Harbor, taxes, and they turned around public opinion in that room.[79]

The 1988 presidential campaign had all the ingredients of ceremonial discourse. It was the politics of personality and pathos, not reason. Patriotism and personal integrity, not policy prescriptions, became the major concerns. And the "artistic" techniques of the Reagan years were everywhere in evidence; "sound bytes" and "one-liners" became the terms in which not only technicians but the candidates themselves characterized the political process.

To present electoral campaigns as representative of the quality of public discourse might seem unfair. There are two reasons, however, why doing so seems to us appropriate. First, elections are the one opportunity for most citizens of mass societies to participate in national politics. Nations of 250 million people cannot be run on the town-meeting model of colonial New England. The sheer scale of the political process closes out the vast majority of the citizenry from nonelectoral forms of participation. Second, there is evidence that postelectoral public discourse since Reagan is assuming the same ceremonial pattern as it did during his time in office.

We observed earlier that the discourses of populism, civil religion, and national security constitute a set of inventional resources that together can operate to foreclose rather than open up the possibilities of public discussion of contemporary issues. One example of this kind of foreclosure is the classification of opposition to one's policies as "heresy" and of opponents as enemies. Consider, in this regard, Congressman Richard Gephardt's recent experience with the Bush administration. In a speech in March 1990, Gephardt proposed the creation of a "Free Enterprise Corporation" and other changes in our international trade practices. Rather than responding to the proposals per se, the Bush administration rebuked Gephardt personally, suggesting that he had "gone bananas." Stung, Gephardt wrote a letter published in many newspapers in which he said:

> I criticized the president's foreign policy. I didn't try to get under his skin. I wanted to debate how best to advance democracy in Eastern Europe and the Soviet Union. This debate should be about policy, not personality. But that's not the state of our discourse these days. Instead of refuting an idea on its merits, politicians too often do what works in campaigns—they shoot the messenger. We are living in a time when all the cliches have come true: we are at a crossroads of history, this is a turning point in the course of human events and the choices we make now will determine the peace and prosperity of generations to come. So we had better discuss and debate those choices. . . . If we continue our ostrich approach, refusing even to debate important policies, we are doomed to be the victims of change.[80]

Gephardt's stake in public discourse is by no means a disinterested one. He was a candidate for the presidency in 1988 and could be again in 1992, when Bush will run for reelection. But the peremptory reaction of the administration to his suggestions is instructive, nevertheless. In attempting to engage Bush in a dialogue about policy, Gephardt was violating the norms of ceremonial discourse. The administration's response was its way of restoring a kind of discursive equilibrium, one we have associated with the legacy of Ronald Reagan.

These examples of presidential discourse in the early years of the Bush administration are suggestive of what we believe Reagan's rhetorical legacy will turn out to be. The persistence of its effects will derive both from the success with which Reagan used ceremonial discourse and from the fact that he was using materials, inventional and noninventional, that were not unique to him but sprang from sources deep within the dynamics of American ideology and technology. The discursive subsystems he mined, the artistic techniques he exploited, were there for his use. But he used them with extraordinary energy, persistence, and skill. The 1980s, then, are likely, in retrospect, to stand out as ceremonial discourse's finest hour, but not its swan song. The Reagan years will serve as a standard of what ceremonial discourse can do, a kind of school for would-be Ronald Reagans, and, indeed, for Michael Deavers and Richard Wirthlins as well. There are and will be many eager students.

2

Rhetoric, Legitimation, and the End of the Cold War

Ronald Reagan at the Moscow Summit, 1988

G. Thomas Goodnight

At 2 P.M. on Sunday, 29 May 1988, Air Force One touched down at Vnukovo Airport. As the plane taxied to a stop, a waiting band struck up the American national anthem, onlookers cheered, and dignitaries snapped to attention. On that sunny spring day in Moscow an American president stepped onto Soviet soil for the first time in fourteen years. For Ronald Reagan, surely one of America's most virulently anti-Communist presidents, the pilgrimage to Moscow was long in the making.

This essay is a study of Ronald Reagan's last and in many respects most masterful international performance, his courtship of the Soviet people and Secretary Gorbachev at the Moscow summit. Spoken in the language neither of détente nor of the cold war, the summit speeches warrant investigation because they set forth the rhetorical complexities of attitude and tone characteristic of United States–Soviet relations in the "post-postwar" world. Thus the inquiry takes up one of the most interesting and controversial aspects of the Reagan legacy, the relationship between the rhetorics of the first and second administrations, and the end of the cold war.

Toward Moscow

American-Soviet relations, never stable throughout the post-war era, had all but ruptured in July 1979. In response to the invasions of

Afghanistan, Jimmy Carter boycotted the Moscow Olympic games; and in response to approaching nuclear parity with the Soviet Union, he crafted an expanded defense budget.[1] What was the Carter foreign policy soon became the Reagan crusade. "The only morality they [the Soviets] recognize is what will further their cause, meaning they reserve unto themselves the right to commit any crime, to lie, to cheat," the newly elected president told the American public, and anti-Soviet themes echoed loudly throughout the first administration.[2] Following the president who called détente a "one-way street," members of the administration spoke of nuclear warning shots, of new, better nuclear weapons for Europe, and of rolling back Communist insurgencies.[3]

In the decade of the 1970s, the international community had become more or less comfortable with the neutral idiom of Mutually Assured Destruction (a nuclear doctrine that recognized the linked interests of the superpowers in the condition of mutual vulnerability and survival).[4] Reagan rhetoric made MAD controversial. Not since the height of the cold war had international relations been articulated in quite so strident a rhetoric of polar opposition, nonaccommodation, and ideological struggle; not since John Foster Dulles spoke unabashedly of "massive retaliation" had the political uses of nuclear weapons been so openly stated.[5]

Reagan unsettled business-as-usual with apocalyptic discourse. The Soviet Union is an "evil empire," he warned a cheering group of right-wing fundamentalists, which will eventually be thrown on the ash heap of history.[6] It was as if Reagan's rhetoric sought to drag down the reigning god terms of Soviet-American relations, *cooperation* and *peaceful coexistence*, and thereby extirpate détente.[7]

While the rhetoric of the first Reagan administration was painted largely in cold-war hues, it also was colored by contrasts. Confronted by European allies made insecure by the prospects of regional nuclear deployments, assailed by an international antinuclear protest movement, and battered repeatedly by a recalcitrant Congress that was, on the whole, loath to engage in Reagan's risky business, the administration offered negotiations with the Soviets as a gesture of benign intentions on its part, and as a way to call the Bear's bluff on peace issues.[8] Thus were the Soviets enjoined in a series of extended negotiations over intermediate nuclear forces (INF), strategic arms limitations (START), and conventional weapons (MBFR). As the steeply one-sided terms of these offers became clear, and as confrontation through proxy wars spread, bilateral relations deteriorated to such an extent that few believed negotiations could be productive.[9] The president's 1984 landslide reelection victory, however, coincided with the emergence of a bold, relatively young Soviet leader, Mikhail Gorbachev, and the terms of international relations soon began to shift because, in the words of Margaret Thatcher, Gorbachev was a man with whom it was possible to "do business."[10]

The rhetoric of the second Reagan administration is peculiarly turbulent. On the one hand, the president pressed forward defense buildups, most visibly the Strategic Defense Initiative (SDI) which to him meant the creation of a purely defensive umbrella dedicated to the protection of the United States and its allies, but to the Soviets signaled a new level of complexity in the arms race and the growing potential of an American nuclear first strike.[11] On the other, Reagan engaged in a series of summits with the Soviet general secretary. From Geneva in 1985, where an agenda for future discussions was set, to Reykjavik in 1986, where a surprise Soviet offer nearly resulted in an agreement to cut ICBMs substantially, to the Washington summit of 1987, where talks concluded on the INF agreement, the dubious president was assaulted by the "charm offensive" of an internationally popular, visionary Soviet leader who spoke the new language of *perestroika* and *glasnost*. In a series of strategic moves, Gorbachev broke up the logjam in nuclear negotiations by conceding to Western demands, crossing what had been called unbridgeable barriers to agreement.[12] Although the Soviet leader was speaking of a new era, some were suspicious and claimed that "only the packaging . . . is changing and not the substance of Soviet foreign policy."[13]

The Great Communicator was no mean student of persuasion either; but, as knowledge of the quasi-legal affairs with Iran and the Contras oozed out of the White House basement, and as a quarrelsome Congress continued to block his more right-leaning political goals, the president moved toward the end of the second administration with the prospect of an ephemeral and tarnished legacy, a distant echo of the "Reagan revolution."[14] To many in the press, the lame-duck president had little to do at the upcoming summit; at seventy-seven, all he had left to accomplish was to stand pat, present the official instruments ratifying the INF agreement (a treaty passed by the Senate just days before departure), collect kudos for making peace, and close shop. Almost everyone expected the Moscow summit to be "longer on symbol than substance."[15]

If only symbols were to be on the table, however, the game was still to be played for high stakes; for it was the future relationship between superpowers that Reagan (and the six hundred officials who accompanied him) hoped to influence, even define, through a dramatic performance of foreign policy discourse. The Moscow summit was thus painstakingly choreographed for an international audience who would see a new relationship between the USSR and the United States displayed in a consummating dialogue between Ronald Reagan and Mikhail Gorbachev. So it is that in the critical investigation of the summit discourses—the speeches, press conferences, photo sessions, and ceremonies of the visit—the legacy of the Reagan administrations can come to be understood.

Symbols and Summitry

An international summit is a meeting between world leaders with their respective teams of advisors, politicians, support staff, and spouses. It provides opportunities for private discussions and public announcements, for formal ceremonies of state and informal get-togethers, for culminating the incremental processes of negotiation and opening possibilities for breakthroughs, for articulating scripted discourses and personal diplomacy, for standing tall in the eyes of domestic audiences and seducing foreign publics. Part diplomatic working session, part media spectacle, and part celebrity party, the modern superpower summit is a drama played large on the international stage, and its chief rhetorical form is courtship.

"By the 'principle of courtship' in rhetoric we mean the use of suasive devices for transcending of social estrangement," Kenneth Burke writes.[16] As persuasion, courtship thrives in the symbolic construction of identification and difference, union and separation, care and indifference. As communication, its motives range from calculated manipulation to pure passion, and in between all reasons are mixed. In the social dramas of courtship, one is never sure whether discourse proceeds more from shared, logical premises than from reciprocal needs, more from realistic necessity than from tantalizing possibility, or more from reasoned conclusions than from disguised opportunities and pitfalls. In its forms of appeal, the rhetoric of courtship frequently stylizes social identification. Garbed in acts of display, ceremony, and ritual, courtship is spoken as pledge, contract, declaration, and toast. Courtship is an unstable and sometimes destabilizing form of symbolic action, though. Discourses of resolution, disclosure, consummation, and desire are evoked by gestures of disguise, discovery, disappointment, and opportunity. In its mixtures of speech acts and gestures, courtship coalesces into an exciting but turbulent rhetoric.

To analyze a superpower summit as courtship may seem frivolous. After all, the fourteen presidential addresses, numerous public appearances, and events were all carefully scripted by the Department of State and the White House staff in Washington (and for Gorbachev by his counterparts in Moscow). Moreover, one finds in the summit's discourses more avowal of realism than romance. However, the summit's swirl of characters—the interaction of Soviet and American politicians, diplomats, generals, intellectuals, protestors (refuseniks), media personages, and stars—set in motion a pageant whose drama of estrangement and identification surely exceeded the anticipated sum of its parts.

Reagan was not unaware of the rhetorical strategies of a romantic hero, and he had experience sharing the stage with another lead; but the situation multiplied constraints and opportunities among varied audiences.[17] Could Reagan court the Soviet people to the virtues of democracy while not being too critical of Gorbachev? Alternatively, could he court Gorbachev while at the same time not distancing himself from his own conservative legacy? Could Gorbachev induce an avowed enemy of Communists to accept the legitimacy of a reformed Soviet Union?[18] More, could the champion of *perestroika* and *glasnost* convince his countrymen of the wisdom of change, in light of Reagan's more than occasional reversions to cold-war rhetoric?[19] The dialogue at the summit would both signal and constitute the durability of a newfound relationship, and a key to its successful performance would be the balancing of the issue of legitimacy among conflicting claims.

Because the Soviet Union and the United States have long engaged in cold-war ideological competition, the degree of estrangement between the nations is great.[20] As a consequence, while courtship among its leaders may be desired to soothe the anxieties of a nuclear age, any claimed identification has to face severe tests of legitimation. Jurgen Habermas writes: "Legitimacy means that there are good arguments for a political order's claim to be recognized as right and just. . . . *Legitimacy means a political order's worthiness to be recognized.*"[21] Claims to legitimacy are always contestable, and any leader who proffers a discourse grounded in the right and just must be prepared to "show how and why existing (or recommended) institutions are fit to employ political power in such a way that the values constitutive of the identity of society will be realized."[22] Cold-war rhetoric had eroded the discursive space for reciprocal grants of legitimacy between the United States and the Soviet Union, figured as capitalist and Communist systems. The argumentative burden around which gestures of courtship were acted out at the summit thus centered upon speaking a language of legitimation that would grant opposing views some "right and just" place for an emerging dialogue between nations, a very difficult task. How could opportunities for legitimation be opened up when, for the preponderance of the cold war in general and the bulk of Reagan's administrations in particular, concerns about the legitimacy of respective systems were at the heart of opposing rhetorics?

The Moscow summit was a courtship that proceeded by the dramatic enactment of a legitimation controversy.[23] Reagan and Gorbachev both worked to dismantle the delegitimizing arguments of the cold war in order to craft a new linguistic horizon for international cooperation and, at the same time, strove to maintain "true and just" positions in the eyes of their own domestic political audiences by questioning each other's rhetoric. A discussion of the controversy provoked through their efforts should provide a richer understanding of the discursive contraints on international relations left to us as a legacy of the Reagan administrations.[24]

The Summit: The First Day

Even before Air Force One had landed, the intentions of the respective leaders were announced to one another and to their respective domestic audiences. Gorbachev, in an unprecedented, extended interview published by the *Washington Post* and *Newsweek*, set high sights for the summit. "I am convinced that positive trends are unfolding in the world. There is a turn from confrontation to coexistence," Gorbachev said, and then added: "The winds of the Cold War are being replaced by the winds of hope."[25] Lest the president appear rushed by the gusts of Soviet politics, his advisors underplayed the potential of new summit accomplishments while insisting that the sturdy, four-part agenda for the summit be pushed forward with modest progress.[26]

Since Geneva the United States had insisted upon a four-part agenda for each superpower meeting. For Moscow the agenda remained the same. Arms control was divided among three issues: an exchange of the instruments of ratification for the successfully concluded INF agreement; the signing of "minor" accords that permitted exchange of scientific information on nuclear testing; and the possibility of further progress on START, a negotiation that found both countries united on the ends of significant nuclear arms cuts but divided over means of counting and inspection. To these concerns were added negotiation over "regional issues," which amounted to reducing or putting an end to indirect confrontation through proxy wars in Angola, Nicaragua, and Afghanistan, and negotiation on bilateral matters concerning trade and cultural exchange. The fourth element of the agenda was human rights, and it was on this matter that Reagan made his opening gambit.[27]

On his way to Moscow, Reagan stopped at Helsinki and delivered an address that prefigured the American position on each agenda item; at its heart was the announcement that "there is no true international security without respect for human rights."[28] From a respect for human rights, the president claimed, all prospects of peaceful cooperation flow: "The [Helsinki] accords have taken root in the conscience of humanity. [There is] increasing realization that the agenda of East-West relations must be comprehensive, that security and human rights must be advanced together or cannot truly be secured at all. . . . The provisions of the Final Act reflect standards that are truly universal in their scope."[29]

A divided agenda permitted the American delegation sufficient breadth of issues to select one or more items for emphasis; yet parts of the agenda could be linked (as in this case, respect for human rights was held as prerequisite to common standards of action, trust, and security). Reagan's Helsinki address signaled that the summit would produce easy reconciliation of differences, for it was on the human rights issue that the administration had been most critical.

At St. George's Hall in the Kremlin the leaders of the two most powerful nations on the planet squared off in an exchange of greetings that would set the tone of the four-day summit. Gorbachev, speaking first, welcomed the Presidential party: "As we see it, long-held dislikes have been weakened; habitual stereotypes stemming from enemy images have been shaken loose. The human features of the other nation are now more clearly visible. This in itself is important, for at the turn of two millenniums, history has objectively bound our two countries by a common responsibility for the destinies of mankind."[30]

He offered the Reagans a Russian proverb: "It is better to see once than to hear a hundred times." "Let me assure you," he added, "that you can look forward to hospitality, warmth and good will" accorded by the Soviet People.[31] The significance of Reagan's presence on Soviet soil was thus made patent. As Robert Kaiser of the *Washington Post* wrote: "With his warm reception for Reagan, Gorbachev was signaling his countrymen that even this one-time ogre of Soviet propaganda had earned a new acceptability," an acceptability cultivated by *perestroika* and *glasnost.*[32]

The greeting served two legitimation functions. First, it undercut the president's long-standing description of the Soviet Union as an alien land by indirectly attributing his views to stereotypes, which can be broken by experience; at the same time, the welcome invited the president to share the highest mutual concern: finding the means to avoid nuclear confrontation. Such a transcendent goal subordinates the question of differences between ideological systems. Gorbachev thus established the issue of nuclear weapons as a key criterion for successful summitry.

Reagan responded adroitly, thanking the general secretary for his kind words of welcome and succinctly stating the four-part agenda: "human rights, regional issues, arms reduction, and our bilateral relations."[33] He then went on to remind everyone that the glass is not half empty but half full.

> We signed a treaty that will reduce the level of nuclear arms for the first time in history by eliminating an entire class of U.S. and Soviet intermediate range missiles. We agreed on the main points of a treaty that will cut in half our arsenals of strategic offensive nuclear arms. We agreed to conduct a joint experiment that would allow us to develop effective ways to verify limits on nuclear testing. We held full and frank discussion that planted the seeds for future progress.[34]

The key to these successes was a step by step, cautious approach. "*Rodilsiya ne toropilsiya*—it was born, it wasn't rushed," Reagan said, and there were "tremendous hurdles yet to be overcome."[35]

Although Reagan refrained from pushing the American position in his words of greeting, human rights soon became a matter of conflict; at the first private meeting of the heads of state, sparks reportedly flew over differences on the issue.[36] For the remainder of the summit, the legitimacy of the Soviet Union's position would be assaulted by its own invited guest, but for the moment the reported clash of wills was offset by a contrasting media event, a spontaneously staged "plunge into the crowd" by President and Mrs. Reagan during a walk through Arbat Mall, an area variously described as a small-merchants' paradise and the Georgetown section of Moscow. This first foray at handshake diplomacy met with uneven success. At first, mixing with the people brought smiles, a contact between Americans and real Russians; but smiles turned to alarm as surprised Soviet security shoved, pushed, and elbowed their way through the crowd to clear a path to the president's limousine, even jostling a few journalists. The pomp and pageantry of the official greeting gave way too quickly to a common touch. "Someone must have spread the word in advance," Reagan quipped, making light of the whole affair.[37] This was just the opening act of a budding romance between the president and the Soviet people.

The Summit: The Second Day

Monday, 30 May, started off with another closed-door session among superpower leaders. If there were lingering consequences of the "barbs" traded on the first day, they were concealed beneath the waves, smiles, and handshakes captured by the press at St.

Catherine's Hall. While the staffs of the four working groups met to sort out negotiations, the president made his way to the Danilov Monastery, a thirteenth-century Russian Orthodox spiritual center that was celebrating the millennial anniversary of Russia's conversion to Christianity. Danilov, turned into a factory after the Bolshevik revolution, had been returned to the Russian Orthodox church in 1983; and on that morning, it was to be transformed yet again into a symbol of common aspirations between the Soviet and American people.

The Danilov Address

In his address at Danilov, Reagan recognized that progress had been made in the Soviet Union on the issue of religious freedom, but instead of lauding the reforms, he spun out a utopian dream: "We don't know if this first thaw will be followed by a resurgent spring of religious liberty—we don't know, but we may hope. We may hope that *perestroika* will be accompanied by a deeper restructuring, a deeper conversion, a *mentanoya*, a change in heart, and that *glasnost*, which means giving voice, will also let loose a new chorus of belief, singing praise to the God that gave us life."[38]

The speech concluded with a lyrical quote from the Soviet dissident writer Aleksandr Solzhenitsyn, reminding the audiences—abroad and at home—of the long spiritual tradition of the Russian people, a spirit symbolized in Russia's country churches:

> When you travel the by-roads of Central Russia, you begin to understand the secret of the passifying Russian countryside. It is in the churches. They lift their belltowers—graceful, shapely, all different—high over mundane timber and thatch. From villages that are cut off and invisible to each other, they soar to the same heaven. People who are always selfish and often unkind—but the evening chimes used to ring out, floating over the villages, fields, and woods, reminding men that they must abandon trivial concerns of this world and give time and thought to eternity.[39]

Repeating a wish he had made at Helsinki, Reagan said that "we" prayed that once again the church bells would ring across the countryside, "clamoring for joy in their new-found freedom." Rhetorically, the passage works in a manner similar to Reagan's dare to tear down the Berlin wall. In the face of all other matters of great moment, his desire to let freedom ring in Russia is an irritatingly strict symbolic threshold for legitimacy and identification, but it does preserve critical distance. Unlike the Berlin wall remark, however, ringing bells is a threshold softened by the good will of good wishes.

The American press reported, rather cynically, that summit planners front-loaded the human rights issue to get controversy out of the way early in the summit.[40] But the Danilov speech was more than a passing moment. By featuring religious freedom as the key to other freedoms, and by connecting the United States and Russia through the transcendent principle of faith and a poeticized depiction of country life, Reagan showed respect for the people who were his hosts, affirmed and broadened the goals the Soviet Union was setting in its newfound commitment to democracy, and—importantly—enacted a language that assured Americans of his first-order commitment to a civil religion, the very wellspring of conservative discourse.[41] The speech thereby purified intentions and initiated a courtship with the Soviet people.[42]

The Spaso House Speech

The human rights agenda was defined and amplified at the residence of the American ambassador. There, in diplomatic sanctum, the president met ninety-eight Soviet dissidents and refuseniks. Progress on freedom of religion, speech, travel, and private enterprise was presented as the summit's working aims. Indeed, the president said, individual rights are correlated with "self-worth" and self-worth to the enterprising spirit. "On the fundamental dignity of the human person, there can be no relenting, for now we must work for more, always more."[43] Implied is the basic premise of the summit: whatever reforms had been made, they were not enough; that the president met with the official outcasts of Soviet society was enough to say that there was a long way to go before the Kremlin could make claims of right and justice.

No strangers to political propaganda, the Soviet media were prepared for this aspect of the legitimation controversy. As often happens in American political broadcasts, clips of the meeting were shown without the controversial discourse, and coverage was "balanced" by a panel discussion with a group of American dissenters who went to the Soviet Union with claims that their own political freedoms were circumscribed in the United States.[44] And it was asserted by the Soviet press that at least one of the so-called dissidents had Nazi connections.[45] Still, for such a meeting to take place at all must have signaled that changes were in the wind.

Nancy Reagan and Raisa Gorbachev

To focus too much on official speeches and working groups distorts the drama, for a major theme of the affair was the courtship between the respective leaders' wives, who were as much in the limelight of international publicity as their celebrated husbands. In one sense, Nancy Reagan, a former dancer and Hollywood actress, played the part of an American abroad, a person whose visits to cultural sites transmitted the glories and mysteries of Russian history. In another, Nancy Reagan, the powerful Washington insider, was emplotted in a social battle with the flashy, intellectual, and primly proletarian Raisa Gorbachev. How would they get along?

As the two emerged from a tour of Assumption Cathedral, holding hands in the Russian custom, initial reports appeared encouraging. However, the press probed for tension. Donald Regan had reported that Nancy once said of Raisa, "Who does that dame think she is?" Was that true? asked an NBC News correspondent at a joint press conference. With her "smile turned to ice and her eyes to fire," Nancy Reagan replied that he was "wrong." As for their previous meetings, Raisa Gorbachev claimed, "We've gotten along very well every other time."[46] What would the next three days of joint tours yield between two of the most powerful women in the world?

Toasts at the Gorbachev Dinner

In the Anglo-American tradition the dinner toast is a wish of good cheer, a congratulation to the host and assembled party, a spontaneous evocation of good will; and a clever toast honors its maker as much as its audience. In the Russian tradition, a toast is all of these things and more, for the toast is the occasion for a speech that offers heartfelt words to guests gathered in the peace of a meal that is part of the long flow of history. More than bon mot, the Russian toast discloses all the thoughts and feelings appropriate to a moment of meeting.

At the televised state dinner in the sumptuous surroundings of the Kremlin's Faceted Chamber, General Secretary Gorbachev rose to address the president and an international audience. "I welcome you in the Moscow Kremlin. For five centuries, it has been the site of events that constituted milestones in the life of our state. Decisions crucial to the fate of our nation were made here. The very environment around us is a call for responsibility to our times and contemporaries, to the present, and to the future."[47]

The historical trappings of power represented by a dinner of state display solemnity amidst sociality. The setting calls for a discourse carefully measured for a situation of rare and great importance. What fate and opportunity would frame the occasion this evening? Gorbachev explains:

> Normal and, indeed, durable Soviet-American relations, which so powerfully affect the world's political climate, are only conceivable within the framework of realism. Thanks to realism, for all our differences, we have succeeded in arriving at a joint conclusion which, though very simple, is of historic importance: a nuclear war cannot be won and must never be fought. Other conclusions follow with inexorable logic. One of them is whether there is any need for weaponry which cannot be used without destroying ourselves and all of mankind.[48]

And the rest of the agenda is laid out as a broad-based concern for "the realities and imperatives of the nuclear and space age, the age of sweeping technological revolution when the human race has turned out to be both omnipotent and mortal."[49] The toast creates a sense of proportion by tying the realization of joint responsibilities to the risks and opportunities of a new era.

But what foreign policy analysts often refer to as "responsible behavior" on the part of the Soviet Union has its price. While speaking of the fruits of new thinking and openness, Gorbachev draws a sharp limit on change:

> We see ourselves even more convinced that our socialist choice was correct, and we cannot conceive of our country developing without socialism based on any other fundamental values. Our program is more democracy, more *glasnost*, more social justice with full prosperity and high moral standards. Our goal is maximum freedom for man, for the individual, and for society. Internationally, we see ourselves as part of an integral civilization, where each has the right to a social and political choice—to a worthy and equal place within the community of nations.[50]

The legitimation argument, then, is clear and complete. To accept the Soviet Union as a responsible international partner is necessary in order to reduce the risk of catastrophe from nuclear war or environmental damage. However, such acceptance requires recognition of the right to difference and tolerance for a system whose institutions, launched by an alternative value system, circle in an expanding orbit of social justice. The speech is coherent, precise, and difficult—a ringing challenge to the cold war's construction of a hegemonic and implacable foe.

The president spoke next. There was much dancing around the

discourse. Of course there are common bonds, he said: a triumph over Hitler, an admiration for the arts, brave "first steps" to disentangle nuclear standoffs, a search for common grounds. He then launched into a rambling digression describing an obscure 1950s film, *Friendly Persuasion*. The retelling of the film was supposed to serve as a vehicle to identify reciprocal positions at the summit, but Reagan's narrative dissolved into incoherent references to Quakerism, the American Civil War, and difficulties of communication. Ironically, the director of the film the president professed so much to admire had been blacklisted as a Communist in the 1950s. This was not the president's finest hour, but the address minimally accomplished its end. "Let us toast the art of friendly persuasion, the hope for peace with freedom" he concluded. With so much pomp and circumstance no one seemed to notice a heartfelt emptiness in the president's toast; at least no one had the bad manners to comment critically on a long-sought display of cordiality. Joseph Whelan, who studied the summit extensively, even calls the speech "a public relations performance of the first magnitude."[51] And it can be said that, all in all, the first two days of the summit were a smashing rhetorical success. On the third, the legitimation controversy would blossom.

The Summit: The Third Day

At the third private meeting between the heads of state, Reagan set aside for the moment the gospel of human rights and listened to a dissertation on *perestroika* and the need for a stable trade relationship between the United States and the Soviet Union. Only six years before the administration had pleaded with allies in western Europe not to extend help on a natural-gas pipeline for Europe. In the face of a flagging Soviet economy and the absence of favored-nation trade standing from the United States, loosening trade restrictions was a most desirable goal for Gorbachev.[52]

The leaders emerged from the private space of negotiation for a stroll in the most public of cold-war spaces, Red Square—the site of many May Day parades and public celebrations of the Bolshevik Revolution. That morning, the displays of state power were replaced by quiet conversations. "We want our children to live in peace," said a woman holding a small child. Gorbachev took the child into his arms and said, "Shake hands with grandfather Reagan."[53] Asked about Reagan's criticism of human rights, the Soviet leader adroitly finessed the controversy: "We are so critical of our own country that even the President's criticisms are weak. We know what our problems are."[54] Added Reagan, affirming newfound communication possibilities: "What we have decided to do is to talk to each other and not about each other, and that's working just fine." Asked as to whether he still believed the Soviet Union was the Evil Empire, he responded, "You are talking about another era."[55] So was spoken a defining moment in history: just as Gorbachev embraced self-criticism of Soviet society, so the president let go into the past the most celebrated moment of his own cold war rhetoric. But there was more. "At one point, as they strolled just a few feet away from Lenin's tomb, Reagan put his arm around Gorbachev's shoulder. That simple and spontaneous gesture made them look like two old friends."[56] For the moment, images and words fused to suggest the possibility that an alternate discourse was being born.

The moment was not lost on the reporters who were broadcasting the walk live and in color. A picture of a politician kissing babies is old hat—it is a gesture of love and renewal that is comforting even if staged—but the picture of the leaders of the free and Communist worlds taking up the attitude of the familial commonplace, chatting with a child in the middle of the storied public space of Red Square, spoke volumes. "The pictures say a new era in U.S.-Soviet relations is not mere illusion," reported Tom Shales.[57] In fact, the whole openness afforded to the coverage of the summit itself bespoke the new relationship. "It is a totally different world," gushed William Lord, ABC News executive producer. "Moscow is right now an open city. It is just as open as Washington, D.C. is."[58] Two speeches, delivered the afternoon of the third day, would both spread and dampen the euphoria.

Remarks at the House of Letters

"As Henry VIII said to each of his six wives, I won't keep you long," quipped the president, beginning a luncheon address to an assembled group of Moscow artists and "cultural leaders." The speech, a brief discussion of the relation of art and politics, made for a comfortable, reflective performance by this actor-turned-president, who chose this moment to reflect on the meaning and success of his life. In it he revealed "two indispensable lessons" taken from his "craft" into "public life."

First, he quoted Eisenstein, who, when making *Ivan the Terrible*, said: "The most important thing is to have the vision. The next is to grasp and hold it. You must see and feel what you are thinking. You must see and grasp it. You must hold and fix it in your memory and senses. And you must do it at once."[59]

The lesson? The "House of Letters Speech" continued: "To grasp and hold a vision, to fix it in your senses—that is the very essence, I believe, of successful leadership not only on the movie set, where I learned about it, but everywhere. And by the way, in my many dealings with him since he became General Secretary, I've found that Mr. Gorbachev has the ability to grasp and hold a vision, and I respect him for that."[60]

The message was plain and compelling. An actor might be held without credentials for politics, or worse, thought a mere stooge for his script writers; but a great actor "fixes" a vision, a truth sensitive to the events around him, and it is this "ability to grasp and hold a vision" that is the locus that permits Reagan, he claims, to "respect" the general secretary (a rhetorical gesture prerequisite to legitimate courtship). Forgotten in this transcendent moment was the fact that *Bed Time for Bonzo* is hardly the equivalent of *Ivan the Terrible*. But the merger of the values of artistic inspiration with political sensibility, mediated through the invocation of Eisenstein and Gorbachev, was sheer magic. For a Moscow audience, where the art of politics finds its most proper expression in the politics of art, such greatness of soul must have been as unexpected as it was welcome.

"The second lesson I carried from acting into public life was more subtle," Reagan continued, quoting a short but beautiful passage from Anna Akhmatova's *Requiem*, where in a prison line, a woman, face twisted and lips blue with cold, came up to Akhmatova and whispered, " 'Could you describe this?' And I answered her, 'I can.' " Akhmatova saw, on the face of the other, the shadow of a smile. Of this encounter, the president said:

> That exchange . . . is at the heart of acting as it is of poetry and of so many of the arts. You get inside a character, a place, and a moment. You come to know the character in that instant not as an abstraction, one of the people, one of the masses, but as a particular person—yearning, hoping, fearing, loving—a face, even what had once been a face, apart from all others; and you convey that knowledge. You describe . . . the face. Pretty soon, at least for me, it becomes harder and harder to force any member of humanity into a straitjacket, into some rigid form in which you all expect to fit. In acting, even as you develop an appreciation for what we call the dramatic, you become in a more intimate way less taken with superficial pomp and circumstance, more attentive to the core of the soul—that part of each of us that God holds in the hollow of his hand and into which he breathes the breath of life.[61]

The attention to the detail of the single face, the breaking of stereotypes, and the appreciation of difference, of course, were not without possibilities for transforming poetic to political discourse, and Reagan was quick to press the point.

> As I see it, political leadership in a democracy requires seeing past the abstractions and embracing the vast diversity of humanity and doing it with humility, listening as best you can not just to those with high positions but to the cacophonous voices of ordinary people and trusting those millions of people, keeping out of their way, not trying to act the all-wise and all-powerful, not letting government act that way. And the word we have for this is freedom.[62]

Reagan wrapped up with a paean to the new freedom in Soviet artistic expression and provided a link between it and his broader concerns for human rights. "The greater the freedoms in other countries the more secure both our own freedoms and peace. And we believe that when the arts in any country are free to blossom the lives of all peoples are richer."[63]

The argument of the speech moved from articulating the pain on a single person's face, to a sensitivity to the vast differences among us, to a duty for universalizing free expression. It was a coherent, if sweeping, statement of concern linking Reagan's training and values as an actor to his political agenda as summit leader. Again, there was comic irony at play, for Reagan rhetoric had not been noted widely for its sensitive attention to the diversity of the Soviet people (even if it was distinguished by its grasp of the local and anecdotal, reserved by and large for common Americans who supported administration views), and his work with blacklists in the 1950s certainly did not bespeak the value hierarchy he avowed this day.[64] Nevertheless the speech did spread subversive good will. By narrowing the focus of political rights, and by magnifying the stature of the arts—defined as "pillars" of universal human rights—the speech centered the Russian arts community itself in a moment of self-liberation.

It is said that even the mighty Homer nods, and during this performance so did Reagan, nodding off to sleep during a harangue by a fellow artist. Graciously covering the incident, his hosts said that his catnap was understandable because the other speeches were "a bore."[65] However, this incident set a theme for the remainder of the conference. Marlin Fitzwater denied that the president's health was limiting his activities; nevertheless, rumors of a "stamina gap" began to spread.[66] Could the seventy-seven-year-old president keep up with the demands of the summit?

Address at Moscow State

The speech to the students and faculty at Moscow State University, judged by both its length and its coverage, was supposed to be the rhetorical centerpiece of the Reagan summit; it was close to a rhetorical disaster. Long, tendentious, void of particular subject matter or interesting prose, the speech reads like discourse worked through the mills of cold-war bureaucracy, with the single but important difference that the Moscow State address took the high road of ersatz philosophical discussion.[67] The speech, premised on the need for freedom during a global information revolution, linked human rights with economic success. Freedom was forwarded as a god term meaning "the right to question and change the established way of doing things."[68] Such questioning, of course, serves as the engine of change that Reagan and his speech writers hoped the young Soviet intellectuals would hop aboard. A cultural exchange was proposed, but one wondered what the president had in mind, since the lesson of freedom was somehow illustrated by *Butch Cassidy and the Sundance Kid*, in an obtuse, rambling digression that caused listeners to shake their heads in puzzlement.[69] The long, exhausting performance ended with a gloss on Tolstoi and the virtues of being young and in the spring—when there are so many opportunities for politics.

There was a moment in the address, a joke, meant to bear away the guilt of difference while lending to the leader a common touch. Explaining that individual "experience" is the greatest teacher of all, Reagan told a tale:

> And that's why it's so hard for government planners, no matter how sophisticated, to ever substitute for millions of individuals working night and day to make their dreams come true. The fact is, bureaucracies are a problem around the world. There's an old story about a town—it could be anywhere—with a bureaucrat who is known to be a good-for-nothing, but he somehow had always hung on to power. So one day, in a town meeting, an old woman got up and said to him, There is a folk legend here where I come from that when a baby is born, an angel comes down from heaven and kisses it on one part of its body. If the angel kisses him on his hand, he becomes a handyman. If he kisses him on his forehead, he becomes bright and clever. And I've been trying to figure out where the angel kissed you so that you should sit there for so long and do nothing."[70]

In socioanagogic readings of tragic drama, differences between groups are reconciled only by creating a scapegoat who is symbolically killed to maintain the identity of the group. The comic counterpart plays down such differences by presenting social evil as the product of accident, not essence, and the result as a rectifiable mistake rather than ineluctable fate.[71] Heaping social sins on "bureaucrats" creates a base of identification between speaker and audience that enunciates mutual confidence in common good will and intelligence without, of course, having to name names.[72] Provoking laughter and applause, the story of a misplaced kiss generated the only warm moment in a turgid, didactic performance.[73]

Worse than the speech was a question-and-answer session where a student decided to test the president's knowledge of and concern for human rights by asking why a group of Native Americans came to the Soviet Union. Why were they not able to see him in Washington?

> Let me tell you just a little something about the American Indian in our land. We have provided millions of acres of land for what are called preservations—or reservations, I should say. They, from the beginning, announced that they wanted to maintain their way of life, as they had always lived there in the desert and the plains and so forth. And we set up these reservations so they could, and have a Bureau of Indian Affairs to help take care of them. At the same time, we provide education for them—schools on the reservations. And they're free also to leave the reservations and be American citizens among the rest of us, and many do. . . . Maybe we should not have humored them in that wanting to stay in that kind of primitive lifestyle. Maybe we should have said, no, come join us; be citizens along with the rest of us. As I say, many have; many have been very successful. . . . Some of them became very wealthy because some of those reservations were overlaying great pools of oil, and you can get very rich pumping oil. And so, I don't know what their complaint might be.[74]

The unfortunate answer abutted soaring concern for human freedom with a grossly insensitive, B-movie script on history and the social problems faced by Native Americans. Of course, the morning papers whipsawed the Moscow State address, carrying replies by outraged spokespersons for Indian groups, on the one hand, while showing Reagan behind a podium dwarfed by a gigantic bust of Vladimir Lenin on the other.[75] Part of leaving a rhetorical legacy involves speaking to the young, imbuing the impressionable with a future discourse, and on this ground the lecture was a justified part of the summit; but the president was not playing on his best court by entering an austere, intellectual arena to engage dialectic on the international state. The demands for critical self-questioning there could turn around much too quickly.

Toasts at the Reagan Dinner

The substance of the work of 31 May was the signing of what were billed as several "small" arms-control agreements, each providing for an exchange of information on nuclear arms and testing; the breakthrough was represented as not of enough moment to involve the main protagonists. Nancy Reagan's tour was extended to Leningrad, with appropriate fanfare, coverage, and sparks.[76] The dinner at Spaso House, the residence of the American ambassador, would provide again an opportunity for an important exchange of feelings and views.

In a sense, Reagan and Gorbachev's toasts were remarkably similar; both emphasized that differences persisted between nations, that the summit process had created an important discussion that would be continued; however, even in the celebration of "communication" itself there were strong notes of difference. Following the religious leitmotif of his summit discourse, Reagan quoted a passage from Boris Pasternak's poem "The Garden of Gethsemane," where Peter cut off the ear of a guard who came to arrest Jesus, and Jesus admonished: "You cannot decide a dispute with weapons; put your sword in its place, O man,"[77]

"This is the imperative, the command," Reagan said somberly. "And so we will work together, that we might forever keep our swords at our sides." The sword is sheathed, but it goes without saying that it still hangs at the side. Picking up the bright cadence of an Irish wish, Reagan raised the glass: "May this lovely home never lack for visitors, and shared meals, and the sounds of spirited conversation, and even the peal of hearty laughter."[78]

Gorbachev was less literary and more visionary.

Soviet and American people want to live in peace and communicate in all areas in which they have a mutual interest. The interest is there, and it is growing. We feel no fear. We are not prejudiced. We believe in the value of communication. I see a future in which the Soviet Union and the United States base their relations on disarmament, a balance of interest, and comprehensive cooperation rather than deterring each other or upgrading their military capabilities.

I see a future in which solutions to real problems are not impeded by problems historically outdated or artificially kept alive, inherited from the times of the cold war, and in which the policies of confrontation give way to joint quest based on reason, mutual benefit, and readiness to compromise.

I see a future in which our two countries, without claiming any special rights in the world, are always mindful of their special responsibility in a community of equal nations. It'll be a world that is safer and more secure, which is so badly needed by all people on earth—by their children and grandchildren—so that they could gain and preserve the basic human rights: the right to life, work, freedom, and the pursuit of happiness. The path toward this future can be neither easy nor short. We may be standing at the threshold of a uniquely interesting period in the history of our two nations. This new meeting between the two of us, Mr. President, confirms that 3 years ago in Geneva, we took the right decision.[79]

Those were portentous words enunciating a code of international behavior that scrupulously dismantled cold war discourse. The speech was an affirmation, even as it was an enactment, of the "dialogue" that characterized the discussion of differences initiated at Geneva. It placed the burden of international relations primarily on the challenge to sustain communication. At that moment, relations between the Soviet Union and the United States had never been better, but would these gestures of communication bear "substance"?

The Summit: The Fourth Day and Farewell

The chief official business for the heads of state on the official agenda of the Moscow summit was a formal exchange of INF Treaty documents at the Kremlin. Debate over the modernization of nuclear weapons in Europe had been long and contentious, igniting the international nuclear-freeze movement and engendering explosive public displays at the bargaining table. At this moment, that debate was brought formally to a close. Whereas the main arms-control measures of the 1970s, SALT I and SALT II, had provided texts rationalizing the growth of reciprocal nuclear stockpiles and armaments, for domestic political reasons the agreements had never been ratified. More than the end of one debate, the act of consummation was taken as a sign of portentous things to come.

Gorbachev's short address at the exchange daringly extended the remarks of the previous evening. Defining the discourse as "big politics," he called each summit, including this one, "a blow at the foundations of the cold war."[80] Supplanting the old divisive discourse was one aimed at a "nuclear-free and nonviolent world" generated through "constant" dialogue. Of course, this meant harder work on the START treaty, an agreement long in negotiation that had advanced but little in Moscow.

Reagan's address was less elevated and more concrete. He reminded his audience that seven years ago his own promise for arms control was defined as "unrealistic" and "dismissed" as a "propaganda ploy or a geopolitical gambit."[81] But the promise for negotiation had proven sincere: "We have dared to hope, and we have been rewarded." He continued:

> For the first time in history, an entire class of U.S.-Soviet nuclear missiles is eliminated. In addition, this treaty provides for the most stringent verification in history. And for the first time, inspection teams are actually in residence in our respective countries. And while this treaty makes possible

a new dimension of cooperation between us, much remains on our agenda. We must not stop here, Mr. General Secretary; there is much more to be done.[82]

No one who covered the event suggested that a reduction of a few hundred nuclear weapons makes little practical difference in a world where there are tens of thousands of such instruments of mass destruction, nor did anyone note that there were many more modernized nuclear weapons in the world after the Reagan administration than before. Coverage was drained of critique, for the "reduction of tensions" brought about and signified by a hearty handshake and hug between the leaders was an embrace long desired by many.

Less visible to the public but of greater significance to institutional arrangements between countries were the "minor treaties" of the summit, signed the day before, and the INF agreement, which together signified the emergence of a new order. More than ending the debate of the 1980s, the mutual agreements put to rest an argument between arms controllers and the military, sustained throughout the length of the cold war, over the value of treaties per se.[83] From now on, the powers would not wall off their experts but would exchange scientists and officials from reciprocal nuclear establishments in order to verify data. With comparatively little political fanfare, the Moscow summit affirmed the rationalizing of nuclear regimes as a separate and self-contained part of diplomatic life. At this moment, the Soviet and American states were joined in a symbolic structure legitimating the rational control of nuclear weapons.[84] This was the first and only treaty with the Soviet Union that Ronald Reagan supported completely.

Joint Statement of 1 June 1988

The work of a summit is encapsulated in a joint statement that expresses the leaders' official views of what has been accomplished and their assessments of the significance of the meeting. Such statements necessarily are written in the stilted, formal language of diplomacy; they suppress oppositional discourse in favor of what is commonly accepted. The "strategic ambiguity" of the discourse thereby permits continued discussion by deflating differences that might otherwise break apart the courtship. Of course, areas where agreement is noncontroversial or views that have shifted to become more compatible can be announced straightforwardly. Given the expansive range of discussion on the agenda of the Moscow summit, there seemed to be room for both strategies.

Reagan's contentious human-rights crusade was summed up in a statement to the effect that "dialogue" should continue with greater exchange of "information" and "contacts." In this manner, each side formally withheld its own opinions about the other's views of its own human rights stand, while maintaining that advances were being made mutually.

Gorbachev's summit strategy, however, did not permit diffusion of differences in the traditional manner. His visionary discourse sought nothing less than the goal of legitimating the reformed Soviet Union in a new world order. On the first day of the summit, he proffered this text to the president on a plain white sheet of paper:

> Proceeding from their understanding of the realities that have taken shape in the world today, the two leaders believe that no problem in dispute can be resolved, nor should it be resolved, by military means. They regard peaceful coexistence as a universal principle of international relations. Equality of all states, noninterference in internal affairs, and freedom of sociopolitical choice must be recognized as the inalienable and mandatory standard of international relations.[85]

The president said he "liked it"; but later, after consulting with American officials—who feared that the terms were too "ambiguous" and smacked of the disappointing language of détente, and who reportedly were concerned that such language might result in the reductions of the defense budget at home—the president told Gorbachev he just couldn't do it. Officials drafted a watered-down statement that placed on "dialogue" the burden of finding means to settle differences. It seems that avowal of realism requires that international norms be set to the minimal common denominator of productive behavior rather than to higher aspirations.[86]

Press Conferences

Occasionally a meeting between leaders will result in a joint press conference to give public visibility and to certify the common hopes and achievements emerging from private discussions. Despite the highly publicized pictorial and verbal pledges of hardy friendship and affiliation, each leader found differences significant enough to define for himself the work of the summit.

Gorbachev's press conference, like his interview with the *Washington Post* and his live discussion on the "MacNeil/Lehrer Hour" the preceding evening, was a virtuoso performance.[87] The press conference worked to recontextualize the American position. Gorbachev affirmed that a sense of realism and a commitment to dialogue are appropriate constraints for sincere communication, and that some summit discussions met these criteria; but he also said that one must expect "propaganda moves, demarches and attempts to score points through propaganda tactics."[88] As to the human rights themes, he claimed, "the American administration has no understanding of the actual situation with respect of human rights or with respect to the processes that are under way in our country in the sphere of democracy." He added, "We too probably don't understand the American side," and proposed a constructive solution, an international seminar that would test the thesis of openness.[89] This gambit diffused American rhetoric as error or worse, even as it proved a commitment to "freedom" by forwarding willing discussion.

Gorbachev did not stop at setting up future forums. He harvested the most important symbolic fruit of the summit, recalling the question in Red Square.

> Yesterday, when the President was talking with our people, where I was present, someone asked, and I think the press printed this: "Mr. President, do you still consider the Soviet Union an 'evil empire'?"
>
> No, he replied. Moreover, he said this at a press conference near the Tsar-Cannon [a fountain], in the Kremlin, in the center of the "evil empire." We take note of that; it means, as the ancient Greeks said, "Everything flows, everything changes."[90]

With this elegant Heraclitean trope Gorbachev pushed the flow of discourse in a positive direction, only to recognize that its course had been impeded. "I think that an opportunity to take a major step in shaping civilized international relations was lost at this meeting," he said with regret; yet "politics is the art of the possible," and he reminded the audience that it was through Soviet willingness to change and compromise that the current situation was constructed. Americans, he said, hold on to the Jackson-Vanik Amendment (prohibiting favored-nation trade status) as a constitutional document of relationship. "One of them is physically dead, and the other is politically dead. Why should they hold us back? After all, that amendment was adopted in a completely different situation, decades ago."[91] Such irony bespoke the frustrated earnest desire of a spurned suitor who offered unconditional affiliation only to be given a tepid response without concern for what might have been.

Showing signs of fatigue, Reagan produced a much shorter, and, according to the press, a much weaker performance; but, even granting its absence of precise dialectical maneuver and fresh thinking, the discourse was not less dramatic and not without consequence.

Reagan said that Gorbachev has gotten the right message; he affirmed his own "consistency of expression as well as purpose."

"At every turn I've tried to state our overwhelming desire for peace, I have also tried to note the existence of fundamental differences. And that's why it's a source of great satisfaction that those differences, in part as a result of these meetings, continue to recede," he said.[92] In fact, the lack of trust had receded so far that Reagan recanted the very premise of his administration: that the Soviets were implacable foes who lied and cheated for the purpose of world domination. He averred that he was not giving an opinion but quoting others. Following on this exchange came the question of the summit and perhaps his entire presidency:

Q. Well, that's what you thought then. Do you still think that, and can you now declare the cold war over?
The President. I think right now, [he paused] of course, as I've said, *dovorey no provorey*—trust but verify.[93]

In the space of a brief silence, where the president appeared to be pondering the end of the cold war, he traveled from his formative political days, the time of the Jackson-Vanik Amendment and poor Soviet-U.S. relations, to the present. That he could not give up the past entirely was not unexpected; that he appeared to be considering it spoke of the dawning possibilities of the future.

The press corps was not tantalized by this play of ambiguity but seemed to desire the usual red-meat, headline-grabbing items that international affairs often bring, and they knew where to probe. Picking up on the theme of bureaucracy, one reporter asked if it was not a policy of the Soviet government, not just its bureaucracy, to keep émigrés from leaving. Reagan responded, confused, "I don't know that much about the system, but it was a question that was presented to me on the basis that it possibly was a bureaucratic bungle." A rambling story on moving a file cabinet followed. A follow-up question was asked: "Don't you think you're letting Mr. Gorbachev off a little easy on just saying it's a bureaucracy?"

The president responded: "No, as I said, I don't. The way the question was framed, I thought that there was a possibility of that. No, but I just have to believe that in any government some of us do find ourselves bound in by bureaucracy, and then sometimes you have to stomp your foot and say, unmistakably, I want it done."[94]

For a week, the president had carried the human rights torch to the Soviets in a dozen speeches, yet this single remark produced an unlikely coalition of critics. Editorials proclaimed that Reagan had gone soft on human rights, and a number of prominent conservatives took to the hustings to make political hay out of this answer by indicting the Soviet "system."[95] The most charitable interpretation of the remark attributed the discourse to a strategic move on Reagan's part, letting Gorbachev save face because the summit had failed to produce breakthroughs.[96]

At work here is a misinterpretation. *Bureaucracy* is a code term which, for Reagan and Gorbachev, meant those officials who oppose their policies, and in fact the Soviet term *bureaucracy* has no positive association, even though it does connote officials with power.[97] To say that the bureaucracy is at fault does not mean the policy is not entrenched; but for the Western journalists who are used to a bifurcation between administrative function and policymaking authority, the answer is a gloss. Conservatives at home complained loudly and long, since in their view, communism has always been and is a single system.[98] The press's interpretation is a product of cultural differences, even as it is a remnant of hegemonic cold-war domestic politics.

Although the president was not making political points at home, he also was trying to salvage his human rights argument abroad. In short, the Soviet press and leaders had found his lectures on human rights patronizing, especially when Reagan's own discourse did not show any special sensitivity or ideas about problems in the United States.[99] Asked about Gorbachev's rejection of this discourse and his professed lack of "admiration" for Reagan on this score, the president tried to explain the standing of his own concerns.

> One out of eight Americans trace their parentage and their heritage, if not their own immigration, to the Eastern bloc.
>
> And so, I have put it this way: that you don't stop loving your mother because you've taken unto yourself a wife. So, the people in America do have a feeling for the countries of their heritage. . . . And so, when we feel that people are being unjustly treated—imprisoned for something that in our country would not be a crime, calling for such a sentence—our people get aroused, and they come to us, and they want help. They want something done.[100]

Neither expressing familial concern for blaming bureaucratic indifference was sufficient, however, to spring the president completely outside the tensions that had arisen. However ardent the expressed desires for reconciliation were that week, it was clear, from the expressions of difference, that reestablishing relationships would take time. The cold war was over and it was not over.

The Last Night in Moscow

The tour was not without its snags for Nancy Reagan, either. A whirlwind rush through the Hermitage, a comment on the ugliness of new high-rise housing projects, and a contretemps over the schedule with Raisa were all duly reported; but the first lady was known at this summit more for what she didn't say than for what she did say. Obviously agitated, she simply declared the relationship between herself and Raisa Gorbachev a "Mexican standoff."[101] The final evening in Moscow was to be entertainment: the Bolshoi Ballet, a dinner at a private dacha outside Moscow, and then a stop to see the moonlight streaming down on Red Square. The time for departure had come. The Reagans left the next day, and after the summit Gorbachev heavily criticized the human rights theme of the conference, even while the president returned home with much of his previous stature as the Great Communicator restored. "Summit Ends with Smiles, Hugs and a Signed Treaty," read a *Los Angeles Times* banner. "If the summit did not fulfill the hopes of the principals, it did fulfill their needs," the papers concluded, speaking well of how the summit played before domestic audiences.[102] The press reported, as a side bar, that Hitler's last command bunker was blown up on 2 June to make way for a new housing project in East Berlin. It might have been the end of an era.

Cold War Rhetoric and the Reagan Legacy

So momentous were events in eastern Europe and the Soviet Union in the year after the Moscow summit that the *Yearbook* of the Stockholm Institute on Peace Research declared the "end of the cold war."[103] Any analysis that seeks to understand the Reagan legacy will have to come to terms with the differences in foreign policy rhetoric between the first administration, which dismantled détente and reinvigorated the language of the cold war, and the second administration, which reopened and gradually broadened lines of communication with the Soviet Union. To be sure, Reagan rhetoric cannot be evaluated apart from the events and changes in the Soviet Union and the emergence of Gorbachev, a leader who played a vital role in creating a space where discussion between superpowers could be pursued. However, Reagan's engagements with Soviet leadership altered American public discourse concerning Russia and the Soviet Union. Thus, as William Hyland observed, Reagan "of all people" contributed significantly to "end the cold war."[104]

Not all critics are convinced that the language of the cold war is entirely moribund. Robert Ivie argues that the staple of cold war rhetoric is the portrayal of Soviet "savagery"—the Soviet Union as an alien, implacable foe kept at bay by nuclear constraints. He observes that the cold war has been characterized by "a cycle of adoption, extension, criticism, and reversal" of this image.[105] Could Reagan's summitry be simply one more spin of the wheel? Ivie seems to think so. He has written: "The central tendencies of Cold War rhetoric remain intact despite Mikhail Gorbachev's celebrated policies of *glasnost* and *perestroika*," and these tendencies are affirmed by Reagan's postsummit oratory.[106] And, as Cori Dauber shows, cold war rhetoric with its bipolarities, militarism, calls for global struggle, and authoritarian impulses still remains a rhetorical resource and a potent bureaucratic tool.[107] But at this summit wasn't there something new added, too?

The most popular phrase of the summit, repeated by both the president and the general secretary, was simply this: "A nuclear war cannot be won and must never be fought." In the early 1980s a rejuvenated cold war rhetoric had pushed nuclear dilemmas to the surface, to the very point of public legitimation.[108] Reagan-Gorbachev diplomacy worked to constrain nuclear utilization theories. True, leaders have avowed before that nuclear weapons are useless; what was new at the summit was joint public investiture in, and legitimation of, a system of rationalization setting in place a regime of data, information, and scientific exchange on core nuclear issues. The INF treaty did more than get rid of a "class" of nuclear weapons; it began to construct the machinery that makes nuclear control feasible. With nuclear anxieties detached from other, independent areas of disagreements, the Reagan-Gorbachev arguments could do more than articulate different stands in a struggle bound by ultimates; they showed that it was possible to have a productive mixture of positive and negative discourses in an international relationship between superpowers.[109] Thus, a superordinate interideological structure was established to contain nuclear development even while reciprocal criticism was featured as a way to advance United States–Soviet relations.[110]

These new "containment" principles are implicit in the summit structure itself; its disjointed agenda vitiates the integrative qualities of cold war rhetoric that amalgamate distinct areas of contact into hardened ideological opposition; that Reagan could criticize human rights while making progress on bilateral relations suggests that the new rhetoric, while carrying forward its own version of criticism in respective areas, structures opportunities for progress. Cold war symbols are still available as signs of difference, but such language does not dominate the horizons of discourse as it once did; and the dialogue of self-criticism spoken by Gorbachev and Reagan continues to prompt unexpected changes.

The speeches of the summit enact a broad range of discourse, articulating a daring mix of affiliation and criticism not characteristic of cold war or détente rhetorics. Reagan spoke about religious concerns (at Danilov), dissent (at the Spaso luncheon), the arts (at the House of Letters). He spoke to the social elite (at the Gorbachev and Reagan dinners), students and intellectuals (at Moscow State), arms controllers (at the INF signing), the military establishment (in the minor accords), the media (in television interviews), and some common folks. His discourse both praised and blamed the Soviet Union, even as it moved away from an implacable anti-Soviet stand and toward a criticism that encourages change. The performance, if too complex and demanding to be seamless, was nonetheless carried off well. To the end, Gorbachev played the anxious suitor pressing to go forward, faster; while Reagan stayed a skeptical listener to "friendly persuasion"—neither swept away by the moments of passionate appeal nor disappointed by the upsets and controversies of the moment.

As always, a key to Reagan's rhetorical performance was the personal touch, anecdotes and stories told in his effort to talk the language of the Russian people. Erickson, a rhetorical critic, complains that "the heroes and villains of Reagan's speeches are far from realistic; they are tools through which the speaker manipulates us by translating our complicated and varied lives into simplified stock, two-dimensional dramatis personae embodying virtue and voice."[111] It must be admitted that during the summit even these two dimensions sometimes collapsed. But are not two dimensions better than one, or none? Reagan personalized his experience, and it is reported that if the Soviet people did not always agree with the ideology, they were impressed with the effort.[112] The president, late in his administration, thus muted his hostility toward the Soviet Union with an avowed concern for its people, its culture, its future. He enacted a model for developing relationships distinct from cold-war constraints or Metternichian balance-of-power designs.

In a summary judgment of Reagan's rhetoric, Kathleen Hall Jamieson writes:

> Although it embodies what passes in an electronic age for eloquence, Reagan's rhetoric leaves much to be desired. A command of history enables a leader to learn from the past; a command of fact enables him to test the legitimacy of his inferences. Reagan's leaps from the single anecdote or personal experience to the general claim are sound only if his anecdote is

accurate, typical, and representative of a larger universe of experience. His subordination of fact to conviction and truth to ideological convenience ultimately cost him his credibility.[113]

Although Reagan showed the same slips of inference and anecdote at the summit that had hurt his "credibility" throughout the second administration, he nonetheless returned home to a 70 percent approval rating, quite near the peak of his popularity in the first administration.[114] Why the gap between rhetorical norm and rhetorical effect? Could it be that the summit was a spectacle that simply awed the American people, that its images were beheld without understanding? Perhaps, but there is an alternative explanation.

Jamieson is correct that Reagan was not an innovator in the sense of staking out bold new territory, but she is wrong that all the Reagan rhetoric did was to "revivify and reinforce" old values while failing to "advance our understanding of ourselves, our country, and its institutions."[115] In the end, Reagan was a controversialist, able to deploy symbols to provoke response. In the summit, for example, his most effective human rights discourse was to engender a dispute in the Soviet Union over whether soviet women were being properly honored. True, the values he uttered were old-fashioned and stuffed with patriotic gore, but the result of this discourse was to invite dialogue in a public discussion of differences. By provoking controversy, Reagan loosed oppositional discourses at home and abroad. His discussions of differences between the United States and the Soviet Union beckoned, yet retarded, a time and discourse beyond the cold war. If his versions of the past and future are not accepted as completely satisfactory, they do reassure some that America is still secure while inviting others to go farther, faster in pursuing realistic change. It is this heady mixture of motives—this simultaneous identification and difference, union and separation, care and indifference—that constitutes the legacy of Ronald Reagan's courtship of Gorbachev, the Soviet people, and the American public. Thus, a flawed and magnificent performance by an aging president made the future production of international discourse as complicated as the task of evaluating the Reagan legacy. Such were the speeches of that Moscow spring.

3

President Reagan at the London Guildhall

A British Interpretation

Robin Carter

On 3 June 1988 President Reagan delivered a thirty-minute speech to the Royal Institute of International Affairs at the London Guildhall; the president had stopped in London en route from Moscow to Washington. The occasion was, ostensibly, to "report" on his five-day visit to Moscow, a visit that had culminated in the ratification of the Treaty on Intermediate Nuclear Forces, with its implications for further nuclear disarmament. The ceremonies were introduced by the Lord Mayor of London, and Prime Minister Margaret Thatcher offered a five-minute reply to President Reagan. The president was accompanied by Mrs. Reagan, and among the invited guests were senior cabinet ministers, the Leader of the Opposition (Neil Kinnock of the Labour party, which at the time was committed to unilateral nuclear disarmament), at least two former prime ministers (Harold Wilson and James Callaghan, both of the Labour party), and twenty-four ambassadors of their respective countries. Fanfares and national anthems were provided by the band of the Grenadier Guards, and the host dignitaries wore their traditional robes and regalia; all involved entered in a formal procession. The event, which commenced around noon GMT, was televised live in the United Kingdom and transmitted by satellite to the United States.

According to Ian Brodie of the *Daily Telegraph* (4 June 1988), "The Guildhall address was compiled with input from the State Department and the American Embassy in London. The final version was crafted by Mr Tony Dolan, a White House speechwriter in his early thirties who had been sketching out draft versions for weeks. But Mr Reagan nagged his staff to get the tone right and even in Moscow they were preoccupied with changing the text in response to his suggestions and margin notes that flowed from the President's hand." In the event, President Reagan spoke with his customary oratorical expertise, using the techniques that habitually served him well: humor, self-deprecation, veneration, recollection, homily, political maxims, exhortation, and near-evangelism were interwoven with his treatment of several interrelated themes. The topics broached began with Britain's historical traditions, World War II, and the Anglo-American alliance. This led to an encomium of the postwar Western alliance, with particular attention being given to the new strategy of the 1980s, its opponents, and its perceived success; special mention was then made of Britain's and Mrs. Thatcher's contribution to that success. Those topics were all dealt with in the first third of the speech and were followed by a five-minute disquisition on the current situation in the Soviet Union. Some reflections on the importance and value of freedom, reinforced by a commentary on the 1944 Battle of Arnhem, prefaced the peroration, which provided a validation of Western beliefs via the Judeo-Christian tradition.

The purposes of this essay are as follows: to identify those elements of President Reagan's oratory that were specifically aimed at a British audience, and to demonstrate the skill with which the president touched the emotions of a definable (and important) sector of British opinion; to place these elements in the context of U.S.–U.K. relations; to indicate, where appropriate, relevance to the wider context of American foreign policy insofar as it pertains to Great Britain (in both contemporary and historical contexts); to show these factors—and others, such as his new attitude to the USSR—within the context of Reagan's rhetorical style; to give detailed attention to two aspects of the speech that seem rhetorically creative and original; and to comment briefly on Mrs. Thatcher's reply, and on immediate local reaction to the event.[1]

The two aspects that invite special consideration are Reagan's use of the positive connotations of the concept of "freedom" and his extensive reference to the Battle of Arnhem. With regard to freedom, I shall show how this concept was linked to almost every topic addressed, thus providing a unifying theme for the entire speech; I shall also suggest that the word *freedom* was useful as a device to represent and justify Reagan's (or his government's) own interests and beliefs, enabling him to discuss East-West differences in an impersonal and conciliatory way. With regard to the Battle of Arnhem, I shall argue that Reagan—very skillfully—explicitly praised Britain's military effort while, at the same time, implicitly reminding us of the shortcomings of British leadership.

I shall interpret the speech by exploring, in turn, the aforementioned extrinsic and intrinsic topics—that is, the historical context, the president's homage to Britain, Britain's place in American foreign policy, the president's references to the USSR, his appeals to religious conviction and to "the cause of freedom," and the Arnhem episode. Since this speech may not be widely known in the United States, I have included several lengthy quotations in order to illustrate the stylistic devices and rhetorical strategies to which I refer.

U.S.–U.K. Relations: The Historical Context

It needs to be stated straightaway that the historical context in which President Reagan offered his remarks, and to which he addressed himself very specifically, namely, the Anglo-American alliance, has rarely been straightforward. There are several conflicting aspects of this context. In the first place, we have the perception of the alliance, shared by many Britons, that is sentimental and reassuring; this view is frequently expressed in the phrase "special relationship." Secondly, there is the common American perception of the alliance, in which it assumes less importance, given the global interests of the United States and the relative insignificance of the United Kingdom within those interests. A third aspect is the political reality, which for over fifty years—according to contemporary historians—has been a function of unsentimental self-interest on both sides. In contrast to that is, fourth, the social reality, which consists of thousands of lifelong friendships between British and American citizens, as well as the "GI brides" of the 1940s (and, one might add, such interesting connections as Winston Churchill's American mother and Margaret Thatcher's American daughter-in-law). These bonds of genuine friendship give rise to another factor, namely, the propensity of politicians (especially on the British side) to try to make such friendships serve the aforementioned political interests. Finally, there was, in the 1980s, the relationship between President Reagan and Prime Minister Thatcher, which was widely perceived—to quote a recent study—as "one of the closest personal bonds in the history of Anglo-American co-operation."[2]

Given the Thatcher-Reagan relationship, it may safely be assumed that, when he spoke, the president could be confident of a sympathetic response from Mrs. Thatcher's constituency (that is, the 32 percent of the electorate who voted Conservative in the 1987 general election). But Reagan's wartime reminiscences reached a wider public than that, for, in Britain, the over-forties have grown up with either direct or transmitted experience of World War II, while succeeding generations have had ample opportunities for vicarious (if often implausible) engagement therein. This last fact is important, because a common British version of World War II—evident in books, comics, films, newspapers, and television—has been, for many years, sentimental and nationalistic: if we can draw inferences from the continuing popularity of these artifacts, many Britons are still susceptible to the emotional entailments of that conflict. Reagan's appeal to this response was therefore well judged (although not without an ironic twist, as I shall show).

There was, however, one contextual factor that might have had an adverse effect on the image of Anglo-American relations that both Reagan and Thatcher were trying to convey. For the seven weeks preceding the speech, BBC Television had been broadcasting a series of one-hour documentaries analyzing the history of Anglo-American relations; entitled "An Ocean Apart," the series was made by BBC-TV in conjunction with KCET-TV of Los Angeles. The emphasis of the entire series had been on the political realities mentioned above, and had seemed to demonstrate the self-interest of both countries in their twentieth-century dealings with each other. The final program in the series was broadcast *the night before* the Guildhall event. This final program dealt with Reagan-Thatcher relations and revealed how, once again, a British military success (the 1982 war in the South Atlantic) had owed a crucial debt to American expertise and hardware—a debt that was collected in April 1986, when U.S. aircraft flew from British bases to bomb Libya. The program ended—only hours before the two leaders were due to speak—with a summing-up to the effect that, however close their peoples might be, the realities of world politics meant that Britain and the United States would remain "an ocean apart."

Reagan's Homage to Britain

Reagan's speech began with acknowledgment of two elements for which many Britons regard themselves as distinctive: the antiquity of their culture and their wartime resolve. He did so with typical good humor, thus:

I wonder if you can imagine what it is for an American to stand in this place. Back in the States, we are terribly proud of anything more than a few hundred years old; some even see my election to the presidency as America's attempt to show our European cousins that we, too, have a regard to antiquity. Guildhall has been here since the 15th century and while it is comforting at my age to be near anything that much older than myself, the venerable age of this institution is hardly all that impresses. Who can come here and not think upon the moments these walls have seen: the many times the people of this city and nation have gathered here in national crisis or national triumph?

Reagan then recalled Ed Murrow's broadcasts from London during the 1940 Blitz, and went on, "From the Marne to El Alamein, to Arnhem, to the Falklands, you have in this century so often remained steadfast for what is right—and against what is wrong. You are a brave people and this land truly is, as your majestic, moving hymn proclaims, a 'land of hope and glory.'" Comments such as these are music to the ears of an older generation of Britons (even those outside Mrs. Thatcher's constituency), and the skill with which Reagan evoked their emotional response probably helped them to overlook the somewhat erroneous impression given of the Guildhall's place in British history.[3]

Reagan then employed one of the devices that gave his oratory a personal, even intimate touch. In an anecdote about his own first visit to Britain (in 1949, apparently), he described a conversation with the owner of a pub who had reminisced fondly about wartime American servicemen: "She was looking off into the distance and there were tears in her eyes. She said, 'Big strapping lads they was, from a place called Ioway.'" This anecdote provided the introduction for the major theme of Reagan's speech—his "forward strategy of freedom"—thus: "From a place called Ioway. And Oregon, California, Texas, New Jersey, Georgia. Here with other young men from Lancaster, Hampshire, Glasgow, and Dorset—all of them caught up in the terrible paradoxes of that time: that young men must wage war to end war; and die for freedom so that freedom itself might live. And it is those same two causes for which they fought and died—the cause of peace, the cause of freedom for all humanity—that still bring us, British and American, together."

References to Britain's past were thereafter absent until the medieval roots of British political institutions were adduced towards the end of the speech, when Reagan began to talk about his "prayerful recognition of what we are about as a civilisation and a people": "I mean, of course, the great civilised ideas that comprise so much of your heritage: the development of law embodied by your constitutional tradition, the idea of restraint on centralized power and individual rights as established in your Magna Carta, the idea of representative government as embodied by the mother of all parliaments."

Finally, at the end of his speech, the president "touched base" again, with a reminder of wartime Britain. This time, though, the reference was neatly linked with the present, uniting the major theme (the success of the global strategy), the minor theme (Britain's contribution thereto), a popular English poet, and God: "More than five decades ago, an American president told his generation that they had a rendezvous with destiny; at almost the same moment, a prime minister asked the British people for their finest hour. This rendezvous, this finest hour, is still upon us. Let us seek to do His will in all things, to stand for freedom, to speak for humanity. 'Come my friends,' as it was said of old by Tennyson, 'It is not too late to seek a newer world.' "

In the circumstances surrounding the oratory of Roosevelt and Churchill, to which Reagan referred, times were more parlous: The British army had just been defeated and evacuated from Dunkirk, and France itself had fallen to Hitler's Germany. Churchill's rallying cry was intended to stiffen the resolve of a people confronting the prelude to an attempted invasion; given that the British response was successful, Reagan's audience would have been disposed to pride and nostalgia at the reminder.

Britain's Contribution to America's Global Strategy

It will be obvious that the above tributes to Britain and its people might be seen as serving the foreign policy of the United States very well. To illustrate this comment, I quote from the concluding pages of *An Ocean Apart*:

> In the past Britain mattered to America as its front line of defence, its entry point to the continent of Europe and its ally in the containment of communism around the world. But Britain's power has waned, America's interests are shifting, and the relationship has become less important to the United States. . . . [America's] transatlantic commitment after World War Two sprang directly from its hostility to the Soviet Union. The decision to defend Western Europe against Russia was taken to ensure American, not European, security, for fear that the resources of one of the most important economic regions of the world would fall under Soviet control or influence. Out of it have come NATO, for ever in crisis but still the most durable peacetime alliance in modern history, and a tight bond with Britain, which serves as a vital base for air, submarine and missile defence, as well as for sophisticated intelligence gathering.
>
> This role of "unsinkable" aircraft carrier, which Britain has played for so long, could change if the process of detente with the Soviet Union continues and America's budget and trade crises are not solved.[4]

Clearly, that is but one interpretation of U.S. policy: others would, by contrast, stress the magnanimity of the Marshall Plan, and emphasize acknowledged historical and cultural ties. Equally clearly, though, the "unsinkable aircraft carrier" remains useful, as does the support of its leaders.

The latter point would constitute at least one good reason why Reagan was careful to give Britain credit for the perceived success of his foreign policy, as he proceeded to outline his perception of that policy. Defining the policy in terms of "the cause of peace, the cause of freedom for all humanity," the president named "Great Britain, the United States, and other allied nations" as he proclaimed that

> for them [the aforementioned causes], we embarked in this decade on a new post-war strategy, a forward strategy of freedom, a strategy of public candour about the moral and fundamental differences between statism and democracy but also a strategy of vigorous diplomatic engagement. A policy that rejects both the inevitability of war or the permanence of totalitarian rule; a policy based on realism that seeks not just treaties for treaties' sake but the recognition and resolution of fundamental differences with our adversaries.
>
> The pursuit of this policy has just now taken me to Moscow, and let me say: I believe this policy is bearing fruit.

Reagan then referred to opponents of these policies as "voices of retreat and hopelessness," and described certain subsequent events as "momentous." These "momentous events" were the INF Treaty, the changes in the Soviet Union, and its withdrawal from Afghanistan—all carefully linked, presumably in order to imply that all three, and not just the first, were the direct results of his policies.

At this point, Reagan gave great prominence to his acknowledgment of the British—particularly Mrs. Thatcher's—contribution:

> And here I want to say that through all the troubles of the last decade, one . . . firm, eloquent voice, a voice that proclaimed loudly the cause of the Western alliance and human freedom, has been heard. . . . So let me discharge my first official duty here today. Prime Minister, the achievements of the Moscow summit as well as the Geneva and Washington summits say much about your valor and strength and, by virtue of the office you hold, that of the British people. So let me say, simply: At this hour in history, Prime Minister, the entire world salutes you and your gallant people and gallant nation. And while your leadership and the vision of the British people have been an inspiration, not just to my own people but to all of those who love freedom and yearn for peace, I know you join me in a deep sense of gratitude toward the leaders and peoples of all the democratic allies.

It should be noted that, in the above paragraph, Reagan *three times* asserted a link between the prime minister and the British people—

a link which, since it implied reference to *all* the British people, was formally illicit: as Reagan himself had just observed, there had been, throughout much of the 1980s, a "crescendo" of opposition to the Reagan-Thatcher policies. Indeed, the Labour party's defeats in the general elections of both 1983 and 1987 were partly attributable to their espousal of contrary policies. Assuming that in 1988 it was still U.S. policy to regard Britain as a useful ally, this policy would obviously be better served if the "British people" could be identified with the Conservative party that formed the government, rather than with the then-antinuclear Labour party.

It will also be apparent that the president's praise of the British people and their leaders (past and present) was intertwined with the aforementioned vindication of his own policies. The rhetorical effect of such a strategy is very powerful: any British listener, whether in the Guildhall or watching television, who—like, say, Opposition Leader Neil Kinnock—might wish to dispute Reagan's analysis, would first have to disentangle strands whose knots are tightly tied. Until one does so, rejecting Reagan's claims (that his, Mrs. Thatcher's, and NATO's policies have enhanced the prospects of peace, freedom, and so on) courts the danger of apparently rejecting Reagan's premises (which include his admiration of British institutions, history, ideology, and heroism). So closely did he identify these several propositions that it would be a brave opponent who tried to dispute them publicly; indeed, Mr. Kinnock himself, in a BBC radio interview less than an hour after the speech, politely gave Reagan credit for the progress in arms reduction and even described the president's praise of Mrs. Thatcher as "a generous and gentlemanly thing to do." It was left to some of the journalists to provide more skeptical comment.

Reagan's Observations on the Soviet Union

The topic of the Soviet Union was introduced after the tribute to Mrs. Thatcher and began with an announcement of the ratification of the INF Treaty. After listing three other items on the agenda, namely, regional conflicts, human rights, and bilateral exchanges, Reagan went on to express his interest and satisfaction concerning the changes in Russia, and recalled the positive and optimistic signs there that had given him encouragement. Again, the style was homey and personal: "To those of us familiar with the post-war era, all of this is cause for shaking the head in wonder. Imagine, the President of the United States and the General Secretary of the Soviet Union walking together in Red Square talking about a growing personal friendship and meeting, together, average citizens, realizing how much our people have in common."

However, Reagan was careful to qualify his pleasure with a warning: "Let us embrace honest change when it occurs; but let us also be wary. And ever vigilant. Let us stay strong." At this point, Reagan began the disquisition on the "crusade for freedom" that is given in detail below. That, in turn, was followed by his reflections on the Battle of Arnhem.

The Importance of Religious Belief

The comments about Arnhem ended with a quotation from one of the battle's surviving veterans (Colonel John Frost), to the effect that he regarded his annual reunion with other Arnhem veterans as a "pilgrimage." This provided an effective link to Reagan's concluding passage, in which he reminded us of the fundamental importance to our civilization of its Judeo-Christian basis and tradition: "As those veterans of Arnhem view their time, so, too, we must view ours; ours is also a pilgrimage, a pilgrimage toward all those things we honour and love: human dignity, the hope of freedom for all peoples and for all nations. And I have always cherished the belief that all of history is such a pilgrimage and that our maker, while never denying us free will, does over time guide us with a wise and provident hand. . . . I cherish, too, the hope that what we have done together throughout this decade and in Moscow this week has helped bring mankind along the road of that pilgrimage."

The closing minutes of the speech were given to affirmations of Judeo-Christian faith—with quotations from Evelyn Waugh and (via the film *Chariots of Fire*) the Scottish athlete Eric Liddell—and the power of prayer. In the final lines of the speech (which have already been quoted in the section on Reagan's homage to Britain, above), all these values were fused with an Anglo-American harmony represented by President Franklin D. Roosevelt and Prime Minister Winston Churchill, which, in turn, was linked with the Divine Will.

"Freedom": The Rhetorical Binding

Although, for the sake of analysis, I have separated the topics broached in the speech, this does a disservice to the skill with which the said topics were all woven together. Reagan moved effortlessly from past points of reference to present points, and, as has been seen, from past quotations to present circumstances. Almost invariably, the connections were made via the invocation of "freedom": the emphasis on the value of liberty provided the unifying theme of the speech and regularly served to bind together events and persons that are separated by space and time. In the thirty-minute speech, the words *free*, *freedom*, and *liberty* were used forty times; in one three-minute section they were used no fewer than eighteen times.

From the information available to me, the invocation of freedom seems to have been a regular stylistic feature of Reagan's public address.[5] Liberty is, of course, a notably difficult concept to define, and a useful rhetorical counter for that reason.[6] In London the tone was set in the second minute of Reagan's speech: "After a long journey, we feel among friends. . . . I have come from Moscow to report to you, for truly the relationship between the United States and Great Britain has been critical to NATO's success and the cause of freedom." As usual, the sense in which the word *freedom* was being used was not defined. However, Reagan often used the word to mean "nontotalitarianism,"[7] and this meaning certainly seems appropriate for many of the contexts denoted in the speech under discussion.

The noticeable features of Reagan's use of *freedom* in this speech are the way in which it was presented as the concept that inspires and justifies virtually every policy and every achievement, and the way in which it was used to link the topics addressed. These functions can be shown by examining the presentation of the topics listed at the beginning of this essay. The first example of both justification and linking can be seen in two passages from which I have already quoted—one about wartime Britain, the other about postwar strategy. The sentences that linked these past and present references bear repeating:

> Young men must wage war to end war; and die for freedom so that freedom itself might live.
>
> And it is those same two causes for which they fought and died—the cause of peace, the cause of freedom for all humanity—that still bring us, British and American, together.
>
> For these causes, the people of Great Britain, the United States, and other allied nations have, for 44 years, made enormous sacrifices to keep our alliance strong and our military ready.

It would not be hard to find objections to these assertions. One might be that it is easy for noncombatants to proclaim that "young men must wage war to end war"; another, that this assertion is manifestly self-contradictory: it could only be sustained if human history had never witnessed more than one war. More striking, though, is the way in which Reagan *reduced* a plethora of complex motives and circumstances to one simple formula: he wished to persuade us that those men fought and died for "the cause of freedom for all humanity"—not, obviously, from fear and loathing, or national self-interest, or because they were conscripted, or ambitious, or patriotic, or unemployed, or the victims of political and/or military incompetence. This simplistic interpretation of cause and effect, by way of a two-word rationale—"for freedom"—was a device that served Reagan well throughout the speech.

Some of the other topics that were linked by reference to "freedom" have already been quoted—for example, the allied nations' "forward strategy of freedom," Mrs. Thatcher's proud proclamation of "the cause of the Western alliance and human freedom," and the British people's "inspiration . . . to all of those who love freedom and yearn for peace." This last reference became the framework within which Reagan began to discuss his visit to Russia: "Rarely in history has any alliance of free nations acted with such firmness and dispatch, and on so many fronts. In a process reaching back to the founding of NATO and the Common Market . . . [the Western Allies created] a startling growth of democratic institutions and free markets all across the globe: in short, an expansion of the frontiers of freedom and a lessening of the chances of war. So it is within this context that I report now on events in Moscow. On Wednesday, at 8.20 Greenwich time, Mr Gorbachev and I exchanged the instruments of ratification of the I.N.F. Treaty." The phrase "within this context" makes it explicit that Reagan wished to connect "freedom" with successful arms control. After referring to "signs of greater individual freedom," he concluded his reminiscences about Russia by recalling "the faces of hope, the hope of a new era in human history, and, hopefully, an era of peace and freedom for all."

The president proceeded to remind us that Western strategy would continue to require wariness, vigilance, and strength, and then presented his reflections on the philosophical context of all these events. This was the section with the most concentrated discourse on liberty, and, in order to convey the full impact of the argument, it needs to be quoted in its entirety. Recalling his words to the British Parliament in 1982,[8] Reagan reaffirmed his opposition to "totalitarian rule," and went on:

> Noting the economic difficulties reaching the critical stage in the Soviet Union and Eastern Europe, I said that at other times in history the ruling elites had faced such situations and, when they encountered resolve and

determination from free nations, decided to loosen their grip. It was then I suggested that the tides of history were running in the cause of liberty, but only if we, as free men and women, joined together in a worldwide movement toward democracy, a crusade for freedom, a crusade that would be not so much a struggle of armed might—not so much a test of bombs and rockets as a test of faith and will.

Well, that crusade for freedom, that crusade for peace, is well under way. We have found the will. We have held fast to the faith. And, whatever happens, whatever triumphs or disappoinments ahead, we must keep to this strategy of strength and candour, this strategy of hope—hope in the eventual triumph of freedom.

But as we move forward, let us not fail to note the lessons we have learned along the way in developing our strategy. We have learned the first objective of the adversaries of freedom is to make free nations question their own faith in freedom, to make us think that adhering to our principles and speaking out against human rights abuses or foreign aggression is somehow an act of belligerence. Over the long run, such inhibitions make free peoples silent and ultimately half-hearted about their cause. This is the first and most important defeat free nations can ever suffer. For when free peoples cease telling the truth about and to their adversaries, they cease telling the truth to themselves. In matters of state, unless the truth be spoken, it ceases to exist.

It is in this sense that the best indicator of how much we care about freedom is what we say about freedom; it is in this sense that words truly are actions. And there is one added and quite extraordinary benefit to this sort of realism and public candour: this is also the best way to avoid war or conflict. Too often in the past, the adversaries of freedom forgot the reserves of strength and resolve among free peoples, too often they interpreted conciliatory words as weakness, too often they miscalculated—and underestimated the willingness of free men and women to resist to the end. Words of freedom remind them otherwise.

It would be easy to criticize the above passage on the grounds that the constant repetition of *free* and *freedom* drains those words of their meaning and reduces their impact. It seems to me, however, that such a criticism would miss the point, which is that those words can sometimes be seen to function here as substitutes for personal pronouns (*I*, *we*, *us*) and possessive adjectives (*my*, *our*): this has the effect of making the discourse less assertive and combative than it might be otherwise. For example, if we compare a statement that Reagan *did* make, namely, "Too often in the past, the adversaries of freedom forgot the reserves of strength and resolve among free peoples," with an equivalent one that he *might* have made, such as, "Too often in the past, our adversaries forgot our reserves of strength and resolve," it will become apparent that the second rendering of the proposition is rather less tactful and diplomatic than the first. In addition, the replacement of first-person forms by third-person forms *whose subject is a philosophical/ political concept that suggests a humanistic idealism*, conveys an altruism and a detachment that might otherwise be lacking. Even Communist ideologies include theories of freedom, albeit different from "Western" concepts.

The reason for this diplomatic tone is, perhaps, to be found in the events of the previous few days in Moscow. One headline-worthy development had been Reagan's withdrawal of his "Evil Empire" stigmatization of the Soviet Union. The retraction was reported on 1 June 1988, amid predictions that the Soviet authorities would be extremely gratified by Reagan's recognition of the changes in Soviet society. Thus, the avoidance of provocative public statements became appropriate, and the sense of historical detachment conveyed by the passage quoted seems to have provided a suitable style: by shifting the burden of the argument from I/we to freedom, the argument seems less self-interested and thus more generally acceptable.

While achieving this softer effect, the passage quoted from Reagan's speech also exemplifies, at length, the aforementioned technique of reducing a complicated world to a simple, but heroic, formula. When considered as such, we can see that Reagan offered a fairly conventional argument whose rationale might well be universally applicable. To test this proposition, it is instructive to imagine the above passage being delivered, perhaps in other times or contexts, by other people using different key words: thus, for example, Philip II of Spain might have conveyed a similar message by substituting the words *Catholic* and *the Faith* for *free* and *freedom*. It would be contentious to suggest specific contemporary examples.

The simple-but-heroic device was repeated yet again in the following section, when, moving on to his reflections on the Battle of Arnhem, Reagan once more connected past and present, World War II and modern Russia, using freedom as a link. Reagan relayed a story of how Cornelius Ryan, while engaged in the writing of *A Bridge Too Far* (the best-known history of the Arnhem episode), had been moved to tears by the heroism of the above-mentioned Colonel Frost: "The writer could only look up and say [to his wife] . . . 'Honestly, what that man went through . . .' A few days ago seated here in Spaso House with Soviet dissidents, I had that same thought, and asked myself, What won't men suffer for freedom?"

It is mere pedantry to point out that, strictly speaking, the latter was not "that same thought"? Colonel Frost and his men may, or may not, have had the defense of freedom on their minds through nine days of battle, slaughter, hunger, and thirst: Ryan does not tell us. The flaw in Reagan's argument linking heroism and freedom is that, as we know to our cost, the "enemies of freedom" are also capable of heroism and suffering—like the German paratroopers defending Monte Cassino in 1944, who similarly endured deprivation and destruction while withstanding three months of land and air bombardment. Reagan's rhetorical skill lay in his suppression of such odious comparisons by his insistence on identifying parallel motives linking Colonel Frost, Soviet dissidents, and himself—all united in the defense of freedom.

To conclude the theme, the "cause of freedom" was overtly fused with Reagan's proclamation of Judeo-Christian belief and inspiration: phrases such as "crusade for freedom" and "His will" reinforced the message. Thus, a quotation from the Book of Isaiah (chapter 40, verses 29–31, via the film *Chariots of Fire*) provided "our formula for completing our crusade for freedom," which was given further religious authority through a reminder that "our faith is in a higher law." As noted earlier in this essay, Reagan ended the speech by once more drawing together Britain in 1940, the United States in 1941, and both countries in the present day, implying divine guidance of his ideology and policies: "Let us seek to do His will in all things, to stand for freedom, to speak for humanity."

Although many of Reagan's speeches included references to freedom and the invocation of liberty as a device to elicit support, none that I have seen offer the consistency of the "freedom" theme that was apparent throughout this London Guildhall speech, much less the skillfully created structure described above. However much Reagan's propositions about freedom may beg questions from logic, philosophy, or politics, as *rhetoric* they seem to me to have been well-placed and effective. To a British audience, exposed to nine years of governmental assertions of increased liberty, they had a familiar ring.

The Battle of Arnhem: An Example of Rhetorical Subtlety

Although later public comment failed to mention them, questions are raised by Reagan's reference to the Battle of Arnhem. It is when we address these questions that we may be able to identify an additional dimension to Reagan's rhetoric, a dimension that suggests considerable subtlety beneath the simple but effective argumentation that I have described thus far.

Reagan devoted a few minutes toward the end of his speech to the Battle of Arnhem, and he used his familiar technique of relating conversations concerning individuals—although, in this case, the source was a newspaper article, rather than Reagan's own experience. The individuals were the aforementioned Colonel John Frost and the American historian Cornelius Ryan. The topic was wartime courage—of the highest order. Reagan showed some awareness of the detail of the Arnhem operation, and specifically cited the title of Ryan's book, *A Bridge Too Far*. It is difficult to imagine that anyone who has read the book, or seen the film, can fail to be riveted, and emotionally affected, by the courage and the carnage that are described and depicted. As Ryan tells us, "Allied forces suffered more casualties in Market Garden [the Arnhem operation] than in the mammoth invasion of Normandy . . . combined losses—airborne and ground forces—in killed, wounded and missing amounted to more than 17,000. British casualties were the highest: 13,226."[9]

The first question is, Why Arnhem? Why did Reagan choose to give such serious attention to an operation that ended in the catastrophic defeat of the allied forces? There were, in World War II, other Anglo-American and Allied operations that were more successful: some that immediately come to mind are D-Day, the North African campaign, the Italian campaign, the Battle of the Atlantic, the tactical bombing campaign against V-weapon sites, and the exchange of vital intelligence. Although there were tensions and conflicts, it was "the closest wartime alliance in modern history."[10] So—why Arnhem?

To answer that question, it might help to ask another: Why did Operation Market Garden go wrong? The answer emerges clearly from Ryan's book: the airborne operation, conceived and (largely) executed by the British, was flawed by serious misjudgments and miscalculations on the part of the British planners. Responsibility for these errors must rest with Field-Marshal Montgomery and Lieutenant-General Frederick Browning, who refused to believe that there would be serious German resistance—even after being shown photographs of Panzer tanks near Arnhem; these photographs, brought back by low-flying reconnaissance Spitfires, confirmed intelligence reports from agents on the ground. General Eisenhower and his chief of staff then had serious reservations about the plan but, for political reasons, felt unable to interfere. The sequence of these events is described in detail in part 2, chapters 1 and 5, of Ryan's book, and it is brought out vividly in the film (which was shown by BBC-TV on 19 September 1987—only nine months before Reagan's visit).

Reagan's reference to Arnhem was thus, perhaps, a reminder to the British, especially to the British "establishment," that they are quite capable of monumental incompetence. Whether we consider cataclysmic events like World Wars I and II, less destructive episodes such as Suez, or mere confidence-breakers such as the betrayal of security, Americans can point to a long record of damaging British (and European) follies—many of which the Americans have had to pay for. Furthermore, genuine bad feeling between Allied leaders persisted long after World War II, particularly between Eisenhower and Montgomery: British cabinet papers reveal that, as late as April 1959, Prime Minister Harold Macmillan felt it necessary to mollify the then-President Eisenhower after Montgomery made some tactless comments in a television interview. It would not be surprising if the White House used the 1988 occasion to remind the British establishment of the *real* nature of the alliance. Of course, it would be idle to pretend that U.S. governments and their armed services cannot compete in folly, and, conversely, we should not detract from Britain's genuine accomplishments, but all that is beside the point: the point is that, in emergencies, Great Britain *needs* American support, and that, on this occasion, Reagan had the floor.

If this interpretation of Reagan's Arnhem reference has validity, it reveals an impressive degree of skill on the part of Reagan and his speechwriters, for he conveyed his message *without ever making it explicit*. In fact, he conveyed *two* quite different messages simultaneously: one, the overt message, told the British audience how wonderful they were; the other, the covert message, suggested that, without American know-how, their policies had far less chance of success. Furthermore, this double message was targeted with unerring accuracy. Those (particularly those in senior positions) who were interested enough to *know* about Arnhem, and who saw the point, had no choice but to accept the implied reminder of their fallibility: they had to "grin and bear it." Those who heard only laudatory remarks—which were probably quite genuine—about their countrymen's valor, would have been happy to accept the tribute at face value. The four-minute passage seems to me to be a dazzling example of oratorical irony.

There are those who would maintain that Reagan lacked the subtlety necessary for this kind of finesse: after all, even Mrs. Thatcher, one of his most fervent public admirers, is quoted as saying of the former president, "Poor dear, there's nothing between his ears."[11] It might also be said that Reagan would not have been aware of all the historical details of the Battle of Arnhem—although, even if he had not read the book *A Bridge Too Far*, he may well have seen the film, wherein the failings of the British high command are vividly presented. The inescapable point, though, is Reagan's direct reference to Cornelius Ryan and his book, and the prominence given them in the speech, which, when carefully pondered, make it difficult not to take the hint. As far as Reagan's abilities are concerned, it should not be forgotten that, as was noted at the beginning of this essay, both the U.S. State Department and the U.S. Embassy in London contributed to the speech. Objections based on Reagan's (perceived) lack of knowledge or subtlety would fail to take account of the fact that Reagan had talented advisers and speechwriters, who obviously could have provided both the stylistic and the historical ironies.

Mrs. Thatcher's Reply

The Guildhall event should not be understood as being only about President Reagan. Although his speech was the main item, its content, and the circumstances, make it clear that the focus of attention was also on Prime Minister Margaret Thatcher; indeed, the president's tribute to her, and the ceremony itself, made the occasion one of the public high points of Mrs. Thatcher's career. Hence, our analysis of the context would not be complete without a brief special reference to her contribution and her relationship with President Reagan.

Mrs. Thatcher's reply—only five minutes in length—took the form of a tribute to Mr. Reagan's achievements as president, and she confirmed the themes of friendship and freedom that Reagan had interwoven. She summed up this message in a quotation: "I recall what Winston Churchill once wrote: 'Where we are able to stand together and to work together for righteous causes, we shall always be thankful and the world will always be free.'" Her personal compliments were no less enthusiastic than Reagan's had been to her: "Your personal courage, your gentle humour and your spirit of optimism are all part of the special quality which you have brought to the presidency. Above all—and in this you have done the greatest possible service not only to your own people but to free people everywhere—you have restored faith in the American dream, a dream of boundless opportunity built on enterprise, individual effort and personal generosity." These comments, and others like them, might have seemed unnecessarily effusive for a transaction between elected heads of state. To so judge them, however, would be to overlook the political context within which Mrs. Thatcher had to operate.

From what we know of Anglo-American relations, aspects of which I have conveyed above, we may safely assume that the prime minister was most anxious to impress her American audience, who, we may recall, were able to watch this by satellite relay. Her interest might have been personal as well as national: the debt owed by Mrs. Thatcher to the United States should not be underestimated. I pointed out at the beginning of this essay that American technical and logistical support was crucial to Britain's defeat of Argentina in 1982. Given the agreement of most analysts that Mrs. Thatcher's convincing general election victory in 1983 was partly attributable to the "Falklands Factor," this is tantamount to saying that the U.S. government was instrumental in her reelection. It is understandable that Mrs. Thatcher should want to show her appreciation even if, to British viewers, her valediction—*pace* Irving Berlin—sounded rather fulsome: "My Lord Mayor, I believe everyone present in this Guildhall will join me in saying: 'Thank you, Mr. President! Thank you for the summit, thank you for your presidency, thank you for your testament of belief and God Bless America!' "

The significance of the American audience was not lost on the *Guardian* reporter (Michael White), who observed in his report the following morning that "Mrs. Thatcher . . . had arranged a magnificent opportunity for her political soul-mate to deliver his personal report on the summit—on time for television breakfast shows at home. It provided the perfect photo-opportunity with its robed aldermen, golden heralds, the ornate Gothic chamber . . . and other quaint British tourist props." That said, it was probably rather unkind of Colin Brown, in the 4 June *Independent*, to describe Mrs. Thatcher's flat-topped wide-brimmed white hat as "a white stetson that was clearly meant to make Reagan feel among friends."

Reactions and Summing Up

To many Britons, Reagan's oration would have been moving and flattering: a population conscious of a long period of relative decline is bound to feel pleasure when complimented by the "leader of the free world."[12] Certainly, the radio and television commentators who covered the speech were impressed by the effusions of respect for Mrs. Thatcher.

Later press comment (4 June) varied according to the newspapers' different affiliations. The secondary headline on the front page of the pro-Conservative *Daily Telegraph* read "Guildhall praise for Britain's role," and the report concluded that "[Reagan's] voice is softer now and there are days when the relentless duties of his great office weary him. But he rose to the occasion." The professedly neutral *Independent*, in an editorial contemplating Mrs. Thatcher's relations with both Mr. Reagan and Mr. Gorbachev, noted that "it is tempting for those who have mourned the post-imperial waning of

Britain's influence in the world to feel a sense of mild euphoria at these developments"—but warned against such optimism. The center-left *Guardian* seemed the least impressed, Michael White going so far as to criticize Reagan's argumentation: "History, as well as Lords Wilson and Callaghan who were in the audience, may dispute their [Reagan's and Thatcher's] verdict that it was they who brought forth Mr. Gorbachev and the opportunities he represents, rather than the Politburo"; for that reporter, the importance of the speech lay in its implications for U.S. politics, in that Reagan, by supporting Mr. Gorbachev, had made it "virtually impossible" for the American Right to revive the cold war. The *Times* noted that "some observers found [Reagan's speech] embarrassingly over-generous," and suggested that the president wanted to reinforce Mrs. Thatcher's status as the U.S. election approached—partly to ensure that continuity of policy was perceived by the Western allies.

The most thoughtful interpretative comments on 4 June were those of Stewart Fleming, in the *Financial Times*, who observed the diplomatic tone of the president's speech: "By once again avoiding comments which could be regarded as harshly critical of Moscow, the President seemed intent on continuing his own 'peace offensive.'" The reasons given for this were "to try to counter M. Gorbachev's popularity in Western Europe and to demonstrate that the U.S.—and not just Moscow—is committed to the improvement of East-West relations"; this improvement would have the added benefit of helping the Republican candidate in November's election. Fleming was also conscious of Reagan's emotive power: "He once again displayed his remarkable ability to seize on the atmosphere of an event and shape it to his own political purposes. In a sometimes moving address he conjured up an atmosphere of common purpose within the West, lavished praise on his British audience and on Mrs. Thatcher." My analysis of the speech shows that the theme of freedom was a key factor in achieving these various political and sentimental effects.

As may be inferred from the public comments quoted above, the stylistic features of the president's speech that I have examined—personal involvement, individual anecdotes, emotional appeals, and the redefinition of most issues in terms of "freedom"—could be as effective in Britain as in the United States; the only significant difference might be that there would not be such a ready audience for the quasi-evangelistic topics as would be found in the United States. As a feature of the U.S. government's foreign policy, it is easy to see how the maintenance of U.S.–U.K. solidarity and cooperation could be a useful byproduct of the president's very positive affirmation of admiration and friendship for Britain and its prime minister.[13]

Part II

The Style of Reagan's Discourse

4

Acting like a President;

or, What Has Ronald Reagan Done to Political Speaking?

J. Jeffery Auer

It is appropriate to title this essay in the two-level style of old-fashioned theatrical melodrama because it will be a melodramatic account of what sometimes has seemed the longest-playing theatrical performance of our times—the primary campaigns, nominations, election campaigns, and presidential tenure of Ronald Reagan.

Amateur and professional theater have contributed greatly to our national culture, but we should be troubled today by the degree to which actors and acting have influenced contemporary political speaking, and by the reality that underlies everyday references to "political theater."

Max Lerner, for example, asserts that "modern politics is shot through with the theatric, and can be understood best only if we view the exchange between political actor and political audience as theater."[1] Perhaps most troubling is that when Lerner made that observation in 1972 the metaphor had an element of cuteness about it, but eighteen years and five presidential campaigns later it is no longer only metaphorical. It now accurately describes reality.

Any longtime student of presidential communication must come reluctantly to the conclusion that Ronald Reagan, while acting as president, has substantially expanded the theatric elements of national campaigning and governing. In doing so he has created a hazard for political discourse, and it's time to shout "Fire!" in the theater of politics.

This essay is intended to perform such a public service by inquiring into the contemporary exercise of presidential power through public communication as it has culminated in the behavior of Ronald Reagan.

We are far removed today from the early national period when presidential power derived largely from personal alliances and small cabals of fellow members of the national legislature; from the post-Jacksonian period and the emergence of national political parties, when presidential power was closely related to the management of party patronage; and from the Roosevelt period, when presidential power became for the first time dependent upon the support of an amalgam of special groups. By 1960 Richard Neustadt could write, in what became a handbook for many generations, that presidential power came from the power to persuade and to bargain with leaders of often diverse constituencies.[2] But after the Kennedy campaign managers exhibited their refinement of techniques for interpreting public opinion poll data, and using it in speechwriting for the candidate, the focus of presidential power shifted to the mobilization of broad public support as against reliance upon coalitions of special interest groups or upon traditional party machinery. Indeed, our last two presidents have been almost as much *outside* of their parties as they have claimed to be "outsiders" in Washington.

As we moved through the 1980s, the chief lesson in all this appears to be that influencing public opinion by broadly based appeals and image building has become not only a way of campaigning but a way of governing. Sidney Blumenthal capsulized this effect by calling it "the permanent campaign," and Kathleen Hall Jamieson described it in a study entitled *Eloquence in an Electronic Age*.[3] This writer's perception is reflected by titling this essay "Acting like a President." The White House, in our day, has become the best-known stage in America, and perhaps the world, because it was occupied for eight years by a consummate actor.

So closely are Washington, the White House, and the presidency associated through the quantity, if not the quality, of the incumbent's rhetoric that when a telemanic Ronald Reagan was succeeded by a lower-keyed George Bush, Washington's dominance of the nation's news focus was decreased. In 1989, for example, the Center for the Media and Public Affairs reported only one-third as many evening network news stories on Bush personally as there were on Reagan in 1981.[4] Alistair Stewart, the British television news anchor in Washington, politely observed that Bush is "a very cautious and reactive president . . . he doesn't like to lead from the front."[5] However the Bush reticence ultimately plays out, the "bully pulpit" is commonly enhanced and the preacher is inspired in times of war. Thus, even his January 1991 State of the Union address earned Bush the judgment of "superb performance" from David Broder,[6] though its success may be due to the reading of Reagan rhetoric by the Bush speechwriters. That the performing arts of the presidency as practiced by Ronald Reagan may live on is not the least of the reasons for reviewing them here.

An Overview of Ronald Reagan as a Political Actor

In his thoughtful book on the 1980 campaign for the presidency, John F. Stacks, *Time* magazine's chief political correspondent, observed that "it is not an accident that a movie actor has become the President of the United States. It was almost necessary. We have so altered the way we choose our presidents that political theater has overwhelmed the more traditional aspects of our leadership selection." Stacks reminds us that once upon a time the "keys to advancement up the ladder of American politics" were "legislation written, favors done, political debts accumulated, friends made in party and government circles, faithful service and devotion to the political party." But "they are no more. Now the key is television."[7]

While Stacks is surely correct in suggesting that political theater has overwhelmed the more traditional aspects of presidential selection, the fact is that American politics, and especially presidential politics, has always had substantial elements of the theatric about it. The influence of television has been to magnify and make more potent what has long since been theatrical. Thus it is, indeed, no accident that a movie actor has become president of the United States. As John O'Toole succinctly reminds us, "Ronald Reagan isn't our first actor president; he's just the first one with screen credits."

The truth is that most politicians spend their lives learning how to be convincing actors. But in Ronald Reagan's case we have an actor who has spent some years learning to be a politician. In him the arts merge.

Reagan recalls in his autobiography that in 1965, some months after his memorable campaign speech for Goldwater, he was asked about entering politics and replied that he didn't want to be in government but "just wanted to keep making speeches about it." "I'm an actor, not a politician . . . I'm in show business."[8] But when the question was put another way, and he was asked whether an actor could become governor of California, it was logical for him to reply, "I don't know—I've never played a governor before."[9] And how far off the pattern was he after trying on the part, when in 1966 he told Stuart Spencer that "politics is just like show business. You need a big opening. Then you coast for a while. Then you need a big finish."[10] Ten years later, during his run at the presidency in 1976, he was joking that "professional politicians like to talk about the value of experience in government. Nuts! The only experience you gain in politics is how to be political."[11] Is it any wonder that on 4 January 1985 a comic-strip character, the army private Beetle Bailey, began a conversation with Sarge by asking, "If you could pick anyone to be president after Reagan, who would you pick?"

"Well, let's see," said Sarge, "maybe Senator. . . ."

"No, no," interrupted Beetle, "I mean what ACTOR?"

Paul Duke, moderator of "Washington Week in Review," reported that Reagan refused to take the bait when asked if he felt that his training as an actor enabled him to become a better politician.[12] But his friend Bob Hope (the ultimate judge?) asserted on a 16 May 1988 NBC special broadcast that Reagan is a better actor now than when he was in Hollywood.

Paul Duke notwithstanding, under some circumstances Ronald Reagan confirmed his reliance on his acting ability. In May 1984 he and Nancy visited China, and while there he made a speech to the student body at Fudan University in Shanghai. The students were still talking about it two months later, and there was a detailed report in an English-language magazine. In the question period after his speech the American president was asked what collegiate work had proved helpful to him in his political career. "Well," he replied, "of course I was an economics major. But you'd be surprised at how helpful my courses in acting have been."[13] Four years later in a final television interview (22 December 1988) he responded to David Brinkley's question about how anyone could handle the presidency by wondering "how you could do this job and not be an actor."

During the summit meeting in Moscow, President Reagan's staff created an opportunity for him to address a group of intellectuals and artists. "In looking back," he told them, "I believe that acting did help me prepare for the work I do now." He said he learned "two indispensable lessons." Most important is "to have the vision. . . . You must grasp and hold it. You must see and feel what you are thinking. . . . You must hold and fix it in your memory and senses." Then he concluded the point: "To grasp and hold a vision, to fix it in your senses, that is the very essence, I believe, of successful leadership."[14] The second lesson, Reagan went on, in language appropriate to the Actors' Workshop, is, "You get inside a character, a place, and a moment."[15]

It is clear that our last president thinks in visual terms, in graphic symbols, and that he believes a critical role in his political career has been played by his skills as an actor.[16] Is it any wonder that a *New York Times* correspondent, Steven Roberts, reported that in Moscow "the visual images were at least as important as the words that were spoken" (in "14 Reagan speeches," added Jack Nelson), and that it should go down in history as "the photo opportunity summit"?[17] And is it at all surprising that Reagan's press secretary, Marlin Fitzwater, acknowledged that creating the "photo ops" was a primary concern in planning the trip to the Russian capital?

On the informative PBS series of 3 January 1988 weekly "Campaigning on Cue" broadcasts, Britt Hume of ABC News argued that appearances on television by presidents and presidential candidates were important to the viewing public because they permitted, in his words, "a demonstration of indispensable qualities." Among these, we inferred, was how well the speaker could hold up under stress; and we were further to infer that one test was his or her spontaneity. Lest we be misled by this notion, let us understand that our presidents and presidential candidates practice their apparent spontaneity. Much of the time what we see is contrived and rehearsed behavior that is so well perfected that it seems brilliantly spur-of-the-moment. Consider two examples.

The first one is the well-remembered Reagan line, addressed to Carter in their 1980 debate: "There you go again." Although most viewers may have applauded what they thought was an instantaneous put-down, and Ronald Reagan even assured one journalist later that it was impromptu, this was just not true. In fact, it was reported in print by Myles Martel, one of Reagan's debate coaches, and confirmed by reporters who were present, that the line was carefuly crafted and repeatedly rehearsed. In the simulated television studio in the oversized Reagan garage, two days before the debate, when David Stockman and Martin Anderson stood in for Jimmy Carter as they role-played for the real thing, it was decided that some way had to be found for Reagan to turn Carter off if he pressed potentially dangerous charges too hard. At the same time, of course, the deflecting quip had to be respectful and not make Reagan seem strident. The result was "There you go again."[18] It was a "theatrical coup," wrote Elizabeth Drew a few days later, and "has the ring of something thought out in advance."[19]

That same line came up again in the first debate in 1984 between Mondale and Reagan. There is an unwritten and absurd rule that there is to be no direct confrontation or cross-questioning in these so-called debates. But Reagan's phrase backfired, as Mondale picked up on the line, saying, "You used that phrase once before in debating Carter . . . do you remember?" And a grim-faced Reagan nodded assent, "Uh-huh," and thereby created the first and so far the only question-and-answer sequence in a presidential debate. Forget that Mondale then made a major point out of it, and just ask yourself whether you thought the episode was a spontaneous happening. In fact, Mondale and his advisers had guessed ahead of time that Reagan might again try out his successful one-liner, and they planned and practiced exactly how Mondale would jump on it. Recall that, when Mondale asked Reagan the question, he turned from the lectern and faced the president. Even this simple "pivot" maneuver was carefully rehearsed ahead of time with Pat Caddell as the coach.[20]

So much for the performing art of spontaneity.

There is no reason to think that Ronald Reagan has read much of Edmund Burke, but he doubtless would understand and agree with Burke's sentiment, expressed early in the nineteenth century, that the political actor is obliged to support policies that he personally believes to be good for society, but that his real profession is to satisfy his public.

Thus it can be argued that the simple distinction between the basic political act of speechmaking and theatrical play-acting lies in the character of the performance. When the performance is outstanding the message is strengthened; when the performance is pedestrian the message is often judged to be no better.

The reason that Calhoun, Clay, Corwin, and Webster have been remembered for a century and a half is that they were great orators, politicians who combined matter and manner in impassioned performances that aroused their listeners. As much could be said of William Jennings Bryan, often of Robert LaFollette, and sometimes of Woodrow Wilson.

In our own day Franklin Roosevelt and Adlai Stevenson may have come closest to the earlier platform giants, but they were both obliged to adapt to the electronic age—Roosevelt to radio and Stevenson to television—and in both cases the medium itself dictated the most suitable performing style. In a paraphrase of Marshall McLuhan's classic statement "The medium is the message," Joe McGinnis summed up the effect. "Style becomes the substance," he wrote about the 1968 Nixon campaign. "The medium is the massage and the masseur gets the votes."[21] From the kind of speechmaking suitable for great halls and outdoor arenas, Roosevelt found it possible to adapt his style of performance to what McLuhan called the "hot medium." But Stevenson in the campaigns of 1952 and 1956, no matter how significant his message, could never bring his listeners to feel that he was comfortable in "the cool, participant medium" of television.

Though the distinguishing characteristics of old-fashioned political oratory may have been muted by the advent of radio and television, the omnipresent political speech remains the ultimate political act. Jeff Greenfield calls the speech "the centerpiece of political life."[22] Those who campaign for the office and those who hold it must, like the actor, secure applause from the political groundlings. A president may not operate always in that theater described by Aristotle, where there is "willing suspension of disbelief," but even with skeptics in the box seats, he must work for a union of esthetic creation and social reality. His purposes, mental and physical resources, and techniques must be those of any public performer.

The Performing Arts of the Presidency

It is important that we do not become entangled in the concepts of *performance* and *performer*. Whether it is President Harry Truman making a political speech or James Whitmore on the stage impersonating Harry Truman making a political speech, it is a performance. And its quality will be determined by at least four factors. First, the performer must have a sensitivity to his audience, its makeup, mood, and expectations. Second, he must project a positive public persona, an identifiable personality, character, and image. Third, he must generously display his distinctive abilities, whether dynamic delivery, wit, or personal style. Fourth, he must get a good script, merging the message with the moment. Each of these four performing arts of the presidency deserves fuller consideration.

A PRESIDENT MUST HAVE A SENSITIVITY TO HIS AUDIENCE, ITS MAKEUP, MOOD, AND EXPECTATIONS. Surely the successful reading of audience moods and expectations can be cultivated, but some distinguished speakers have testified that it came to them like a flash. William Jennings Bryan was a presidential candidate in 1896, 1900, and 1908. A dozen years before his first nomination, his wife reported, he came home late one night after a speechmaking trip to western Nebraska, and awakened her. "Mary," he said, "I have had a strange experience. Last night I found that I had power over the audience. I could move them as I chose. I have more than usual power as a speaker. I know it. God grant I may use it wisely."[23]

Late in his life President Woodrow Wilson spoke of the ends of oratory as pleasing the ear, governing the emotions, giving a sense of definiteness, and imparting a sense of reality. These he pursued in a lifetime of what he called exhilarating experiences in public speaking. "I enjoy it," he wrote to his fiancée, "because it sets my mind—all my faculties—aglow; and I suppose that this very excitement gives my manner an appearance of confidence and self-command which arrests attention. However that may be, I feel a transformation—and it's hard to go to sleep afterwards." Especially, he wrote to her on another occasion, he found "absolute joy in facing and conquering a hostile audience . . . or thawing out a cold one."[24]

Franklin D. Roosevelt was "admirably equipped to make use of the new medium" of radio, concluded Earnest Brandenburg and Waldo W. Braden. "Each listener received the impression that Roosevelt was talking directly to him. Millions of Americans sat at their radios and agreed that they 'could practically feel him physically in the room.' His voice communicated his expansive personality; it registered what was in him and what he wanted other people to grasp—conviction, sympathy, humility, gravity, humor—in harmony with situations as he saw them."[25] Roosevelt's audience sensitivity was no unconscious effect: Walter Winchell once reported that Roosevelt deliberately tried to keep in his mind while speaking a hypothetical audience that included a farmer, a housewife, a teacher, a stevedore, and a businessman. By creating this mental image he stayed in touch with his unseen audience.

But now consider former president Reagan. In the Dixon, Illinois, high school, young Ronald played not only football but also stage roles. Reagan wrote, "I learned almost all of what I know about acting today" in B. J. Fraser's drama classes. He practiced the art in high school as Ricky in Philip Barry's *You and I* and as the villain in George Bernard Shaw's *Captain Applejack*.[26]

Then, in his first few months as a freshman at Eureka College, he tried the speaker's platform. A strike committee proposed that the student body sit out classes until the college president and trustees reversed a decision to balance the budget by canceling some upperclass courses and firing their teachers. Reagan represented his class on the committee and was assigned to "sell the idea" of a strike at a student rally. He had only his high school acting experience to call upon, but here is how he described what happened: "I discovered that night that an audience has a feel to it, and, in the parlance of the theater, that audience and I were together. When I came to actually presenting the motion there was no need for parliamentary procedure; they came to their feet with a roar—even the faculty members present voted by acclamation. It was a heady wine. Hell, with two more lines I could have had them riding through 'every Middlesex village and farm'—without horses yet!"[27] In his 1990 recollection, Reagan wrote that "giving that speech—my first—was as exciting as any I ever gave. For the first time in my life, I felt my words reach out and grab an audience, and it was exhilarating."[28]

Like Bryan, Wilson, and Roosevelt, Ronald Reagan's speechmaking is a performing art, and it begins with a highly developed sensitivity to his audience. However well each of these candidates succeeded in finding his real audience, we must remember that only Reagan was totally subjected to the great leveling influence of television. It homogenizes the demographic variety of any candidate's real audience. Moreover, even though we sometimes can't abide a particular candidate, we seldom get to the television set fast enough to turn off his paid political message. So, one size, like the modern sportsman's cap, fits all.

A PRESIDENT MUST PROJECT A POSITIVE PUBLIC PERSONA, AN IDENTIFIABLE PERSONALITY, CHARACTER, AND IMAGE. Here we ask not what a president is, but what he is thought to be. What arts of performance contribute to the creation of a favorable image, one that will win support in campaigning and in governing? In recent years many presidents and presidential aspirants have had difficulty maintaining a consistent persona. In several campaigns we saw a "New Nixon," and in the campaign of 1988 control over persona was a problem for Mondale, Hart, and Jackson.

But Ronald Reagan, as governor, General Electric pitchman, presidential campaigner, and two-term president, has always been a model for "what you see is what you get." His is a clearly identifiable personality, and a steady image. He was the answer to what the electorate seemed to need in the past several decades. Critical to that appraisal, of course, is an understanding that Reagan's message was one America wanted to hear, or at least Americans decided that that was so after they heard it. If part of the decline of eloquence is attributable, as Kathleen Hall Jamieson and others have argued, to the fact that the public lacks a shared collective history, or even national mythology, Ronald Reagan was the man to restore it, in his own way. Bemoaning national "malaise" was not his way, but preaching upbeat optimism was. Unlike his most recent predecessors, he could be earthy, simple, and direct in patriotic appeals that cloaked his favorite social and economic proposals. The image was bigger than the man.

In his landmark 1953 book *The Lonely Crowd*, David Riesman concluded that although the voters were losing confidence in their ability to judge technical competence, they believed that they could judge such "image" qualities as sincerity, and thus they would be relieved of the necessity of making an emotional response to a speaker's argument. Certainly this was the forerunner of Daniel Boorstin's significant treatise on image and the influence of pseudo-events.[29] Despite these judgments, the editor of a volume entitled

The Rhetoric of Our Times argued that in fact the decade of the 1960s witnessed a resurgence of popular interest in issues as against personalities.[30]

By 1984, however, Theodore Sorenson, who had been chief speechwriter for Jack Kennedy, was lamenting the diminution in importance of both issues and true personality:

> Today choosing policy advisers is insignificant compared to lining up the right pollster, media advisers, direct mail operator, fund-raiser and makeup artist. Today policy advisers are not comprehensively articulated but condensed into bumper-sticker slogans and clever TV debate ripostes that will please everyone and offend no one. Today experience and intellect are no more crucial to the multimedia campaign than the candidate's hair, teeth, smile and dog. Today volunteers have been replaced by computerized mail, automated telephone banks and other marvels of technology in an industry that has shifted from labor intensive to capital intensive. Today the news media rarely report what the candidates are saying on the issues. They report instead on a horse race—which horse is ahead, which one has the most physical stamina, which one is lame and which one is attracting the big money.[31]

But if image-making politics has proven successful in Reagan's campaigns, in the Reagan administration it has become standard operating procedure in governing as well. Sidney Blumenthal described it well as "the permanent campaign." Dick Wirthlin told Hedrick Smith how Reagan's presidential speeches were tested to develop "power phrases" and "hot moments" that could be repeated in speech after speech. The technique was to chart graphs of audience response, and then superimpose them upon videotapes of the president's speech. In general, "these reaction graphs suggest that people are often responding as much to the speaker's facial expression, tone of voice, and body language as to content."[32] As Paul Taylor learned, "this enabled Wirthlin to know—down to the sentence, the gesture, the word—precisely what had worked and what hadn't, and with whom. The information would be filed away for use in preparing the next speech."[33]

When these power phrases are discovered, of course, they are timed to fourteen and one-half seconds to ensure getting them on the evening news shows. Indeed, President Reagan has been known to make press statements that are just that long, just right for a "sound bite" that the reporter must use, or use nothing. Thus the television news bite has become the fast food of politics, and the visual image is the handy cola that the viewer grabs on the run.

In the Nixon administration Bob Haldeman argued that the visual always beats the verbal, and in the Reagan administration it was

Mike Deaver who became the Vicar of Visuals. "You get only forty to eighty seconds on any given night on the network news," he says, "and unless you can find a visual that explains your message you can't make it stick."[34] His book *Behind the Scenes* is a veritable catalogue of the carefully created and controlled "photo opportunities" of the Oval Office, the Rose Garden, and the cabinet room at home, and the Halls of Parliament, the cemetery at Bitburg, et cetera, abroad.

In short, the sound bite has a life of its own, and it creates the occasion for carefully controlled visual images in the process that Murray Edelman in his latest book calls "constructing the political spectacle."[35]

Does this seem to be overemphasis upon the degree to which the Ronald Reagan's image has been controlled? Then read Donald Regan's book for the description of staged pictures in the naval hospital after President Reagan's cancer surgery. In every picture permitted, Nancy saw to it that the tubes in her husband's nose and mouth were hidden by the camera angle. Later she arranged to be at his side as he waved to the press from the hospital window and, she said, if the two of them were photographed waving from the balcony when he returned to the White House, "that would make a great picture!"[36]

Or read the chief of staff's account of the prolonged debate that took place among the president's advisers just before Reagan was to step outside his villa in Geneva to welcome Gorbachev, who was arriving by limousine. What were Secretary of State Shultz, National Security Council Chairman Bud McFarlane, Chief of Staff Donald Regan, and Reagan's personal assistants Bill Henkel and Jim Kuhn debating about? Whether the president should wear his overcoat when greeting the Soviet leader! What would look best to the cameramen?[37]

Anecdotal support on matters of projecting and protecting the Reagan image may be found in a shelf of books written by his former aides. In addition to those already cited by Michael Deaver and Donald Regan are others by Reagan's press secretary, Larry Speakes, his chief adviser on domestic and economic policy, Martin Anderson, and his budget director, David Stockman.[38] In all fairness, it must be reported that Anthony Dolan, who spent "eight years off and on as [Reagan's] speechwriting chief," found Reagan to be "engaged, persistent and intellectually formidable," and has written an account of those years "showing Mr. Reagan smarter than his critics or his staff, smarter even than I."[39]

On the record Ronald Reagan was uninterested in these works, saying about Stockman's that "I don't have too much time for fiction," and that Deaver and Regan wrote "kiss and tell books." Of course, if he or Nancy took time to read these books they would doubtless be pleased to find each of the authors respectful and loyal, and (surely not surprisingly) on balance portraying their boss as a good influence on America, even predicting, as Donald Regan did, that Reagan "will be a great figure in history," compared favorably to Franklin Roosevelt and Harry Truman.[40]

From this commentary on the Reagan image it is easy to understand why Washington became known as Hollywood East. "In the Golden West," wrote newspaper columnist Marquis Childs, "you do not run for office, you pose."[41] During his years as candidate and president, Reagan has been seen and heard in every voter's home, posing and performing. As Deaver wrote of him, "He is, after all, a performer. The voice is pleasant, the confidence, the timing sharpened by thousands of speeches and scripts. And there is a sense entirely his own of what the moment may require."[42]

"In the presidential realm," concluded Robert Schmuhl, "the Reagan years provide a rich case study in the evolving techniques of leadership through mass media. What earlier presidents began and tried to foster, Reagan raised to a popular art form. The presidency itself was not only rhetorical but theatrical, with daily performances part of the strategy." Thus, after studying "American political life in the age of personality," Schmuhl titled his 1990 book *Statecraft and Stagecraft*.[43]

In the waning days of his presidency, Reagan promised to lend his image to George Bush, one day a week in September, two in October. His push for Bush was in character, a running sit-com, as the president played the role of his old friend Gary Cooper in *High Noon*, or the hero in the shootout at the old corral, wearing a white hat while dueling with the black-hat Dukakis Democrats, the liberal disbelievers, and the Sam Donaldsons.

For this man and his image, politics was television. The only question, as in the Nixon administration, was, How will it play in Peoria?

A PRESIDENT MUST GENEROUSLY DISPLAY HIS DISTINCTIVE ABILITIES, WHETHER DYNAMIC DELIVERY, WIT, OR PERSONAL STYLE. Even presidents no doubt feel most comfortable "doing what comes naturally," but to get to the White House all recent ones have had to "accentuate the positive and eliminate the negative." Some—Nixon, Johnson, and Reagan—have had academic instruction in speechmaking. Others have called upon specialists for help in various aspects of public performance. Robert Montgomery, film star of the forties and fifties, joined the White House staff of Dwight

Eisenhower as the first presidential television adviser ever. Robert Orben, the professional joke writer, tried to add humor to Jerry Ford's personality, and the stand-up comic Don Penny tried to loosen up the Ford delivery. Myles Martel, a former college debate coach, helped direct Ronald Reagan's rehearsals for his 1980 confrontations with John Anderson and Jimmy Carter. And David Stockman and Martin Anderson played the role of Jimmy Carter in those rehearsals.

For some presidents their performing strength has included an outgoing personality expressed in a vibrant voice, like Franklin Delano Roosevelt's; or a grandfatherly and reassuring manner, like Dwight Eisenhower's; or a feisty and forthright disposition, reflected perfectly in a staccato delivery in flat midwestern accents, like Harry Truman's. What gave special platform appeal to each of these three was, first of all, a reflection of their natural personalities. But these characteristics became more marked as their speechwriters gave them materials that suited the particular dynamics of their style of delivery, and this in turn encouraged them to stay with it.

Surely this deliberate nourishing of distinctive abilities has been apparent in the last half-dozen presidents. Jack Kennedy brought youth, buoyancy, native wit, and imagination to his speaking: it was in his press conferences, the first ever to be televised live by any president, that he sparkled, displayed a quick ex-tempore wit, and seemed thoroughly to enjoy himself.

Lyndon Johnson, as Hart has put it succinctly, "preferred the corridor to the coliseum."[44] But anyone who heard his televised address on civil rights knows that Johnson could speak with great effect. His clear preference, however, and his potent best, was for one-on-one persuasion, in which he overwhelmed recalcitrant legislators with what became known as the Johnson "treatment."

About Richard Nixon enough has already been said and about Jerry Ford there is little to say. His bluff, beetle-browed delivery was pedestrian, and even the ministrations of a series of speech coaches had little effect.

Jimmy Carter's presidency was not without virtues, but effective public communication was not among them. Somehow he combined self-confidence, Baptist zeal, and intensity of conviction, with an apparent indifference to the speaking style with which his speech substance was presented. It is, sadly, not unfair to say that as a public speaker he was the very *least* artful man to hold the office since Warren Gamaliel Harding.

At the other extreme stands Ronald Reagan. Surely he is the *most* artful presidential communicator since Roosevelt, or, in the television era, since Kennedy—and one cannot say that he is inferior to either of them. Before his first race for public office, after all, his career was that of actor and professional public speaker, and he had a repertoire of well-honed skills. These made it possible, no matter how divisive his rhetoric, to win personal approval. Sometimes avuncular, sometimes the Irish wit, always anecdotal, he was, in the performing arts aspect of the term, the "Great Communicator" that his White House cronies called him.

Reagan faltered momentarily, of course, in the first 1984 debate with Mondale. But if there was blame to be assigned, probably only a little could be ascribed to his age, his mental agility, or his state of fatigue. Rather, his pathetic confusions and rambling conclusion made an object lesson for all of us: like horses, a candidate's life-long style should not be changed in the middle of a campaign. Reagan talks best in abstract terms, using anecdotal material, laced with glory words and gentle put-downs. He has never done well in handling strict factual information. And when his debate trainers tried to program him to use an entirely different style of persuasion, it was, in the honored idiom, like teaching an old dog new tricks. The crisp, factual, statistical style is just not the real Ronald Reagan, and it was a mistake for him to try it.

Speaking off the cuff, too, Reagan has problems. Who else but Ronald Reagan would tell a group of reporters one day that Libyan Colonel Qaddafi is "flaky," and the next day have an aide telephone William Safire to ask what "flaky" means?[45]

A more serious gaffe was in a telephone interview from the California ranch, when Reagan defended the Botha administration in South Africa by saying "they have eliminated the segregation that we once had in our own country." By ignoring the denial of free movement, citizenship, and voting under apartheid laws, the president set off a fire-storm of opposition. This was how Michael Deaver explained the catastrophe to Hedrick Smith: "No one should have let him get on the phone to do that interview. . . . You never let Ronald Reagan do an interview from his ranch. . . . If he's going to be making policy pronouncements . . . he should deliver it in a more formal setting. He should be standing up." Then the man who knows him best added: "You know, it's funny about Reagan. The way he thinks changes when he sits down. . . . I would never let him be sitting down. He's too relaxed when he's sitting. He's not careful. He's conversational, not presidential."[46]

A sense of humor and a knack for storytelling is a Reagan strength. Not only does it mellow his associates, but, like Abraham Lincoln, he apparently finds humor a help in coping with stress. Incidentally, in this Reagan idiosyncrasy lies one answer to the question "Where was George?" Nearly every day of his vice-presidency George Bush tried to provide a new joke for the president, and he importuned members of the vice-president's staff, and the staffs of his colleagues to give him a story to take into the Oval Office.

There have been times, of course, when Reagan has been a poor judge of humor. Who else would think it funny to test a live mike by announcing in the presence of reporters that "in five minutes we will begin the bombing of Russia"? Or to try to make a joke to reporters during the 1988 campaign by referring to Michael Dukakis as an "invalid"? But at his best, he is the best. As James Reston wrote after the 1985 Gridiron dinner, "He didn't invent pretense, but has practiced it longer and more successfully than anybody up front . . . he has not only revived vaudeville, but he's the best political storyteller in these parts since Lyndon Johnson. . . . He uses stories not to illustrate the facts but to evade them or poke fun at them, and himself."[47]

There have also been times, as we all know, when Ronald Reagan has been careless of the facts. Early in his presidency, when no gaffes seemed to hurt him and his image sustained him, Congresswoman Pat Schroeder christened Reagan "the Teflon president." Tom Brokaw believed that Reagan got "a more positive press than he deserves. . . . In part it goes back to who he is, and his strong belief in who he is. He's not trying to reinvent himself every day as Jimmy Carter was."[48]

Reagan, however, often asserted that he knew what he was talking about. For example, on the 3 April 1980 "CBS Evening News" with Walter Cronkite, Bill Plante asked if Reagan had been thorough in checking his facts. Reagan replied, "On the mashed potato circuit for a great many years . . . I learned very early that you should check them. I didn't at first, I guess, like any other speaker. I'd see something and think, hey, that's great, and use it. And I just learned from being rebutted a couple of times that I'd better be sure of my facts." But in more recent years, when the "straight arrow" image was tarnished by truth stretching, Donald Regan could allow the press to quote him as saying that "some of us are like a shovel brigade that follows a parade down Main Street cleaning up."[49] Some people thought the remark was in poor taste, though probably accurate, but apparently only Nancy Reagan thought it to be outrageous.

When he was a cowboy actor in Hollywood, Reagan may have been at home on the range. But as a politician in Washington, Ronald Reagan was most at home on television. Given a good script, projected on his TelePrompTer, he could squeeze every nuance of emotion out of it, as he almost always did, and so superbly in his five-minute televised presentation of Peggy Noonan's tribute to the crew of the ill-fated space shuttle *Challenger*. As Robert Schmuhl said of the man, "His personality dominated public occasions. Others might have written the words or designed the stage, but he made the appearances *his*."[50]

In sum, by using his distinctive abilities, Ronald Reagan the public orator far surpasses any of his contemporaries. Indeed, in 1988 it was apparent that some Republican campaign strategists feared that if Bush and Reagan appeared together, the latter's far superior presentation of self would further enfeeble George Bush's audience impact.

A PRESIDENT MUST GET A GOOD SCRIPT, MERGING THE MESSAGE WITH THE MOMENT. Early in the 1984 campaign, David Broder, a distinguished *Washington Post* political analyst, wrote an essay in praise of amateurism. The immediate occasion was his weekend exposure to NCAA basketball, participated in, for the most part, by amateur athletes, and to the annual Gridiron Club performance where professional journalists who are amateur actors and singers put on a "roast" of politics and politicians. The ultimate point of the essay was to lament that "there is precious little room for amateurism in presidential campaigns these days, and politics is the worse for it."[51]

What we do have, of course, in both campaigning and governing, is technology rampant. An army of technicians has intruded into presidential politics—and for the primary purpose of enhancing presidential communication. The era of the cathode-ray tube and the microchip has brought pollsters who can advise a candidate or a president where to position himself on issues in order to win maximum support, political consultants to help him adapt to trends and opposition tactics, media consultants who can prescribe costume and tonsorial treatments, telephone-bank experts who can call the party regulars, public relations experts to fashion television commercials to reach everyone else, and direct mail experts who will help raise the millions of dollars necessary to finance the whole enterprise.

By no means ranked last in this army of experts is the professional speechwriter. Some communication scholars believe that the writing of a president's speeches should be the amateur's last stand. There are also those who consider reliance on speechwriters to be unethical. Ernest Bormann and this writer have twice publicly debated these issues at conventions of the Speech Communication Association and have continued the debate, joined by Franklyn Haiman, in the pages of *Communication Education*.[52] This is not the place to replay those arguments. The fact of the matter is simply that today professional speechwriters are indispensable assistants for nearly all government officials, most business and industrial executives, many university presidents, and at least a few high-ranked ecclesiastics. Who would begrudge a little help for presidents and presidential candidates?

The names of the chief speechwriters for recent presidents are doubtless well known: Ted Sorenson for Kennedy, Jack Valenti and Harry McPherson for Johnson, Raymond Price and William Safire for Nixon, Robert Hartmann and Bob Orben for Ford, James Fallows and Hendrick Hertzberg for Carter, and for Reagan a succession of Peter Hannaford, Kenneth Khachigian, Aram Bakshian, Tony Dolan, David Gergen, Pat Buchanan, William Gavin, Bentley Elliott, and others, worthy and otherwise.

Before the Republican convention that nominated George Bush, all of the major speakers had their texts written for them by members of the Bush staff, so that they would be properly coordinated and no one would steal another's thunder.[53] The nominee's speech, far better than his regular primary campaign efforts, was widely understood to have been written by Peggy Noonan.[54] Noonan had been an outstanding writer for Reagan, and she apparently went with the territory.

Ronald Reagan reports that his views on the propriety of using speechwriters were formed when he was riding the General Electric circuit with his traveling manager, Earl Dunckel. On one of the first occasions when Reagan was going to talk on political issues, instead of about Hollywood pictures and their stars, Dunckel offered to write the new speech for him. Reagan later explained that he refused the offer because "I couldn't be the mouthpiece for someone else's thoughts."[55] It should not be inappropriate to observe that creating a character by mouthing someone else's thoughts and words is exactly how every actor earns his living.

But Reagan continues his little conceit. When asked if he liked being president better than being a movie actor, he replied: "Yes, because here I get to write the script, too."[56] Thus, a standard photo opportunity before every major Reagan address has been a news picture of him, studiously writing something on a piece of paper. On at least one occasion this practice proved embarrassing. In anticipation of his January 1986 State of the Union address, such a picture was widely published, conveying the impression that Reagan wrote the speech. Because of the *Challenger* disaster the address was postponed, but before it was rescheduled (weeks after the president had been photographed supposedly completing the text) the newspapers were ironically filled with reports of in-house arguments among Reagan's advisers over which script should be chosen from among those submitted by two different teams of speechwriters who were competing for Reagan's mind.

So the truth is that while he may spend time pondering speech drafts prepared by others, and in fact may perform some editorial revisions, insert favorite "Reaganisms," and put language he regards as stilted into his best "aw shucks" idiom, the president no more writes his own speeches now than he wrote the shooting scripts for his Hollywood movies.

That being true, one might have anticipated revelations about speechwriters in the acknowledgments in Reagan's first postpresidency publication, a collection of fifty-three speeches he delivered, from remarks at a Kiwanis International Convention on 21 June 1951 to his "farewell address" on 11 January 1989.[57] There is none. He thanks an editorial assistant, a researcher, the publisher's staff, his literary agent, and his typist. He does thank Landon Parvin "whose [unspecified] contribution has been immeasurable"; Parvin was, indeed, a presidential speechwriter early in the term and later wrote for Nancy Reagan. One can only assume that Parvin helped select the fifty-three speeches from among the hundreds delivered by the president.

Each speech is prefaced with a note about the occasion or some anecdotal background; but never is there an acknowledgment of speechwriting assistance, and the index lists no member of his speechwriting staff. In the foreword Reagan observes that "some of my critics over the years have said that I became president because I was an actor who knew how to give a good speech. I suppose that's not too far wrong." However, he adds, "I don't believe my speeches took me as far as they did merely because of my rhetoric or delivery, but because there were certain basic truths in them that the average American citizen recognized."[58] If rhetoric to Mr. Reagan means substance or script as against delivery, then his statement clearly claims authorship, sans speechwriters. His publishers support him: the book's jacket reads, "Here, in his own words, is the record of Ronald Reagan's remarkable political career and historic eight-year presidency."

In the absence of any acknowledgments to speechwriters in the 1989 volume of speech texts, one might hope for more candor in his 1990 autobiography. Reagan does acknowledge that he had "a great deal of help" with his memoirs, and that "Robert Lindsey, a talented writer, was with me every step of the way."[59] But when referring to his first gubernatorial campaign he vigorously asserts that he wrote his own speeches; in the postgubernatorial years he certainly implies authorship of "a newspaper column and regular radio spot that gave me a chance to continue speaking out." After losing the 1976 nomination to Ford, he claims, "I began writing my newspaper column and radio scripts again." He concludes a chronology with "Until I got to the White House, I wrote all of my own speeches" until other obligations prevented it, and then "I continued to write my more important speeches, but much of the time I would sit down with White House speech writers [they were never identified, nor were their names indexed] and go over the points that I wanted to make during an upcoming talk, and then they would present me with a draft to edit."[60]

Ronald Reagan's claim to creativity has received family support. Nancy Reagan says that Reagan's daughter Maureen "has the best [and sharpest] memory of anyone I've ever known,"[61] but the two Reagan women encompass different time periods in their testimony. Maureen Reagan recalls that allegations about her father's having "professional speechwriters" followed him throughout his career, "but in fact Dad has always written every speech himself."[62] In contrast, Nancy Reagan says only that "until he became president he wrote his own speeches. . . . Because he wrote his own material he could deliver it with total conviction." But "when he became president there was simply no time to write his own speeches."[63] Technically, since he was not yet president at the time, Mrs. Reagan could be right about the first inaugural address, "which he had written by himself,"[64] except that there is ample evidence that the principal author was Kenneth Khachigian.[65] Khachigian, by the way, is not mentioned in Mrs. Reagan's memoirs, nor is any other of her husband's speechwriters, except that she does credit Landon Parvin, whom she called "one of our best speech-writers," with writing the lyrics sung by Mrs. Reagan at the 1982 Gridiron Club dinner.[66] (The possessive term she used is interesting because Parvin had been borrowed on a more or less permanent basis from her husband's staff).[67]

It would be unfair to conclude a discussion of Reagan as speechwriter without noting that every president in our history, beginning with George Washington, has had some kind of assistance in the preparation of his speeches. This help has ranged from suggesting ideas, approaches, and some phrasing, to 100 percent creation of content and style. The extent to which the speechwriting is total, of course, may lead some speech critics to rechristen "The Great Communicator" and call him by the more limited term "The Great Articulator."

On the PBS "Campaigning on Cue" program, ABC's Britt Hume argued that appearances of presidents and presidential candidates on television are important to the viewing public because they permit, in his words, "a demonstration of indispensable qualities." Among these, we were to infer, was how well the speaker could hold up under stress, and further that one test was his or her spontaneity. But it has already been noted that much of the time what we see in presidents and those who aspire to the office is in fact such contrived and rehearsed behavior that it not only conceals any stress, but seems brilliantly spur-of-the moment.

There is no need here to describe in detail the ideal process of speech composition involving both principal and aide. Peggy Noonan's 1990 description of how it was done by a sizable staff in the Reagan White House can now take its contrasting place beside the previous best insight, William Safire's report on those who wrote for Nixon.[68] Presidents with a good sense of communication work closely with their speechwriters. Good speechwriters, under optimum conditions, reflect Safire's job description of his relationship with Richard Nixon: "to help refine a point of view, to fit into a framework, making allowance for political compromise; then to clothe that point of view in the most dramatic and persuasive words that came to mind; and then to help promote, project, and advance the man and the Administration that I was a proud part of."[69] Noonan reflected her own style in describing her role to Hugh Sidey: "Government is words. Thoughts are reduced to paper for speeches which become policy. Poetry has everything to do with speeches—cadence, rhythm, imagery, sweep, a knowledge that words are magic, that words like children have the power to make dance the dullest beanbag of a heart."[70]

In any event, to use the words of Ben Wattenberg, one of Lyndon Johnson's speechwriters, "When the president walks up to that podium with that black ring-binder notebook, it doesn't make a damn bit of difference who wrote what paragraph—it's his speech. The speechwriter is a creature of the president, not the other way around."

The Wattenberg view is, of course, satisfactory for a study of presidential rhetoric as a genre, when authorial credit goes with the office. So it is, for example, in the focus of the recent volume by Campbell and Jamieson, where they "treat the presidency as an aggregate of people, as a corporate entity. . . . a syndicate generating the actions associated with a head of state, including those deeds done in words."[71] But in a volume about Ronald Wilson Reagan and *his* politicial discourse we properly ask who the members of the syndicate are and what contributions they make.

It would be a mistake to conclude from anything we have said that the candidate with the best speechwriter always wins, but it is certain that no presidential candidate or president can be effective without having a good script, just as it is certain that no president can consistently create his own. In preparing his "Farewell Address" George Washington relied heavily on Alexander Hamilton, and to a modest extent on James Madison. But modern presidents have come to creating structured speechwriting support groups.

What are the ingredients of an effective presidential speechwriting organization? Here is a review of Ronald Reagan's.

FIRST, REAGAN HAD A GOOD TEAM OF WRITERS. Martin Anderson, who watched from the inside and occasionally himself wrote for the president, asserted that "Reagan's speechwriters were probably the most talented group assembled since the days of Kennedy and Nixon."[72] Foremost among them was Kenneth Khachigian, who had written earlier for Nixon. Then came a libertarian, Dana Rohrabacher; the ideologically right Bentley Elliott; Anthony Dolan, whose brother was head of the American Conservative Union; and one of the first women in the role, Mari Meseng. Anderson described them as "policy clones of Reagan. . . . They knew what he wanted written without asking." (Anderson left the president's staff before Peggy Noonan joined it, replacing Mari Meseng as "the woman speechwriter," a classification that vexed Noonan.)

Even with the excellent team that Donald Regan, Martin Anderson, and Peggy Noonan describe in some detail, there was a constant script-review process conducted by a dozen or so senior staff members, such as George Bush, Ed Meese, Jim Baker, Michael Deaver, David Gergen, Lyn Nofziger, David Stockman, and Murray Weidenbaum. In their offices copies of the speech draft were photocopies for the next-level staff members, and thus there was a veritable army of guardians of the official policy, on the alert for deviations from it. Peggy Noonan rather less charitably called these staffers and substaffers (about twenty for an unimportant speech and as many as fifty for an important one) the "mice" who nibbled away at her speech drafts and often chewed out the heart if not the guts.[73]

SECOND, RONALD REAGAN HAD AN IN-HOUSE CRITIC IN HIS WIFE, NANCY. Donald Regan records a conversation with her during the preparation of the 1987 State of the Union address, when she displayed her frequent role as "guardian":

> She said she was glad that Ken Khachigian, not Pat [Buchanan] was writing the State of the Union speech. She liked the second draft, and she and the President had invited Khachigian and his wife to Camp David over the weekend so that they could work on the speech together.
>
> "The parts about abortion have got to come out," she said. I pointed out that the President had particularly wanted some language on the subject included in his address.

"I don't give a damn about the right-to-lifers," Mrs. Reagan retorted. "I'm cutting back on the Iran stuff, too. It's too long and it's not appropriate. Ronald Reagan's got to be shown to be in charge."[74]

THIRD, THE PRESIDENT NEEDED TO KNOW AND TRUST HIMSELF SO THAT HE COULD CONTRIBUTE TO IF NOT CONTROL THE FLOW OF ARGUMENT IN HIS SPEECHES. It may be that Ronald Reagan seldom displayed that quality, but when he did, Donald Regan testified, he was effective. Here is what Regan said when the president outlined his ideas for structuring his speech for 27 February 1987, after the Tower Commission report was released: "Listening to this fluent and persuasive summary, and watching the conviction shining in the President's eyes, I could only wonder how anyone could imagine that it was a good thing to silence this extraordinary man who could speak for himself so much more eloquently than anyone else could speak for him."[75] That claim echoed assurances in early interviews with Kenneth Khachigian, Peter Hannaford, and other speechwriters, who told this writer that Ronald Reagan was perfectly capable of writing his own speeches without any help if he just weren't so busy running the country.[76] Anthony Dolan, who headed Reagan's team of speechwriters, paid his tribute somewhat differently when he called writing for Reagan "a constant exercise in humiliation," because the president "is his own best speechwriter—and always has been."[77] And surely Reagan appreciated Dolan's one-liner: "Speech-writing in the White House is plagiarizing Ronald Reagan."

What Political Speakers May Learn from Ronald Reagan

One way of summing up Ronald Reagan's communication behavior is to review it as it is likely to influence future presidents and other political speakers. "No man who becomes president," wrote William Leuchtenburg, "can evade a confrontation with his forerunners, since they have the most palpable effect on the way he will ultimately be regarded."[78] He was considering, of course, such matters as perceptions of the presidency, domestic and foreign policies, and campaign styles and strategies, as well as public discourse, and all these elements were reflected in the *New York Times* headline that summed up Reagan's 1980 acceptance speech, replete with quotations and paraphrases from FDR: "Franklin Delano Reagan."[79] What Reagan's influence may be on his successors can be suggested by isolating salient characteristics that distinguished his public speaking. They are presented here, without attaching good or bad labels, or arranging them in any hierarchical order.

First, Ronald Reagan came to a White House that had survived a series of failed presidencies. As Haynes Johnson, Pulitzer Prize–winning reporter for the *Washington Post,* recalled them: "Richard Nixon was always looking at his navel and telling how he felt. Lyndon Johnson was profane and paranoid. Jimmy Carter was so earnest and self-righteous, but couldn't command respect." But Reagan, says Johnson, "in a public, ceremonial sense, has all the attributes that make people feel there is someone in control."[80] It was clearly Reagan's training as an actor that made him "comfortable with himself," and his successors will surely note Reagan's audience appeal when he was "standing tall" and proclaiming that "it's morning in America."

Second, Ronald Reagan demonstrated that folksy and especially self-deprecating humor can still be effective with contemporary audiences. Indeed, when it is presented in an avuncular attitude, with a cocked head, a calculated smile, and an innocent manner, Reagan's humor carries far better than did Adlai Stevenson's more sophisticated and sometimes acerbic wit. In interpersonal communication, as with a group of senators in the Oval Office, however, he does not fare as well. Senator Robert Packwood, objecting to Reagan's storytelling instead of discussing the purpose of a visit, said, "He's on a different track," and "We just shake our heads."[81] The same point was made, in another way, when Gerald Ford told a reporter that "he is one of the few political leaders I have ever met whose public speeches revealed more than his private conversations."[82] But humor, even one-liners recycled from long-ago movies, were a stock in trade for Reagan. Sometimes they were used to make a political point; other times they seemed avenues for releasing tension or dealing with stress. Occasionally the Reagan wit seemed quite spontaneous, and at other times rehearsed and stored to quip away an anticipated press-conference question.

Third, Ronald Reagan, like any good actor, has demonstrated the critical value of adapting one's style to the medium. This is a thesis of Kathleen Hall Jamieson's brilliant book, *Eloquence in an Electronic Age: The Transformation of Political Speechmaking.* Reagan knows that he is addressing individuals—perhaps two or three in a family group around the television set, and not a vast audience in a great hall. Consequently, his style becomes conversational and intimate, dramatizing and storytelling, self-disclosing, personal and intense, and keying upon memorable phrases, uttered whenever possible in memorable political settings. Not all of his successors who try the same approach will find it working well for them. After all, Reagan was an actor, reading lines.

Fourth, Ronald Reagan the actor, directed by Michael Deaver, created a new political art form, the visual press release. It centralized the president's own role in making the news. Announcing a new welfare program in the setting of a nursing home clearly conveys to those watching the six o'clock news that their president is personally stringing up a "safety net." Hedrick Smith of PBS's "Washington Week in Review" says the visual press release "is not just some artful makeup—it is the effective communication of a political message." Politics, he adds, "is a lot closer to acting than a lot of us want to admit and having been an actor is not a liability."[83] Other politicians will not fail to note Reagan's apparent success as a continued and engineered presence in the news.

Fifth, Ronald Reagan had a fondness for political language that was often divorced from political reality. Sometimes this was loose handling of the facts or ignoring them altogether, and substituting anecdote for analysis or wishful thinking for wisdom. Sometimes there were factual errors or outright gaffes that may have come from an uncritical acceptance of a *Reader's Digest* tidbit, of the scripts handed to him, or even of lines from his old movies (reflecting what has been called "an uncanny slippage between life and film").[84]

A different problem is presented by Reagan's deceptive television speeches on the Iran-Contra affair, when he denied having traded arms for hostages. An investigation of the administration's role was made by a special review board, chaired by Senator John Tower, with Edmund Muskie and Brent Scowcroft as members. In its report the board concluded that the president's management style was too lax and that he didn't ask the right questions of the principals. Senator Tower obligingly explained in a news conference that the president did not "wittingly mislead the American people" but that "there was a deliberate effort to mislead by those who presented these materials" and prepared his TelePrompTer texts.[85] In his recent volume of memoirs, Tower is more forthcoming in reporting how Reagan changed his own testimony to the board in two separate meetings. In the second one he reversed his earlier statement, and then picked up and read from a supportive document prepared by White House counsel Peter Wallison, after saying, "This is what I am supposed to say." It could not be said in the board's report, but Tower concludes that the about-face was part of Donald Regan's effort to cover up his own involvement in the affair, and that "it was obvious that the president had been prepped by Wallison and words were put into his mouth."[86] Unhappily, it may be concluded either that Reagan was not capable of checking for accuracy the Wallison draft and the texts for his Iran-Contra speeches, or that he preferred not to know what he was really saying. His credibility on these counts continues to suffer even after his eight hours of testimony were videotaped in March 1990 for the trial of John Poindexter.

Sixth, Ronald Reagan has always been known as "a quick study," and with an old actor's ability to "milk the script." But he has never enjoyed the experience of working without a script, unless an old joke or a wisecrack would suffice. Thus his greatest talents do little to prepare him for ex-tempore speech situations, such as presidential press conferences and truly confrontational debates. He clearly demonstrated that there is little substitute for a wide knowledge base, analytical ability, and mental alertness in preparing for publicly televised press conferences. By contrast, Bush is more at ease and more effective in press conferences, and seemingly less comfortable with formal presentations.

Seventh, Ronald Reagan, like most actors, is always willing to take one more bow, ready to make one more speech. Indeed, he appears to revel in the opportunity to monopolize the media, even with what the *Wall Street Journal* labeled as "Rose Garden rubbish,"[87] and when Reagan had held office less than a year Hugh Sidey asserted that "the Reagan Administration talks too much,"[88] though this complaint was not directed at press conferences, which were as scarce as formal speeches were plentiful.

In these respects Reagan differed from his predecessors. Kennedy, Johnson, and Carter all complained about demands for speeches on incidental occasions and topics. Speechwriter Harry McPherson reflected Johnson's feelings when he concluded that "presidents are called upon to speak too often, manufacturing words of lasting significance for gatherings of little consequence to them." Although Ronald Reagan and Margaret Thatcher appeared to be ideological soulmates (in fact, of course, she said of him, "Poor dear, there's nothing between his ears"),[89] they disagreed on this point. It was the former prime minister who said, "The trouble with politicians is we have to speak more often than we have something to say."[90] Reagan would speak anyway.

Eighth, Ronald Reagan has always enjoyed public speaking and is so good at it that he sets an almost impossible standard for those who aspire to replace him. Most significantly, his abilities have been a special threat to George Bush. The vice-president's advisers seemed less than enthusiastic about joint appearances with the president during the 1988 campaign. They felt, with some justice, that many voters would be measuring Bush's performances not against Dukakis but against Reagan. Since Bush's election one often senses public feeling that comparisons continue to be made. Even abroad, the Paris *Libération* regretted his "cautious and vigilant approach, less muscled than the Reagan rhetoric," and *La Repub-*

blica of Rome, after the 1989 conference of the seven major industrialized nations, contrasted earlier "memorable television spectacles" by Reagan with the "unremarkable and nearly secondary role" that Bush presented alongside the other world leaders.[91] And even when David Broder, who had previously called Bush's televised speeches "few and forgettable," judged the Bush 1991 State of the Union address "an important political breakthrough," he attributed its success to the "model" found in Reagan's 1982 speech on the same occasion, not only "transplanted themes" and emotional symbols, but rhetoric in "bold circus type."[92]

Ninth, Ronald Reagan's presidential career must surely send a note of warning to all aspiring presidential candidates that effectiveness in political communication is a critical ingredient for success, and that if they are not naturally gifted, they should seek professional help in developing competence.

By Reagan's own testimony, he was no born orator. Indeed, the early pages of his autobiographies[93] emphasize his childhood feelings of insecurity and his desire to avoid an audience. But equally featured is his account of dedicated efforts to improve. What later was noted as his facile storytelling surely had its beginning in his days as a sports announcer, sitting in a studio and weaving full narratives of action taking place hundreds of miles away, and about which he knew only a few cryptic facts from Morse-coded telegraphic bulletins. For this he practiced, and accepted guidance, even reading commercials to get the "right rhythm and cadence and give my words more emotion." Building upon his ability to ad lib, his ease in memorization, and his sense of timing, he developed his skills by acting in fifty-three Hollywood movies. In those he became known as a "script doctor" because of his ability to write in quick script changes.[94]

In a later career as a General Electric spokesman he learned about adaptation to live audiences, and such aides as Peter Hannaford and Michael Deaver saw him through the postgubernatorial career of political speeches to almost anyone who could pay a fee. In 1979, for example, Reagan's income tax report included $380,500 from speeches, $58,453 from radio broadcasts, and $26,757 from writing for newspapers and magazines. Even after paying his expenses, doubtless including salaries of speechwriters, his net income from these sources was $289,977.[95]

Thus Ronald Reagan came to the presidency with an extensive

background of formal instruction and informal experience that was instrumental in his success. As Anthony Lewis wrote after the 1986 *Challenger* explosion, "People waited: Not for an answer . . . but for words of consolation. They came, with rare grace, from President Reagan . . . in a few words, simple and direct . . . he expressed our inchoate feelings. He was touching without being mawkish. He was dignified. Listening to Mr. Reagan, I thought I understood better than ever before the mystery of his enormous popularity as President. . . . the main reason for public affection lies, as always, in Mr. Reagan's personality and his ability to communicate it. . . . In cold print the next day his words seemed flat. But when he spoke, there was tangible emotion in them, resonating with his listeners."[96]

Tenth, Ronald Reagan's heavy reliance upon speechwriters certainly provides a lesson that will impress all presidential aspirants and other political speakers: *Even if you fancy your own way with words, hire the best speechwriters you can afford, and preferably ones who will be comfortable with your ideology and philosophy.* The significance of this lesson was most apparent on 11 February 1991, when it became known that President Bush's chief of staff, John Sununu, and his chief pollster, Robert Teeter, were hunting a new and better chief speechwriter for him. William Safire, an old hand as presidential speechwriter, addressed a column of advice to the new prospects. *"Don't be haunted by Reagan's Ghost,"* he wrote.

> Mr. Bush is fearful of comparison with Reagan speechifying, and instead shows off his ad-libbing advantage over his predecessor—perversely boasting of his inadequate speech delivery. Enough of that; you've been brought aboard not for sound bites and snippets, but for serious meat and potatoes. As the Ides of March approach, in prime time, the President will solemnly come before us to explain his war's endgame. He has the sense of history in his head but not on the tip of his tongue. It's your enviable job to help him make weapons out of his words.[97]

Conclusion

This essay has focused upon the eight-year presidency of Ronald Reagan, whose political speaking was the performance of a consummate actor, playing a political role. In one sense, of course, all rhetorical as well as dramatic performances are role playing. Moreover, rhetoricians have always recognized that the performing arts are amoral, neither good nor evil in themselves. These arts of the public forum are equally accessible to demagogues and charlatans and to men and women of good will. Thus the use of the performing arts by those seeking or holding office should not be deprecated out of hand. But it is reasonable to insist that the performing arts be used responsibly, not just effectively.

The argument here is that there is a danger of being lulled by the effectiveness of Ronald Reagan's speaking (judged by his election victories, legislative successes, and standing in popularity polls) into assuming that it has also been responsible speaking and that it is thus acceptable as a political speaker's model. The argument has also been made that Reagan's effectiveness in the performing arts tended to conceal basic inadequacies in the substantive arts of rhetorical invention and disposition.

In effect, this essay reaffirms the classic proposition that gives substance priority over style; and it reminds us of Plato's charge that rhetoricians and public orators were more concerned with winning than with searching for truth, even if it meant making the worse appear the better reason. It was Aristotle, of course, who replied that persuasion is a skill that can be used for worthy or unworthy ends. "What makes a man a sophist," he said, "is not his skill, but his moral purpose." He believed that if the better side of the argument is ably presented, even a skilled persuader cannot long make the worse appear the better reason.

Citizens always have it within their power, through diligent use of the democratic process, to choose men and women of good moral purpose as presidents and presidential candidates. Such persons, it must be believed, will employ the performing arts knowing, as Adlai Stevenson put it, that "the hardest thing about any political campaign is how to win without proving that you are unworthy of winning."

5

Antithesis and Oxymoron

Ronald Reagan's Figurative Rhetorical Structure

James Jasinski

Ronald Reagan related the following anecdote to a group of Republican officials in 1988:

> It's like the story about a Congressman sitting in his office one day when a constituent comes by to tell him why he must vote for a certain piece of legislation. The Congressman sat back, listened, and when he was done he said, "You're right. You know, you're absolutely right." The fellow left happy. A few minutes later, another constituent came by, and this one wanted him to vote against the bill. The Congressman listened to his reasons, sat back, and said, "You know, you're right. You're right. You're absolutely right." Well, the second constituent left happy. The Congressman's wife had dropped by and was sitting outside the office when she heard these two conversations. When the second man left, she went in and said, "That first man wanted you to vote for the bill, and you said he was right. And the second one wanted you to vote against it, and you said he was right, too. You can't run your affairs that way." And the Congressman said, "You know, you're right. You're right. You're absolutely right."[1]

Since his election in 1980, Ronald Reagan has been the subject of sustained scholarly attention. One line of inquiry has sought to uncover, in Robert Ivie's terms, the source of the "coherence and rhetorical force" of Reagan's public discourse. Some scholars point to Reagan's use of narrative story lines and anecdotes—like the one contained in the epigraph of this essay—and attribute his oratorical success to these rhetorical forms. Other scholars focus on Reagan's use of argumentative resources and strategies and other techniques of modern political communication.[2]

While not wanting to deny the value of, or discredit the analyses contained in, these studies, I do want to suggest in this essay that there are neglected aspects of Reagan's rhetorical arsenal that merit critical attention. One such overlooked dimension—Reagan's use of rhetorical tropes—is the focus of this study.[3] Specifically, I identify two tropes—antithesis and oxymoron—as fundamental to the force and coherence of Reagan's rhetoric. Reagan's concern with logical contradiction, manifest in his anecdote about the congressman, is not merely a way of attacking Democrats; in this essay I show how Reagan achieves rhetorical coherence and force through manipulating the resources of antithesis and, most important, oxymoron.

The description and classification of tropes has long been a central aspect of rhetorical theory. Recently, theorists and critics interested in the nature of tropes have rejected taxonomic approaches in favor of explanatory conceptualizations. As Hans Kellner puts it: "Tropes in particular have become a kind of *lingua franca* bridging linguistics, rhetoric, poetics, philosophy, criticism, and intellectual history; the descriptive power of tropes has extended itself into an explanatory power, and the solvent capability of these dual processes—description and explanation—seems to offer virtually limitless possibilities of analytic power in many different directions."[4] Contemporary tropological studies are no longer restricted to descriptive categorization; scholars have come to recognize the cognitive and conceptual potency of tropes.

Nonetheless, the ongoing valorization of tropes poses certain problems for the rhetorical critic. Like medieval rhetorical theory, there is a tendency for contemporary tropology to turn in on itself and exhaust its critical energies in efforts to establish one "master" trope or reveal the inherent—but hidden—meaning of specific figurative forms. In this essay I attempt to harness the resources of current expansive tropological theory in the service of the more particular and practical concerns of rhetorical criticism. As such, the analytic focus of the essay is on the tropological structures and patterns that emerge at the discursive level of a message (as opposed to the specific devices that constitute a text's linguistic level). The primary emphasis is not on specific linguistic antitheses or oxymora ("jumbo shrimp"); rather, critical attention is devoted to describing Reagan's more elaborate antithetical and oxymoronic structures (for example, the American need to want but also not want—or at least not want *too much*—arms control). By focusing on the discursive level, I show in the body of the essay how antithesis and oxymoron function as a basic figurative structure in Reagan's discourse. Reagan typically assesses situations, and defines their fundamental problem, by recourse to antithesis. This figurative strategy renders a complex world tractable. Solutions, on the other hand, are constructed oxymoronically; this figurative strategy tames the contradictions that mark our social existence. In the conclusion of the essay, I argue that this problem-solution, or antithesis-oxymoron, structure generates important conceptual tensions (what I would term latent ironic consequences) that have the potential to stimulate more reflective public engagement of the contingencies that are inherent to public life.

A basic assumption in much recent rhetorical criticism maintains that public discourse is crafted as a response to situational exigencies. However, critics disagree over the extent to which advocates *create* the exigencies their discourse seeks to modify.[5] One need not endorse some sort of radical ontological relativism in order to appreciate that the very idea of a situation—and in turn specific types of situations—is produced through collaborative interaction. When a message imposes a name on a certain domain of empirical reality, it "brings into being that very situation to which it is also, at one and the same time, a reaction."[6] As Burke makes clear, names are never neutral. Naming a situation imparts to it a specific form which is, quite often, figurative in nature. Tropes and figures are a common resource through which individuals and social groups "form"—name and/or define—situations.[7]

When messages emphasize an exigence, an "imperfection," they either implicitly or explicitly claim that the "encompassed" situation is problematic. Advocates often rely on tropes, and other figurative resources, as a way of organizing, forming, or encompassing the situation at hand. In the case of Ronald Reagan, situations, and their attendant problems, are organized antithetically.

In antithesis, discursive elements (ideas, events, people, actions, and so forth) are arranged in relationships of contrast and/or opposition.[8] Antithetical contrasts commonly function as descriptions. For example, in an effort to describe one's feelings for another person, one might contrast those feelings with one's feelings for someone else (if the elements contrasted are very unlike, we classify the figurative result as a metaphor). In antithetical contrast, elements illuminate each other without introducing a subordinating relationship. In antithetical opposition, elements are juxtaposed so as to introduce radical difference and evaluation. One element in the juxtaposed pair is subordinated to the other element on the basis of an implied value scale. For example, an advocate might juxtapose one course of action (full disclosure of secret government practices) with another (a cover-up) in such a way as to render them mutually exclusive (an either/or question).[9] Once radical difference is established, evaluation can proceed, as "cover-up" is subordinated to "full disclosure" on the basis of the value of honesty (conversely, full disclosure could be subordinated to cover-up based on the value of effectiveness). Frequently, advocates define a situation as problematic when audiences appear to be supporting the subordinated element in the antithetical opposition.

Numerous critics have pointed out Reagan's reliance on relationships of opposition.[10] It is certainly true that Reagan employs the "master" antithesis—good *versus* evil—with regularity. Yet an exclusive preoccupation with this moral-religious dichotomy obscures other central oppositions around which Reagan constructs his vision of the social world. If we consider samples of Reagan's domestic affairs and foreign policy discourse, a series of antithetical oppositions can be uncovered. In turn, these oppositions interact with each other creating one dominant antithetical "cluster" for domestic affairs (government *versus* people) and for foreign policy (totalitarianism *versus* freedom).

Reagan's campaign for the presidency in 1980 was built largely around the opposition of government *versus* people. In 1988, despite the fact that Reagan was the government's chief executive, this opposition remained prevalent. In his 1980 acceptance speech, Reagan described his "view of government" as one which "places trust not in one person or one party, but in those values that transcend persons and parties. The trust is where it belongs—in the people." The task before the American people in 1980 was to reign in government, "to bring our government back under control." In order to do this, it would be necessary to thwart the Democrats' plan for "more Government tinkering, more meddling and more control." In 1988 this antithetical pattern remained central in Reagan's vision. Speaking about the American Revolution to an audience in Massachusetts, Reagan claimed that "the insight that was at the heart of the revolution begun here two centuries ago [was:] Trust the people, let government get out of the way, and leave unharnessed the energy and dynamism of free men and women." Later in the same address, Reagan returned to this theme: "It is governments, after all, not people, who put obstacles up and cause misunderstandings."[11]

The basic government-*versus*-people antithesis is developed in richer detail in other addresses. Reagan constructs an elaborate vision of the qualities and characteristics that mark the people as superior to the established government. One secondary antithesis focuses on the intellectual virtues of each element in the dominant opposition. The people embody the virtues of the "heart" while government represents the "virtues," which turn out to be the limitations, of abstract theory. The people demonstrate "unquestioning faith" while government agents and bureaucrats cling to abstract theories that commit them to "intellectual skepticism."[12] Faith allows the people to understand the simple truths of the world; skeptics view the world as devoid of truth and, therefore, terribly complex.

One of Reagan's favorite heroes is Knute Rockne. In a 1988 address at Notre Dame dedicating the Rockne commemorative stamp, Reagan used the Rockne legend to elaborate the opposition between faith and skepticism. Some of Rockne's contemporaries apparently did not believe in the legend, did not believe in the Rockne magic. In reality, Reagan tells his audience, "there was little room for skepticism or cynicism; we knew the legend was based on fact." Some people always seem to doubt the truth embodied in a legend. And what is the truth of the Rockne legend? "Most of all, the Rockne legend meant this—when you think about it, it's what's been taught here at Notre Dame since her founding: that on or off the field, it is faith that makes the difference, it is faith that makes great things happen."[13]

Because of their cynicism and skepticism, government representatives and employees are pessimists; the people, bolstered by their faith, remain optimistic. Reagan frequently reinforced the opposition between faith and skepticism, optimism and pessimism, through the use of chiasmus.[14] In one of his weekly radio addresses during the 1984 campaign, Reagan summarized the differences between himself and Walter Mondale: "The difference between us is that we look at a problem and see opportunity, and he looks at opportunity and sees a problem."[15] Skeptics allow the problems of the world to induce paralysis (problems overcome skeptics); people of faith see problems as a challenge to be overcome (people of faith overcome problems). The syntactical inversion reinforces the conceptual opposition that Reagan posits.

The intellectual virtues of the people, as opposed to the apparent virtues of government leaders (especially Democrats in Congress), entail markedly different social programs. Consider two examples. Addressing the National Rifle Association in 1983, Reagan distinguishes two different attitudes toward environmental management. The skeptics in Washington have an "elitist approach." Elitists do not trust the people. They are pessimistic about the public's ability to both enjoy and manage nature's bounty without supervision. Consequently, elitists believe "that vast natural resource areas must be locked up to save the planet from mankind." Elitist policies, Reagan implies, treat the public like children. Another opportunity is turned into a problem.

While vague about the specifics, Reagan counters the elitist attitude with the principle, legitimated by his faith in the people, of "stewardship." Stewardship entails "respect for both man and nature . . . caring for the resources we have for the benefit of mankind." Our model stewards, "the backbone of our conservation efforts," are "American sportsmen."[16] Sportsmen, Reagan implies, instinctively know the appropriate balance between using and abusing natural resources. All Americans need to do is have faith in sportsmen; they are, after all, part of the American "people."

In Reagan's vision of government there appears to be one thing that the skeptics are not skeptical about: the need for greater government intervention in the lives of the American people. Despite their "good intentions," the skeptics' commitment to more and more government intrusion has resulted in the creation of "a growing army of professionals . . . [whose] economic self-interest lay in extending dependency, not in ending it."[17] The common-sense virtues of the people yield policies and programs that will accomplish what they set out to do; all the government skeptics can do is (ironically) perpetuate the very conditions they purport to remedy.[18]

Reagan's welfare policy discourse illuminates another important secondary antithesis within the government-*versus*-people opposition. In addition to intellectual virtues, Reagan's antithetical antagonism is built out of oppositions between the constituent elements of the larger categories. In answering the questions What kind of people are we? and What kind of people run our government? Reagan adds detail to the dominant antithesis. Reagan employs the term *people* in an inclusive manner. The people are the families, the parents, the business community, the sportsmen, and the many others who reside in the private sphere. Reagan recognizes two legitimate categories for describing the roles people occupy in the private sphere: the people exist in either their narrow role (as "individuals") or their collective role (as "Americans"). In Reagan's words: "I don't look at people as members of groups; I look at them as individuals and as Americans."[19]

The government—the public sphere—is populated by "Washington-based bureaucrats and social engineers" with a "big brother" mentality.[20] Government agents invade the private sphere by projecting racial, ethnic, and other essentially artificial categories onto the American people. In Reagan's rhetoric, the American people are homogeneous; no significant occupational, social-class, racial, ethnic, or gender differences are recognized. The only valid distinction is Reagan's opposition of government (individuals who are incapable of recognizing the essential indivisibility of the people) and the general public.

Reagan's foreign policy discourse develops a variation of the government-people antithesis. The principal opposition that Reagan employs to frame and define the world of foreign affairs is totalitarianism and freedom. "We're approaching," Reagan told the British Parliament in 1982, "the end of a bloody century plagued by a terrible political invention—totalitarianism." The signs of totalitarianism include "political control taking precedence over free economic growth, secret police, mindless bureaucracy, all combining to stifle individual excellence and personal freedom."[21] Totalitarianism exists in many guises: "brutal dictatorships," revolutionary guerrillas and thugs out for personal power, and messianic religious leaders.[22]

The opposition between totalitarianism and freedom (liberty, democracy) is total. Reagan establishes the antithetical relationship through a series of questions in his 1982 address to Parliament: "Who would voluntarily choose not to have a right to vote, decide to purchase government propaganda handouts instead of independent newspapers, prefer government to worker-controlled unions, opt for land to be owned by the state instead of those who till it, want government repression of religious liberty, a single political party instead of a free choice, a rigid cultural orthodoxy instead of democratic tolerance and diversity?"[23] As this passage makes clear, Reagan holds this opposition to be absolute.

As is the case with the antithetical constructions used to define the world of domestic affairs, Reagan's oppositions in foreign policy discourse envision a dualistic world of mutually exclusive, either/or relationships. In domestic affairs discourse, Reagan's antitheses render any opposition other than government-people as illegitimate. Antithetical oppositions such as labor-management or consumer-corporation depend on spurious categories fabricated by interventionist government agents (and their supporters). Reagan's foreign policy rhetoric works much the same way. Alternative definitions are eliminated through the antithetical construction. Nothing exists "in between" totalitarian bondage and democratic freedom. Land is either "owned by the state" or "those who till it" (Reagan evidently believes that banks and large corporations have not developed a modern form of tenant farming). Reagan's use of antithesis renders the complexities of the modern world manageable by framing the world into a set of relatively simple oppositions and easily understood problems.

Given the manner in which Reagan constructs problems through antithetical oppositions, identifying solutions would seem to be a simple affair. Reagan's antithetical discourse exhorts audiences to affiliate themselves with the articulated God-terms and utterly reject the subordinated devil-terms.[24] This simplicity is, I believe, deceptive; through tropological analysis we can see that Reagan's proposed solutions display a complex conceptual and rhetorical structure. Reagan's practical and ideological solutions to the exigencies of the modern world are oxymora: paradoxical and/or contradictory prescriptions, with supporting rationalizations, for action and belief that are taken to be valid and legitimate. Hugh Heclo points to this aspect of Reagan's politics: "Reaganism achieves political coherence precisely because it shuns abstract consistency."[25] A developed critique of Reagan's rhetoric needs to describe not only the antitheses but the oxymora that structure, and give force to, the discourse.

Critics have begun attending to Reagan's use of oxymora. Both Goodnight and Rushing imply that the central idea contained in the SDI or "star wars" missile defense proposal, to save ourselves from technology by more technology, is ironic and ultimately paradoxical. Rushing argues that Reagan reconstructs our sense of time in order to provide scientists with an opportunity to redeem themselves for their original "sin." The sense of time that Rushing finds developed in Reagan's "Star Wars" address—a " 'progressive' regression" to the past—is, as Rushing's use of quotation marks suggests, an oxymoron.[26] Reagan's use of oxymora is not a random occurrence, nor is it limited to discussion of strategic weapons. Repeatedly, Reagan escapes the dichotomous logic his antithetical oppositions create by resorting to oxymora. Three categories of oxymora are discussed below: oxymora relating to domestic policy, foreign affairs, and Reagan's persona or character.

One central paradox in Reagan's domestic policy discourse is his claim that the problems of the future are best met by the wisdom of the past. Addressing an audience of youngsters at Epcot Center in 1983, Reagan affirmed that "like millions of Americans, I'm a firm believer in the back-to-basics movement, because it is the basics that will best prepare us for the future." Speaking at Notre Dame in 1988, Reagan maintained that "our program has been to foster innovation and to keep our country in the forefront of change." Nevertheless, Americans must remain committed to the "lasting values and principles that are the heart of our civilization and on which all human progress is built."[27] Reagan's contention, his solution to the problem of complexity in the modern world, was that we must adhere to the traditions and fundamental ideas of the past; in other words, complexity is managed most effectively through simplicity. Reagan figuratively depicted his general approach to problem solving as a kind of "simple complexity."

Oxymora play an important role in solving specific domestic

problems. Reagan's January 1982 speech to the New York Partnership demonstrates the power of oxymora on a number of issues. Early in the speech, Reagan wondered what values are essential in rebuilding America. Responding to his rhetorical question, Reagan asserted that Americans need "initiative, ingenuity, and audacity" as well as "honor, integrity, kindness, and courage."[28] By no means are these values antithetical to each other. Nevertheless, Reagan's juxtaposition (these passages are separated by two paragraphs) implies a harmony among values that many would, I believe, see as potentially discordant. Reagan slipped over the way in which audacity, an arrogant disregard for established practices and customary restraints, is consistent with honor or integrity. Integrity implies that one will remain firmly committed to a code of conduct; audacity implies almost the opposite (the audacious are typically contemptuous of restrictive moral codes). But as Heclo notes, logical coherence is not essential; Reagan's paradoxical oxymoron taps into that dimension of the public's imagination which pictures itself in these dissonant terms.

Elsewhere in the same address, Reagan confronted the question of how to rebuild community. His response: "The key to rebuilding communities is individual initiative, leadership, personal responsibility. If we encourage these qualities in our people—and especially in our young people—then our freedoms will not wither and die." Heclo aptly describes Reagan's vision of community as "communitarian individualism." In the "Partnership" address, the question of community is situated between two passages testifying to the glories of "individual initiative." This long opening section concludes with a common Reagan strategy: invoking the language from the preamble of the Constitution ("We the people") as a vehicle for extolling the virtues of individual action. Reagan manages to transform the collective, collaborative, and very contentious act of founding into an unambiguous act of individual enterprise.[29] In short, the problem of building community is solved by, paradoxically, emphasizing individualism.[30]

Reagan addresses a third problem in the "Partnership" speech: how to stimulate economic recovery. Reagan's program is based on saving: "If America can increase its savings rate by just 2 percentage points, we can add nearly $60 billion a year to our capital pool to fight high interest rates, finance new investments, new mortgages, and new jobs. I believe a country that licked the Great Depression and turned the tide in World War II can increase its savings rate by 2 percentage points—and will." Yet this is not the only approach Reagan advocated. Taking a page out of Franklin Roosevelt's recovery program, Reagan urged a massive tax cut to pump money into the economy and allow consumers to spend. As Reagan put it in a radio address from August 1984, "Our tax cuts have meant more money for you to spend and invest and for you to save and use as you wish, more money to create jobs and expand the economy."[31] Reagan's formula for economic recovery calls for Americans to save *and* spend; it is never clear how this is to be done simultaneously.

Finally, Reagan's approach to social welfare programs demonstrates the persuasive potential of oxymora. As noted in the previous section, Reagan's long-range policy goal is to end the dependency relationship fostered by welfare programs. Michael Weiler suggests that Reagan analogically conceptualizes welfare programs as a habit-forming drug.[32] Given such a view, welfare recipients become "addicts" who need assistance in overcoming their welfare "addiction." What type of help is appropriate? Reagan typically endorses a cold-turkey solution; the addict must get off "the stuff" completely. Initially, this might cause the addict to suffer and make the person responsible seem cruel. But, in the long run, the benefits of breaking the habit outweigh the initial pain. Reagan's social welfare rhetoric, constructed along the lines of a drug analogy, develops a central oxymoron: sometimes (as the popular song lyrics go) "you've gotta be cruel to be kind." Put another way, it appears that there are times when compassion for the plight of our fellow citizens is best shown by curtailing the programs on which they depend.

A similar oxymoronic pattern occupies a central place in Reagan's foreign policy discourse. A staple theme in Reagan's rhetoric is that peace can be secured only through strength. Throughout two terms of office, Reagan has emphasized this basic cold war principle to such an extent, and with considerable success, that it now appears to be an element of our fund of "social knowledge."[33] Reagan's successful fusion of peace and strength has helped displace a mythic understanding, common in the persona of the gunfighter, that no amount of firepower or strength is enough because there is always someone out there who wants to challenge the fastest, or biggest, gun in town.[34]

An oxymoronic logic structures a substantial amount of Reagan's discourse on U.S.-Soviet relations. Reagan describes America's policy toward the Russians as "constructive cooperation."[35] Operationally, this policy entails that America be both cooperative and adversarial; America must challenge the Soviets but at the same time not challenge them. Reagan seems to recognize the inherent tension in this directive through his attempted dissociation strategy, which distinguishes challenge from threat.[36] The policy requirement, as Reagan portrays it, is to be both a cooperative partner as well as a formidable adversary.

In a slightly different form, this oxymoron reappears in Reagan's insistence that America must be willing not only to find peaceful solutions to world problems but to respond with force. "Strength and dialogue," Reagan asserts, "go hand in hand." Reagan's rhetoric builds an association between two attributes—fighting and talking—which, on the surface, appear unrelated. The implication of Reagan's oxymoronic association is that only fighters are good (effective, successful) talkers.[37]

A similar pattern works its way into Reagan's response to the freeze movement and the question of arms control. Addressing the World Affairs Council in 1983, Reagan acknowledged that "we Americans are sometimes an impatient people. I guess it is a symptom of our traditional optimism, energy and spirit. . . . [But] any of you who have been involved in labor-management negotiations or any kind of bargaining know that patience strengthens your bargaining position. If one side seems too eager or desperate, the other side has no reason to offer a compromise and every reason to hold back, expecting that the more eager side will cave in first." Reagan then applies this lesson to the problem of arms-control negotiations. "It is vital," Reagan asserted, "that we show patience, determination, and above all, national unity. . . . That is why I have been concerned about the nuclear freeze proposals . . . however, well-intentioned they are, these freeze proposals would do more harm than good."[38] What is the fundamental problem with a nuclear freeze? Reagan argues that an actual freeze, or even significant support for a freeze, makes America look too eager for an arms-control agreement. In order to be effective at the bargaining table, two antithetical tendencies must be balanced. Americans, Reagan claims, must want but at the same time not want (or not seem to care greatly either way) an arms-control accord.

Finally, Reagan's persona or character exhibits an oxymoronic structure. Broadly considered, Reagan has combined essential features of the Nixon (distance) and Carter (intimacy) presidencies. The result is a persona that is at once larger than life (part hero, part prophet) and down to earth (the guy next door); this combination of characteristics can be described as "intimate distance."[39] Reagan is at once removed from, yet an intimate part of, our daily lives.

Reagan's idea of leadership provides a more concrete example of how the oxymoron figure shapes his public character. In his 1980 acceptance speech Reagan claimed that "we need a rebirth of the American tradition of leadership at every level of government and in private life as well." What is the American tradition of leadership? Historically, Reagan maintained, "America is unique in world history because it has a genius for leaders—many leaders—on many levels." Eight years later, Reagan urged Congress and the nation as a whole "to let a thousand sparks of genius in the States and communities around this country catch fire and become guiding lights."[40] We are, it seems, a nation of leaders. But this places us in a paradoxical situation, for we need, but at the same time do not need, leaders in order to confront our pressing problems.[41] Something like this same paradox applies to Reagan; he is a leader of a significant movement in American public life, yet at the same time he is just an ordinary person articulating basic common sense (Reagan is only doing what anyone would do in his place). Reagan the leader is essential to the "Reagan revolution," but at the same time Reagan the leader effaces himself into the fabric of the unfolding mythology of the conservative "revolution."

Reagan's solutions to the dilemmas and pressing problems of public life involve more than a blind commitment to ambiguous God-terms. Reagan's rhetorical success is based on an antithetical and oxymoronic structure, or "logic," that renders paradoxes palatable and neutralizes contradictions. The implications of this rhetorical form require further attention.

Reagan is not unique in employing the rhetorical resources of the oxymoron. Robert Gunderson observes that there is a significant "oxymoron strain" in American public discourse. In their attempts "to please a majority in a badly divided society," public advocates continually "resort to a rhetoric of negative opposites."[42] The oxymoronic rhetoric of negative opposites has the potential to encourage reflection and enhance the quality of public deliberation. As Karlyn Campbell demonstrates in her study of women's-liberation rhetoric, the oxymoron is an essential resource for advocates confronting value conflicts, moral dilemmas, and "a vortex of contradiction and paradox."[43]

The commonplace distinction between live and dead metaphors can help explain the differences between the oxymora of women's-liberation rhetoric and the oxymoronic structure of Reagan's public discourse. The oxymora of the women's movement, as Campbell helps us see, are extremely potent. For example, by violating the "reality structure," the oxymora of women's-liberation rhetoric "are attempts at radical affirmation of new identities for women."[44] The oxymora of women's-liberation rhetoric are live; their force extends outward from the immediate tasks of the rhetorical occasion by provoking reflection on the stereotypes, contradictory values, and moral dilemmas that exist in contemporary society.

Reagan's oxymora, on the other hand, are typically dead, or sedimented. They do not challenge the hegemonic reality structure but rather artfully conceal its seams or gaps. Live oxymora exploit the conceptual tensions that emerge in the linguistic space established between the figure's terms. Sedimented oxymora evade the conceptual tensions arising from their figurative structure and, as a consequence, repress reflection. The tensions between "peace" and "strength" or the "individual" and "community" remain dormant. Reagan's oxymora are potent in the context of the immediate rhetorical situation—they are instrumentally effective and have the capacity to constrain subsequent controversies—but they are not constructed or employed in a way that stimulates conceptual reflection.

What additional consequences result from Reagan's sedimented oxymora? Any answer to this question at present must remain tentative. Two negative consequences, along with one that is potentially productive, merit discussion. The both/and logic of the oxymoron has the ring of egalitarianism and democratic pluralism. Politics, the pluralist might tell us, is not about confrontations and oppositions between social groups. Politics unites superficially divergent groups into a unified whole; it is inclusive. The both/and logic of the oxymoron is an appropriate vehicle for pluralist practice.

Pluralism continually seems to falter as one set of ideas emerges as "more equal" than the rest. But there is an additional problem: pluralism (as theory) and sedimented oxymora (as practice) can be conceptually compelling but practically disastrous. Rhetoric, as a practice and an art, has traditionally been about making specific choices in contingent, but very real, situations. Discrimination among possible courses of action functions as a prelude to audience decision and judgment. Sedimented oxymora diminish an audience's ability to make discriminating choices. One lesson that emerges from the history of rhetorical theory, which Reagan's oxymora obscures, is that we need not divide the world into a series of either/or relationships, but we do need to make either/or discriminations in order to act and render judgments.[45]

A second consequence of Reagan's oxymoronic discourse is its ability to generate irony. Reagan's rhetoric is rife with "miniature" ironies. Consider the following example. In remarks to local Republican officials in 1988, Reagan reviewed his administration's progress in reestablishing the true principles and practices of federalism. Reagan was particularly proud of Executive Order 12612 (he quipped that "everything in this town has a number attached to it—usually a large one"): "We took a lesson from the environmental movement, and now when any agency in the executive branch takes an action that significantly affects State or local governments, it has to prepare a federalism impact statement, which only seems proper to me."[46] Under the weight of his many oxymoronic constructions,

Reagan fails to note the irony in his action. Executive Order 12612 "solves" the problem of Washington bureaucracy and red tape by creating more bureaucratic regulations and procedures.

There is a larger sense of irony that emerges from Reagan's discourse. In the realm of domestic policy, Reagan's principal goal has been to move people from dependence to independence. Reagan promises people more control over their lives. But do the workers at factories that have closed and relocated to a different state have control over their lives? Do the many people with jobs paying the minimum wage have control over their lives? Reagan's program was designed to give power back to the people, but Reagan seems to misunderstand the nature of social power. The irony of Reagan's program is that many people, instead of having greater control of their lives, are subjected to forms of power that Reaganism appears incapable of understanding.

A similar argument can be made regarding Reagan's foreign policy rhetoric. Reagan argues that peace is attained through strength but does not recognize the potential for his program to produce destabilization. While the superpowers' arms race is currently in remission, continued Third World terrorism and bloody civil wars illustrate the underlying irony in Reagan's foreign policy rhetoric: strength often produces only the illusion of peace and security. Regarding national autonomy, Reagan argues that nations must be free to determine their own destiny. But in many respects, this freedom amounts to an ability to choose one's masters or keepers. Ironically, extending freedom becomes a new form of exerting control.

Reagan's rhetorical structure produces discourse that is incredibly seamless. There is little ambiguity in Reagan's rhetoric, given the definitional force of antithesis, and, thanks to the potency of oxymora, there are no contradictions. Where can one turn in an effort to oppose this discourse? One place is to the unintended by-products of the discourse: the experience of irony.[47] As people come to experience the ironies implicit in Reagan's rhetorical vision, gaps in the discursive structure appear and openings for rhetorical controversy develop. Alternative discourse that explores, and widens, these gaps and exploits these openings has the potential to counteract the coherence and force of Reagan's figurative rhetorical structure.

6

The Transformation of Actor to Scene

Some Strategic Grounds of the Reagan Legacy

Jane Blankenship
Janette Kenner Muir

Ronald Reagan left the presidency with the highest approval rating given any president at the end of his tenure in office since World War II. The precise nature of the legacy left behind so widely popular a president has been, and will continue to be, usefully discussed for some time. In this essay we trace a number of transformations, typically, but not always, upward, that strategically facilitate Ronald Reagan's establishment of a legacy. We do not address this legacy directly; we do, however, examine some strategic underpinnings, for to divorce strategy from substance is to do disservice to both terms. Style, on both strategic and linguistic levels, is a revelation of (sub)stance, in this case, a revelation of Ronald Reagan's self-perception, and of his perception of "the people" and the "ground" they would share with him in situating a vision of America.

That major figures often undergo upward and downward transformations in public esteem is readily apparent to observers of the political process. Sometimes such a transformation takes place quickly and dramatically for all the public to see. Such a transformation prompts or is prompted by what is frequently referred to as a "critical event." At other times, a transformation takes place slowly, less dramatically, and we often do not notice it until we have passed through it and look back.

Despite the everyday use of the term *transformation*, relatively few treatments exist that very fully explore the loci and nature of transformations which occur in political life. In this analysis, Kenneth Burke's "dramatistic pentad" provides us with a useful grammar. He asks, "What is involved, when we say what people are doing and why they are doing it?"[1] Burke suggests that five terms can be central "generating principles," the key "grammatical resources" for an answer to that question: act (what was done), scene (when or where it was done), agent (who did it), agency (how the agent did it), and purpose (why).

The pentad has been used for over forty years to help critics analyze single speeches, debates, a cluster of speeches on the same issue, texts over time such as constitutions or literary texts.[2] At first glance there is a simplicity about the terms of the pentad. Yet they may be developed into "considerable complexity" that at once yields a rich heuristic and still allows that very simplicity to be "discovered beneath its elaborations."[3] There are a variety of ways the pentad may be extended, but two in particular yield rich insights into public discourse. Burke urges us to develop a series of ratios between and among terms to facilitate the critic's viewing from a variety of vantage points: for example, scene-act, scene-actor, act-purpose, agent-agency. For instance, a speaker, allowing the scene to dominate in a scene-agent ratio, argues, implicitly at least, "*I* am a victim of the scene."

The ratios are ubiquitous to be sure, but it is the "pliancy among [the] terms," their "range of permutations and combinations," that makes them particularly useful "when considering their possibilities of transformation."[4] The notion of *transformation* is central to this essay. By focusing on Ronald Reagan's campaign to gain and keep the presidency, we will deal with the upward *and* downward transformations underlying the Reagan legacy.[5]

Below, we briefly recall five rhetorical "moments" of the Reagan presidency. We suggest that these moments are illustrative of the symbolic strategies that facilitate Ronald Reagan's transformation from actor to increasingly grand notions of scene. It is fitting, with so carefully scripted a president, that we choose full texts as our critical objects for, as Richard Darman observes, "speech writing in the Reagan White House was where the philosophical, ideological, and political tension of the Administration got worked on."[6] And as Peggy Noonan comments: "Speech writing was where the [Reagan] Administration got invented everyday."[7]

It would, of course, be inaccurate to assume that the transformations of Ronald Reagan were only upward. During his first term there was a recession with an unemployment rate of over 10 percent. At that point his public approval plummeted to just over 40 percent. Only several months later, however, his public approval rating took a turn upward and through the bombing of the Marine barracks in

Beirut, sending the troops to Grenada, and a 2 percent drop in the unemployment rate, the president's popularity remained on a relatively stable upward course.[8] Even when the public did not approve of Ronald Reagan's policies, they continued to like him, prompting his popular nickname "the Teflon President."

There *were,* to be sure, several other temporary concerns about the president's performance. For example, during the 1984 general election campaign Reagan's incoherent responses in the first presidential debate with Walter Mondale prompted his staff to take blame for "overbriefing" their boss and caused the *Wall Street Journal* to use the "s-word" when it wondered directly on the first page: "Was the President of the United States approaching senility?"[9] It took no more than a jaunty one-liner in the second presidential debate—"I refuse to hold my opponent's youth against him"—for large numbers of people to be reassured that *if* the Gipper had ever been out of it, he was now back. Humor had again served the president well, as it had throughout much of his career. Still, it took the Iran-Contra affair to decrease his approval rating to anything approximating his lowest rating during the economic recession of his first term.

In addition to several upward transformations, in this paper we examine two "moments" of downward transformation—one outwardly imposed when Reagan's credibility was seriously in question during the Irangate period; the other self-imposed, as he proceeded to leave the scene in his farewell address.

The 1980 GOP Primary Debates

The first "moment" we have selected, an upward transformation, is illustrated by examining the set of GOP primary debates in 1980.[10] On the national scene, these debates allowed us to see close up and in detail a microcosm of that year's primary campaign, during which Ronald Reagan was transformed from one among other GOP actors (candidates) to the *Republican Scene.* The symbolic features of his performance in the debates clearly presage those that were to become the staples of the Reagan rhetoric. Here, we will point to specific frames before, during, and after, the six GOP primary debates to show how Reagan's presence in the debates placed him, both verbally and visually, in the foreground of each scene and eventually facilitated his move to become the ground on which other actors played out the 1980 primary campaign.[11]

When we use the old term *presence* we are, of course, referring to the actor's presence in voice, facial demeanor, and disciplined body, coupled with the assurance that comes from knowing how to use the tools of the trade—the husky voice and somewhat breathy delivery to denote emotion; the whole series of nonverbal behaviors we equate with reflectiveness; the easy, relaxed appearance of a person used to audiences; humor to defuse issues and arguments; and the aura of ready confidence. But the presence of Reagan in the 1980 debates was much more. Presence is related to attention; people who have presence are accorded a different place in our consciousness of events. Persons who have presence assume the foreground in our consciousness; others play out their parts in relationship to that ground.

The press, the other candidates, and Reagan himself contributed to his foregrounding. The transformation of Reagan from actor to scene through frames created by the press is demonstrated best by press coverage of the now-famous Nashua, New Hampshire, debate. When Reagan grabbed his microphone and snapped at the moderator, "I paid for this microphone Mr. Green" [his name was Jon Breen], the media had the perfect frame. Replayed numerous times on television news programs, that particular picture of Reagan came to symbolize all the events in the hall that night. Reagan was not simply in the foreground; he became the scene.

But the press did far more than direct attention. The members of the press who served as moderators and panelists also focused attention on Reagan. In Iowa, when Reagan was not present, Mary McGrory introduced him into the debate when she asked John Anderson, "Why are none of your rivals, present and absent, acceptable?" Not satisfied with this implicit reference to Reagan, the next panelist asked John Connally how he differed from Governor Reagan. Led to comment by the press, the candidates mentioned Reagan's absence eight times and directly referred to Reagan policies six times.

When Reagan *is* present, as in the Houston debate, the foregrounding is even more evident: Early in the debate, moderator Howard K. Smith asked Reagan to respond to charges that he often misused statistics. The question was lengthy, and Smith cited numerous examples of Reagan's alleged errors. As Smith asked the question, he allowed Reagan to respond to each example, and on the television screen was a series of frames of Reagan and Smith in an animated, and often funny, dialogue. After Smith concluded the question, he still allowed Reagan to respond further.

Bush was noticeably absent during the entire exchange until the audience heard him say, "Let me, uh . . ." and Smith asked him if he wanted to respond. Smith's word choice was again significant. Directing the sequence so that Bush was nearly irrelevant, Smith chose not to call on Bush to respond; instead he asked if Bush *wanted* to respond.

Bush, too, recognized that he was out of sight. When Smith asked Reagan about his call for a blockade of Cuba, Reagan responded, the audience applauded, and Bush was forced to interrupt, "May I respond to that question?" It is no surprise, therefore, that when the evening was over, Reagan asked, "Was there a debate?"

The Reagan presence is also enlarged by Reagan himself. Drawing on his experiences as both actor and political campaigner, Reagan seizes every opportunity to subsume the other candidates and feature himself. On the platform in Manchester, New Hampshire, the random draw placed Reagan in the last speaker's position. He began, "I agree with everything that's been said." Answering the second question, he went overtime, and moderator Howard K. Smith interrupted him. Reagan quipped, "I just stopped," and the audience laughed, punctuating the sequence and making John Anderson's reply an afterthought. By the third question Reagan was no longer last, yet he began by saying that there was "not much left at the end of the line."

The frames that so feature Reagan are the result of his luck and skill; the cooperation of other actor/agents in the campaign (the press, the television crews, the other candidates, the public); and, in some cases, the ineptitude of the other actors. If Ronald Reagan's strategy of coopting frame after frame[12] allows him to operate as foreground in the audience's attention, so his second strategy, the strategy of creative circumferencing,[13] allows him to thematically envelope both the other GOP candidates and his audience. Moreover, this strategy of creative circumferencing allows him not only to envelope ("contain") the themes of the other candidates,[14] but also to enlarge the scene from scene 1 (a scene *within* a particular debate), to scene 2 (a *particular* debate), to the much larger GOP primary scene.

Of the four dominant themes in the 1980 primary debates, three are especially worth noting here because they point us forward to the next move in the transformation upward:

1. The source of strength of the American people has always derived from their generosity, courage, independence, goodness, and capacity to get things done, and, not insignificantly, from God's divine plan.

2. The strength of America lies in her people, not in her government; that strength can be "unleashed" by a president and party who say, "Follow yourselves."

3. Because of these endowments, the American people are en-

titled to "it all" and "it" is within our power to acquire if we recall our own special destiny.

Reagan's broad brushstrokes, indeed, paint a grand future that goes beyond November 1980 and a Republican victory. In the Nashua debate: "We have it in our power to begin the world over again" and in the Houston debate: "It is all here. It is all possible to the American people."

What larger circumference could any candidate sketch for this nation? The American people, after all, have a right to be "liberated" so they can dream their "glittering" dreams precisely because of their fundamentally heroic nature; for example, Reagan proclaims in Columbia, South Carolina:

> They [Massachusetts Bay colonists] went ashore. . . . They went out and across the mountains. . . . We became a great melting pot. . . . [This] most independent of people . . . a new breed, called an American—the most generous, the finest people in the world.

and in Manchester, New Hampshire:

> I've always believed it was some kind of divine plan that put this continent here . . . to be found by a particular kind of people, people with a love of freedom and the courage to go with it. . . . [Americans are] a new kind of human . . . the most independent, the most generous.

By summoning up the tiny Arabella, Thomas Paine, Abraham Lincoln, past triumphs over fires and earthquakes, Jeremiah Denton and the POWs, Reagan, in fact, enlarges the circumference of the scene to include the whole sweep of American history. The 1980 primary campaign became yet another test of American greatness, of whether we will again dream grand dreams or settle for the smaller dreams of lesser candidates.

In the 1980 heroic vision Ronald Reagan is a modest liberator. He "merely" makes it possible for the American people to dream *their* dreams. We, after all, are the "elect"; Reagan, in 1980, only our temporarily elected Vessel. As the Vessel that contains us, our dream, "The American Dream," Reagan frames the 1980 political season, so that we can play out our dreams through his election. The Reagan strategy of containment, of envelopment, first of the other candidates and then of the voters, is thus complete and clearly points the way to the larger combat with an incumbent president who had repeatedly asked the nation to dream smaller and less glittering dreams. Thus, Ronald Reagan had already laid solid groundwork for the move from Republican scene to presidential

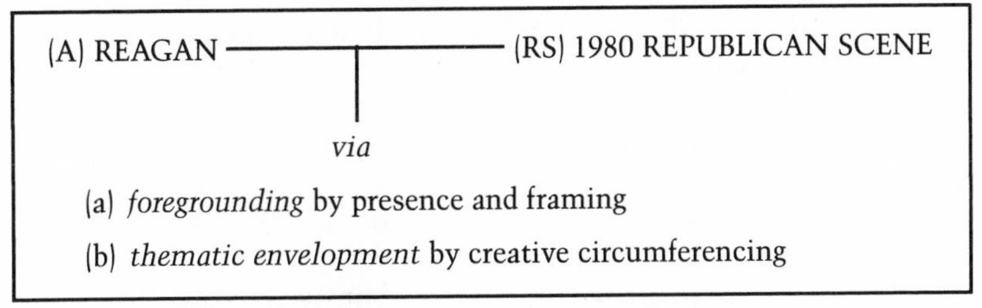

Figure 1. Strategic Moves to the Republican Scene

scene and ultimately to American scene. This transformation is illustrated accordingly in figure 1.

The Reagan Film at the 1980 GOP Convention

In June 1980 Richard Wirthlin, pollster for the Reagan campaign, wrote a confidential strategic memo surveying the political landscape. He observed: "There is a tendency in our increasingly complex and highly technological society to forget that American Democracy is less a form of government than a romantic preference for a particular value structure."[15] Wirthlin goes on to point out that "confidence is essential to legitimacy in government." He argues that only a few of our presidents, such as Franklin Roosevelt and John Kennedy, have been especially good at instilling confidence. "Their leadership," Wirthlin writes, "had a morale-building element to it. They conveyed a sense of hope. They communicated optimism."[16] Finally, the pollster suggests, "the sense of personal normlessness" has reached "its extreme" and that one response to the feelings of personal normlessness is the longing for an authority figure—a leader.[17] "Leadership," Wirthlin continues, "is the ability to enlarge men's vision about the future and give them expectations of a less uncertain and more gratifying future."[18] That memo could have been a blueprint for the film about Ronald Reagan at the 1980 GOP convention—a film that, in microcosm, shows us Reagan's transformation from actor to GOP primary scene, to American *presidential* scene.

The move from Republican scene (RS) to presidential scene (PS) is made with relative ease and no little "strategic" grace. Here, there is time for noting only two kinds of strategic "warrants"[19] or *ways* of moving from actor to scene.

First, recall that in the primary debates Reagan spoke consistently of Americans as a strong people who need to be "unleashed" (other variants were "unhobbled" and "unshackled") to take control of their own destinies. This allowed Ronald Reagan to be a "liberator." In the 1980 convention film, the "salvation effort" is escalated, or should we say elevated, to another, higher plain. In short, Reagan moves from a liberator to a "savior." The savior theme is very clear throughout the film.

Even at a very early age, the film explains, young "Dutch" Reagan was training for his future role.[20] The narrator of the film tells us that Reagan began in a small town, was family oriented, and had a "deep sense of values." He always had to work hard for his spending money, and one way he managed was with his job as a lifeguard at a small pool. While only one line in the film mentions this job, it is significant because, unsurprisingly, it is Reagan's favorite job as a youth; also it indicates that even at an early age, Dutch is in the business of saving people.

Through a series of jobs that took Ronald Reagan from sports announcer to actor to president of the Screen Actors Guild, we learn that Reagan has neatly prepared himself to be a leader. As an actor, people like him because he was "so clearly one of them"; as a union leader his popularity was such that he was elected six times to the presidency. The film narrator reminds us of Reagan's success in this role:

> He *single-handedly* took on the job of keeping union members in control of the union and prevented a takeover of the film industry by organized crime. . . .
>
> This *crusade*, to preserve the freedom of America's working men and women, kindled in Mr. Reagan a deep, abiding interest in the processes of our unique Republican form of government.

The savior image becomes clear as we see that Reagan "single-handedly" saved the union for union members, saved the entire film industry from organized crime, and took on the crusade to preserve the freedom of all working men and women.

The transition moves Ronald Reagan into the political arena with swelling music and the roll of drums, and the narrator begins a story about Reagan's shift from the Democratic party to Democrats for Eisenhower, to officially joining the Republican party. In 1966 he answers the "call of his party" and runs for governor of the state of California. He is elected, inheriting a state on the brink of bankruptcy. With "strong, creative leadership" he gets the state back on track and we are told of the acclamation Reagan is given for his success:

> The *San Francisco Chronicle* said this about Governor Reagan: "We exaggerate very little when we say that Governor Reagan has saved the state from bankruptcy." . . .

When Governor Reagan left office, a $194 million deficit was *transformed* into a $554 million surplus. He saved taxpayers hundreds of millions of dollars; he improved the quality of life for the people of California.

As with his union job, Reagan is given credit for his "saving" qualities as governor of California. He saves the state from financial disaster, transforming California into one of the strongest states in the union or, as the narrator describes, into a "model" state for the nation.

The jobs of lifeguard, union president, and governor of California are all significant for Reagan because they prepare him for the most important job to which he will be "called," the presidency of the United States. It is at this point that the language of the film begins to shift from a secular image of saving to one that is more mystical, a revelation of what the savior will do for America. The narrator tells us:

> This is the man who believes our nation can regain the stature it has lost in the eyes of the world. . . .
>
> He believes in America and the kind of strong, *creative* [special emphasis by the narrator] leadership that can restore our faith in the future.

So, one of the ways for facilitating the transformation to a larger, "higher" scene is the move from liberating to saving; from the more clearly secular to the somewhat more mystical.

Another of the warrants—the *ways* of making the move from actor to scene—is through what is *already known* by the audience. There are several ways in which this is done in the 1980 film.

First, the film's makers bet that we have at the very least some nostalgic residue of *communal* meanings summoned up by the "traditional values" with which "Dutch," with which "Ron," is so clearly allied in the film. For example: close-knit family relationships make up for having "few luxuries"; hard work pays off, not only in the case of student loans but in the long run as well; "deep love" of one's work plus the "help of a friend" plus "a bit of good luck" would pay off in—what else? The making of a star! Not merely a star, but a star whose luminosity is not only his but ours. What was Reagan's appeal in the movies? The narrator tells us precisely: "He became a star who appealed to audiences because he was so clearly one of them."

Second, the film points to what we already know—ourselves and our own traditional values—and reminds us that we have always known, from his early movie days, that Ronald Reagan has shared common values with us. Yet the film also tells us (to the accompaniment of drums, so we won't miss the point!): "In 1952 and 1956, following his own conscience, and with a clear view of the dangers facing America, Ronald Reagan volunteered as a Democrat for Dwight Eisenhower. In 1962 he officially joined the Republican party. The rest is history." History! Something else we already know and have known, at least since 1962, in terms of Ronald Reagan.

So now there are two major aspects of ourselves (our shared substance) "constructed" in terms of the GOP candidate: our values, as embodied *in* Ronald Reagan; and our history, as illuminated by the rise of one of us, Ronald Reagan. But there is also a third aspect of this point. Ronald Reagan is the "natural" choice for president because he is familiar to us and we, together, are familiar with our principles. He is: "a man whose *principles* have been familiar to Americans for thirty years." He does not merely hold our principles. He is confident in them, "confident in the triumph of the principles he has consistently defended." He is so confident, because he understands them: "he is a man who understands the principles that make America great."

In sum, Ronald Reagan is our "natural choice" because we have known and shared with him, values, history and principles for "over thirty years." We have known the interloper Jimmy Carter for only four! What Americans could vote against themselves—their own values, their own history, their own principles? Who could better lead us in "a confident advance into the future" than an idealized version of ourselves—successful, confident, and strong? We are surely ready by now for the last equational structure of the 1980 film. Ronald Reagan's "inaugural day" equals the "day America comes together" for a "new beginning." One can hardly be more explicit than that. A brief account of the main strategic moves can be sketched accordingly in figure 2.

After the initial honeymoon period, Ronald Reagan's first term in office was spent maintaining and enhancing his legitimacy. The first three years of his presidency were marked by critical events, such as the assassination attempt, recession and unemployment at an all-time high, and the U.S. invasion of Grenada. By January 1983 Reagan's popularity with the American public had sunk to its lowest point, with only a 41 percent public approval rating.[21] However, as campaign time approached, Ronald Reagan finally admitted that he would run for reelection, troops were sent to Grenada, unemployment fell, and there was a resurgence in the president's popularity. These events paved the way for Reagan's victory in the 1984 presidential campaign.

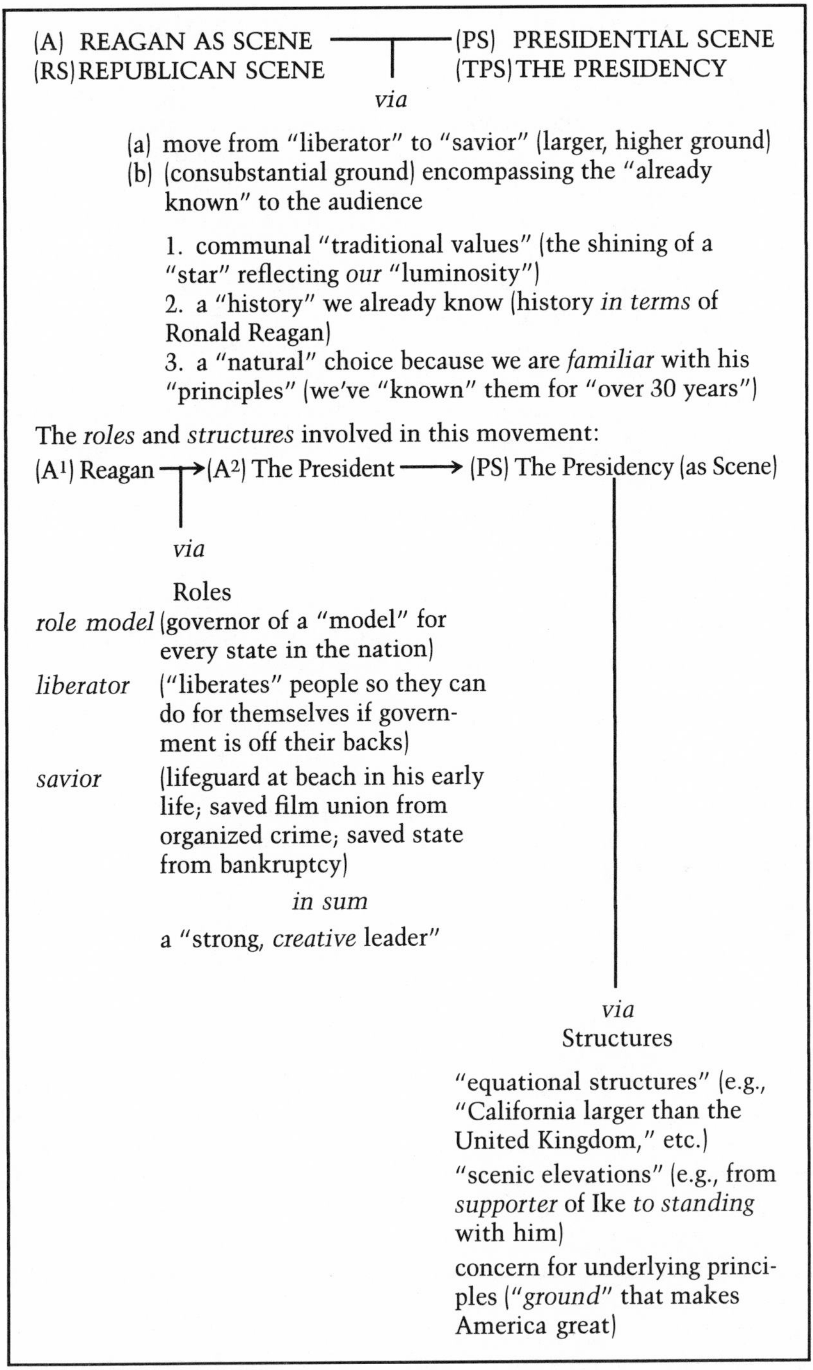

(A) REAGAN AS SCENE ——— (PS) PRESIDENTIAL SCENE
(RS) REPUBLICAN SCENE | (TPS) THE PRESIDENCY

via

(a) move from "liberator" to "savior" (larger, higher ground)
(b) (consubstantial ground) encompassing the "already known" to the audience

1. communal "traditional values" (the shining of a "star" reflecting *our* "luminosity")
2. a "history" we already know (history *in terms* of Ronald Reagan)
3. a "natural" choice because we are *familiar* with his "principles" (we've "known" them for "over 30 years")

The *roles* and *structures* involved in this movement:

(A^1) Reagan → (A^2) The President → (PS) The Presidency (as Scene)

via

Roles

role model (governor of a "model" for every state in the nation)

liberator ("liberates" people so they can do for themselves if government is off their backs)

savior (lifeguard at beach in his early life; saved film union from organized crime; saved state from bankruptcy)

in sum

a "strong, *creative* leader"

via

Structures

"equational structures" (e.g., "California larger than the United Kingdom," etc.)

"scenic elevations" (e.g., from *supporter* of Ike *to standing* with him)

concern for underlying principles ("*ground*" that makes America great)

Figure 2. Transformation Upward to the Presidential Scene

The Reagan Film at the 1984 GOP Convention

In June 1984 Richard Darman, a senior aide to Ronald Reagan, advised the campaign team: "We [the Reagan campaign] should continue to lay claim to the future" and, he counseled, paint Ronald Reagan as "the personification of all that is right with, or heroized by, America. Leave Mondale in a position where an attack on Reagan is tantamount to an attack on America's idealized image of itself—where a vote against Ronald Reagan is, in some subliminal sense, a vote against a mythic 'AMERICA.'"[22]

The Reagan film at the 1984 GOP convention illustrates in microcosm the equation Ronald Reagan equals mythic American scene (RR = MAS) and continues the transformation from actor to scene that began at least as far back as the 1980 GOP primary debates. The incredible richness of this film has been chronicled in detail elsewhere.[23] Here, we will merely indicate several of the key strategies played out in vividly lush visual and audio detail.

Storytelling has often been considered the primary form of communication—highly evocative and deeply personal.[24] In this film Ronald Reagan tells us both his own story and ours. For example, he narrates scenes of his attempted assassination:

> *Reagan* [*voice over filmed scenes of the assassination attempt*]: I didn't know I was shot. The—in fact, I was still asking "What was that noise?" I thought it was fire crackers and the next thing I knew Jerry, the Secret Service, grabbed me here and threw me into the car and then he dived in on top of me and it was only then that I felt a paralyzing pain and I learned that the bullet hit me up here.
>
> *Reagan* [*in close up, talking*]: When I walked in they were just concluding a meeting in the hospital of all the doctors associated with the hospital. When I saw those doctors around me I said I hoped they were all Republicans.
>
> [*Laughter in background.*]

Clearly, Ronald Reagan is a physically strong (his recovery was quick, with relatively few complications) and an emotionally strong man. He can even joke about a funny one-liner to doctors before the operation to remove the bullet from him. Not only is he a physically and emotionally strong man, but by most accounts he was an especially lucky man. Luck was not so much a lady as a supernatural power, according to no less a man of God than Cardinal Cooke: "God must have been sitting on your shoulder." Long accustomed to talking about America's divine mission, now, clearly, it is time for Ronald Reagan to repay good deeds for *that* good deed—to be God's man on earth.

As commander-in-chief, admiringly surrounded by our young troops in Korea, standing tall on the Great Wall of China, applauded by those brave old men who had survived the fire of Normandy Beach, we are not surprised when the film shows Ronald Reagan, chief among the elect, riding a white horse (in soft-focus lens) across an idyllic scene, quoting Scripture: "I will look unto the hills from whence cometh my help."

From the outset of the film, viewers are invited to equate Ronald Reagan's taking the presidential oath with American morning scenes.

[*Visual: Green fields, farm scene*]
Chief Justice Burger: Raise your right hand and repeat after me: "I, Ronald Reagan, do solemnly swear . . ."
Reagan: I, Ronald Reagan, do solemnly swear . . .
[*Visuals: Cowboy in blue jeans moving horses along; construction worker pointing upward*]
Burger: "That I will faithfully execute the office to the best of my ability . . ."
Reagan: That I will faithfully execute the office to the best of my ability . . .
[*Visuals: Newspaperboy riding by on a bicycle throwing paper onto porch; man in suit getting into a car with other commuters; blue-collar men walking to work.*]
Burger: "to serve, protect and defend the Constitution. So help me God."
Reagan: to serve, protect and defend the Constitution. So help me God.
[*Visuals: Raising of flag, children looking up; capitol dome*]

Reagan's "Yes, it was quite a day—a new beginning" summarizes the intertwinings as if they were all *one* narrative.

The commingling of Ronald Reagan's presidency with pristine, even archetypal, scenes continue to make him appear as part of the scene. In this context the scene is used merely as landscape. More broadly conceived is the tying of a vote for Ronald Reagan to the new beginnings of our lives, and an extended notion of family landscape. Clips from television advertisements running during the primary period (when Reagan had *no* challenger from his own party), appear in this film, helping us re-view already-seen television narratives and recapture the new beginnings of our own lives—getting married, moving into a new house, and the like. Here *we* have cowritten the script, for we also know the sights, sounds, smells, and touches of new beginnings. At this level the Reagan presidency equals the metaphysical landscape—warm, optimistic, and trouble-free. It is a landscape made even brighter when viewers juxtapose it with that of his opponent, Walter Mondale. Children in the Mondale commercials are every bit as clean, well dressed, and cherubic as those in the

Reagan ones, but the landscapes they inhabit are vastly different. In one Mondale commercial children are not so much cheerfully playing in their backyard as watching their dad dig a grave/bombshelter. In another, playful children are drawn to a school window outside which nuclear missiles are rising from the ground. Bad times are possible, even for children, in the Mondale landscape.

In the 1984 film the Reagan landscape holds not only smiling, undisturbed children but serene, happy senior citizens strolling along a beach or walking down a friendly neighborhood sidewalk. Not so, in the Mondale landscape; there, elderly women tell us they can barely get by on their social security payments. Life for them is hard and uncertain. But the Mondale commercials pull an even dirtier trick on us: lovely children talking about what they want to be when they grow up, juxtaposed with elderly women talking about a less-than-lovely life—wondering how, or even if, they are just going to get by. The Mondale commercials, by implication, suggest that the Reagan commercials are the stuff of myths. But the Reagan commercials are, more powerfully, the stuff of Myth—of heroes riding white horses on meadows of bright green, of clear waters shimmering in the sunlight, of small, personal, rice-throwing rituals, and of great, lavishly costumed, trumpet-blowing rituals of foreign lands.

The most emotionally charged moment of the film comes as Ronald Reagan shares with us a "representative anecdote" on a windswept beach at Normandy. There, fully understanding the need to let the words come not from the narrator but from one closer to those about whom the Normandy story has been told for forty years, the camera pans to the young daughter of one of the "boys" who had died of cancer eight years before this reunion of World War II veterans. One may, after all, question the authenticity of the choked voice of the storyteller; it is more difficult to doubt the real tears of a real child for her dad. Ronald Reagan reads her letter to her father: "I'm going there, Dad, and I'll see the beaches and the barricades and the monuments. I'll see the graves and I'll put flowers there just like you wanted to do. I'll feel all the things you made me feel through your stories and your eyes. I'll never forget what you went through, Dad, nor will I let anyone else forget. And, Dad, I'll always be proud."

Reagan, a master at "summoning up" real people (recall, for example, his recognition of everyday heroes present in the House gallery during his State of the Union addresses) can summon up the dead soldier through the living presence of his daughter. It is precisely at this moment that audiences tend to, literally, lean into the film—to more clearly become a part of it, to "reach out and touch" it, and they directly associate this emotional impact with the Reagan presidency.

And if the emotional impact of the moment at Bitburg is not enough to engage the audience, then perhaps Lee Greenwood's "I'm Proud to Be an American" provides the final clincher. Panoramic vistas of wheat fields and mountaintops, of the Statue of Liberty and American flags, fill the screen as the song resonates, instilling a pride in the principles that America stands for and, in turn, connecting those principles with Ronald Reagan. In sum, the kinds of strategic vehicles that transport Ronald Reagan from the presidency scene to the mythic America scene can be identified in figure 3.

(A²) The Presidency (PS) (*A New Beginning*—tagline to the 1980 film) → Mythic American Scene (MAS) ("It's Morning in America": an even stronger, more idyllic, and more detailed "temporizing of essence")

via

Shared narratives: (a) *series of romantic vignettes illustrating "morning"*: newly marrieds emerging from a church; young couple moving into their own home; young, happy children at play; happy senior citizens; "upward mobility" everywhere at hand, e.g., homebuilding, skyscrapers, flags being raised; (b) representative anecdotes (real people such as the daughter of a Normandy veteran) who "tell" war stories; (c) Reagan as narrator.

The careful *identification*, e.g., of *Reagan* taking the oath of office, and *idyllic American scenes* (landscapes and "events"). Oath, idyllic scene, oath, idyllic scene.

The careful identification of *church and God* with *Reagan*; e.g., Cardinal Cooke tells Reagan at hospital after the assassination attempt, "*God* must have been sitting on *your* shoulder" and Reagan on white horse, "I will look unto the hills from whence cometh my help." ("Embodiment" in an "authority figure" upon a field of the "elect.")

The "formal" techniques of the film: panoramic vistas of wheatfields and mountains; American flags and the Statue of Liberty; a song about pride in America.

Figure 3. Transformation to Mythic American Scene

Responses to Stories of an Arms-for-Hostages Swap

In July 1986 the president's public approval rating was very high. Well over 60 percent of the American people answered affirmatively the question: "Do you approve or disapprove of the way Ronald Reagan is handling his job as president?"[25] By November it had plummeted by more than fifteen points. Stories of arms sales to Iran to encourage the return of the American hostages buzzed furiously in both the American and the world press. Then, on 13 November, the president spoke to the nation on the Iranian arms sales, assuring the American public that he would give us "the facts" rather than a "lot of stories [over] the past several days attributed to Danish sailors, unnamed observers at Italian ports and Spanish harbors, and especially unnamed Government officials of [his] Administration" about shipping "weapons to Iran as ransom payment for the release of American hostages in Lebanon"—transactions during which "the United States undercut its allies and secretly violated American policy against trafficking with terrorists."[26]

The highly charged atmosphere surrounding the possibility of swapping arms for the American hostages in Lebanon is further charged by the President's language choices. Matters are not merely sensitive; they are "extremely sensitive." Talk is not speculative; it is "merely speculative." There is not merely speculation; there is "unprecedented speculation." There are not merely many reports in the air; there are "countless reports." Reports about a possible ransom payment and secret violation of U.S. policy against trafficking with terrorists are "quite exciting" but "none of them is true." Not only are these reports wrong, but they are "potentially dangerous." Risk is, and will remain, in the air. The "sensitive undertaking," "those with whom we were in contact," and, indeed, the "entire initiative" is at risk or "great risk." Still, if we do not persevere in restoring a relationship with Iran, in ending the war in the gulf, in halting state-supported terror in the Middle East, and in effecting a safe return of the hostages from Lebanon, there are even "greater risks" ahead. In this time of risk, the president urged us not to listen to others (unnamed sources) who tell "stories," spread "rumors," and level "false charges." Rather, we should listen to the president, Ronald Reagan by name, who tells us the "facts" and who speaks the "truth." Once we've heard his side of the story, once he's talked to us "directly—to tell [us] first hand" about the dealings with Iran, the truth will "stay put" (to quote Will Rogers). Ronald Reagan thus remains confident that the American people, given the facts, will make the "right decisions." In just one brief section of the speech, where President Reagan compactly compares "rumors" with "facts," we are asked to distinguish between:

Rumors Denied	*Facts Admitted*
• The United States swapped boatloads or plane loads of weapons.	• The United States sent weapons that could fit into one cargo plane.
• The United States sent spare parts and weapons for aircraft.	• The United States sent spare parts for defensive weapons (Reagan does not specify whether for aircraft).
• The United States has made "concessions."	• The United States did authorize a transfer of weapons.
• The United States swapped arms for hostages.	• The United States would *never* directly trade arms for hostages.

Reagan believes in the American people's capacity to distinguish fact from rumor, the true from the false, the ethos of the named source from the suspiciousness of unnamed sources. Confidence should beget confidence.

The president, however, was operating in a rebuttalist stance, and that is not his best posture. To operate refutatively is to admit there is at least one other (partly) credible view of the way things are. Reagan, a supremely good affirmative speaker, seems less suited to the job of answering charges. Moreover, after a life full of speeches proclaiming that Americans are "special people" and that we have it in "our power to make the world over again," this presentation must open even true believers to modest doubts. If we are so special, why can't we accomplish the four goals the president laid out at the beginning of the speech. Must we, indeed, eventually stoop to bargaining with those he calls "terrorists," "radicals," and the "scourge" of the world?

Reagan *does* work to maintain a strong "I" in the speech, the capable leader making tough decisions and assuming responsibility, for example, *he* authorizes the national security adviser, Robert McFarlane, to go to Iran and to make "stark and clear our basic objectives and disagreements." Although he does urge those who "think that we have 'gone soft' on terrorism" to "take up the question with Colonel Qaddafi," there is no "damn the torpedoes, full speed ahead" tone in this speech. He seeks, rather, our "patience and understanding." We are urged not so much to stand tall but to stand by, and to trust our leader to make the right decisions.

Less than a week later, on 19 November 1986, President Reagan called a news conference to answer questions on what he called "a secret initiative to the Islamic Republic of Iran."[27] Here, we would offer four exhibits that help us chart the downward transformation of the president. "Ronald Reagan" no longer equaled the "*mythic* American scene." His credibility was seriously questioned; he confirmed his own not knowing and misjudgments; he acknowledged his lack of success; and he now talks about diminished possibilities and increased risk-taking.

QUESTIONED CREDIBILITY

Exhibit 1

Question: Mr. President, in the recent past there was an administration whose byword was "Watch what we do, not what we say." How would you assess the credibility of your own administration in the light of the prolonged deception of Congress and the public in terms of your secret dealings with Iran, the disinformation, the trading of Sacharoff for Daniloff?
Question (follow up): You don't think your credibility has been damaged?

Exhibit 2

Question: Mr. President, you have stated flatly, and you stated flatly again tonight, that you did not trade weapons for hostages. And yet the record shows that every time an American hostage was released . . . there had been a major shipment of arms just before that. Are we to believe that was just a coincidence?

Exhibit 3

Question: Mr. President, when you had the arms embargo on, you were asking other nations, our allies in particular, to observe it publicly. But at the same time privately you concede you were authorizing a breaking of that embargo by the United States. How can you justify this duplicity?
Question (follow-up): Sir, if I may, the polls show that a lot of American people just simply don't believe you. That the one thing you've had going for you more than anything else in your Presidency—your credibility—has been severely damaged. Can you repair it?

Exhibit 4

Question: If your arms shipments had no effect on the release of the hostages, then how do you explain the release of the hostages at the same time that the shipments were coming in?

SELF-CONFIRMED NOT KNOWING AND MISJUDGMENTS

I didn't know anything about that till I saw the press on it. . . .

No, I've never heard Mr. Regan say that. And I'll ask him about that. . . .

I don't know whether it's called for or whether I have to wait until we've reported to Congress. . . .

Well, now if I have been misinformed, then I will yield on that.

ACKNOWLEDGED LACK OF SUCCESS

DIMINISHED POSSIBILITIES AND INCREASED RISK-TAKING

We knew the undertaking was risky. . . . This undertaking was a matter of considerable debate within administration circles. . . . Several top advisors opposed the sale. . . . I understand the decision is deeply controversial and some profoundly disagree. . . . Even some who support our secret initiative believe it was a mistake to sell any weapons. . . .

There was the knowledge that we were embarking on something that could be of great risk to the people we were talking to, great risk to our hostages. . . .

There was a risk involved.

In brief, the 19 November news conference allows us to illustrate the downward move from mythic American scene of unlimited possibilities to a far less than mythic America, an America of *limited* possibilities. Ronald Reagan is no longer an actor synonymous with scene, certainly not with MAS and increasingly even less with PS; rather, he has become an actor trying to survive *in* that scene, an actor no longer viewed as a romantic hero "standing tall in the saddle" but an actor with a challenged (if not failed) rhetorical vision, one who frequently lacks even key day-to-day successes in his battles with Congress and who is, increasingly, surrounded by a staff functioning at least partly without his control (as illustrated in figure 4).

The Farewell Address: A Stronger City, a Freer City

On 11 January 1989, in his thirty-fourth speech from the Oval Office, Ronald Reagan bade his formal farewell to the nation. There

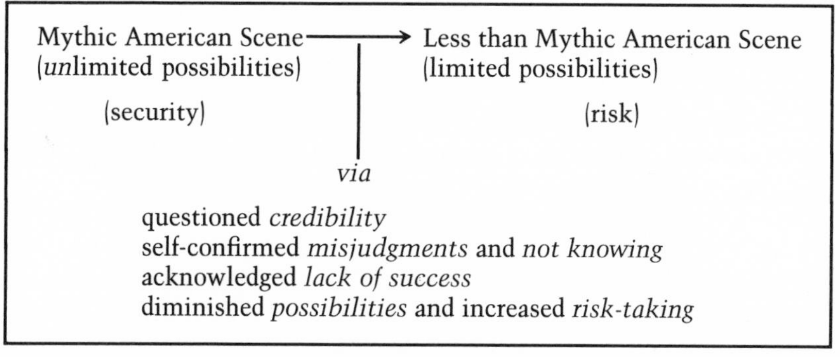

Figure 4. The Transformation Downward to Less than Mythic America

he recalled, as he often did, the words of "an early 'freedom man,'" John Winthrop, to the Puritans: "The past few days I've thought [about] the shining city upon the hill. . . . I've spoken of the shining city all my political life. . . . in my mind it was a tall, proud city built on rocks stronger than oceans, windswept. God-blessed and teeming with people of all kinds living in harmony and peace. A city with ports that hummed with commerce and creativity. And if there had to be city walls, the walls had doors, and the doors were open to anyone with the heart to get here."[28]

In this moment of farewell, the actor Ronald Reagan no longer equaled even the scene (a mythic America still, despite an occasional dimming of her luster). Rather, the scene contained the actor as a kind of concerned spokesman for the scene, the vessel or agency of the presidency, determined to sustain the luster of his shining city on a hill. Always tall, proud, and built on a rock "stronger than oceans" (1980), with the help of Ronald Reagan and "Reagan's regiments" the city was now (1988) still strong and true, but even more prosperous, more secure, happier, its morale higher. Its glow, always steady, had now "changed the world." The shining city was grounded on shared values (for example, love of freedom), shared experiences and narratives (for example, of the "eight-year journey" together, of past history, of "battles" fought and won together), and shared vision (broadly, in 1980 we had it within our power to make the world over again; in 1988 "the future is ours").

The outgoing president was at once proud ("We meant to change a nation, and instead we changed the world") and humble. For example, he points to a long list of the successes of his administrations: an economic program that brought about the longest peacetime expansion in our history; American industry now more competitive; protectionist walls abroad knocked down; superpowers beginning to reduce nuclear stockpiles; regional conflicts beginning to cease; the Persian Gulf no longer a war zone; the Soviets leaving Afghanistan; the Vietnamese preparing to pull out of Cambodia; an American-mediated accord soon to send fifty thousand Cuban troops home from Angola; a "satisfying new closeness" forged with the Soviet Union.

Still, he attributes these successes not to any special capacity of his leadership but to common sense, which told us that when "you put a big tax on something, the people will produce less of it. So we cut the people's tax rates. . . . Common sense also told us that to preserve the peace we'd have to become strong again after years of weakness and confusion. So we rebuilt our defenses."

The president does not even claim credit for the battles fought and victories won to secure those successes. Here, also, he is at once humble ("I never won anything you didn't win for me") and magnanimous ("You won every battle"). He at once embraces the war metaphor and shies away from it: "They called it the 'Reagan revolution,' and I'll accept it, but for me it always seemed more like the 'great rediscovery.'" Indeed, even his most typical appellation, that of "The Great Communicator," is at once accepted and rejected: "In all that time I won a nickname: 'the great communicator.' But I never thought it was my style or the words I used that made a difference: it was the content. I wasn't a great communicator, but I communicated great things and they didn't spring full blown from my brow, they came from the heart of a great nation; they came from our experiences, our wisdom, and our belief in the principles that have guided us for two centuries." Once again he is the spokesman rather than the scriptwriter, the vessel (agency) that prepares us for the future rather than the actor who controls the scene.

The values, the national pride, the shared stories were there all along. Through Ronald Reagan, we "found" them again. He, merely, but not unimportantly, has reminded us of the Puritans' "shining city," of Jimmie Doolittle's "sixty seconds over Tokyo," of the "father down the street who fought in Korea," of the carrier USS *Midway*, which stopped in the South China Sea to pick up Indochinese refugees "crammed inside" a "leaky little boat." By using us as representative anecdotes, he both spotlighted individuals among us (the young woman and her father who fought on Omaha Beach) and illuminated the larger human lessons they stand for.

All was not well, however; this voice of history was worried. "Our spirit is back, but we haven't reinstitutionalized it." Other voices must now do that, such as parents and children "in the kitchen." "All great change in America," Reagan reminded us, begins not in the White House, but "at the dinner table." Others must give voice to his vision.

Reagan is no longer the container (scene); he is, rather, the vessel (agency) for what should follow. Indeed, as he "walks into the city streets" (surely a presidential gloss for "riding into the sunset"), he was soon to be not even the chief actor (president) or the chief spokesperson through which others speak (agency). Yet in this speech he must help us remember the past, set the record straight about the present, and prepare us for the future. He must juggle memory with anticipation. He must at once magnify himself by talking about eight years of triumph and diminish himself so the nation can carry on without him. This was a particularly hard task for someone who worked so diligently toward Richard Darman's equation: a vote for Ronald Reagan is a vote for a mythic America. Now America must remain mythic, but Ronald Reagan must become merely mortal once again. Among the ways he chose to do this was a series of interwoven narratives. First, a narrative of *his* life: "I never meant to go into politics . . . but I ultimately [did] because I wanted to protect something precious. . . . [Back] in the 1960's when I began, it seemed to me that we had begun reversing the order of things, that through more and more rules and regulations and confiscatory taxes, the government was taking more of our money, more of our options, and more of our freedom." Ronald Reagan, a modern "freedom man," went "into politics in part to put up [his] hand and say 'Stop!'" "I was," he recalls, "a citizen politician, and it seemed the right thing for a citizen to do."

Citizen Reagan grew up to be President Reagan. Despite his new elevation to high office, he felt "like the new kid in school." During a formal dinner at the 1981 economic summit with François Mitterand and Helmut Schmidt, he "sort of leaned in and said 'My name's Ron.'" By 1983, no longer the new kid, "all of a sudden, just for a moment [he] saw that everyone was sitting there looking at [him]." Then "one of them broke the silence. 'Tell us about the American miracle.'"

During this growing up period and thereafter, he relentlessly fought gulags, expansionist states, and those who waged "proxy wars" and then, as the Soviets effected internal reforms and withdrew troops from Afghanistan, he came to relish a "new closeness" with his former enemies. Urging vigilance toward his new friends, he did not regret his new attitude.

Regrets were to be saved for an enemy unconquered (the Big Spenders). But, since this isn't a night for arguments, he vowed, "I'm going to hold my tongue." "They," like all the "theys" among the Reagan debunkers, lacked sight or vision. "Reagan's regiments," the American people, understood and shared his vision of the future.

The second narrative, the story of America, wove in and out of the first narrative. Ronald Reagan, Reagan's Regiments, freedom men, We the People, moved as one voice in this larger-than-life narrative—through economic wars (inflation, recession, the deficit) and shooting wars (World War II, Korea, Grenada). His two greatest triumphs were those of the nation—economic recovery (America created and filled nineteen million new jobs) and the recovery of morale (America was respected again in the world and looked to for leadership).

In this "State of the City" message, the outgoing president addressed the answer to his question, "How stands the city on this winter night?" The state of the city in 1989 was set in a "prosperous," "secure" and "happy," nation amid countries across the globe that were "turning away from ideologies of the past," countries that had rediscovered that "the moral way of government is the practical way of government," that "democracy, the profoundly good, is also the profoundly productive." The city, shining on a hill, served during the Reagan administration, as she had for two centuries, as "a beacon, still a magnet for all who must have freedom, for all the pilgrims from all the lost places who are hurtling through darkness, toward home."

This single, compelling image that had served Ronald Reagan so well for so long, now came to serve him in yet another way—still as a scene of near mythic proportions, but a scene of which he no longer was the "representative anecdote." Rather, it was a scene into which he must now walk.

How stands the city on the winter night of 11 January 1989?

> More prosperous, more secure and happier than it was eight years ago. But more than that: after two hundred years, two centuries, she stands strong and true on the granite ride, and her glow has held steady no matter what storm. And she's still a beacon, still a magnet for all who must have freedom, for all the pilgrims from all the lost places who are hurtling through the darkness, toward home.
>
> We've done our part. And as I "walk off into the city streets," a final word to the men and women of the "Reagan revolution," the men and women across America who for eight years did the work that brought America back.
>
> My friends, we did it. We weren't just marking time; we made a difference. We made the city stronger; we made the city freer, and we left her in good hands. All in all, not bad at all.

The transformation downward of Ronald Reagan from scene (container) to actor (the contained) to agency may be schematized in figure 5. All presidents context, at least a part of the scene, for the person who follows after them. The transformation to a "diminished" Ronald Reagan held implications for George Bush, already perceived to be a diminished Ronald Reagan, or at best, a Reagan legatee.

Conclusion

In this essay we have examined some of the fundamental strategies used to help Ronald Reagan establish a legacy. A useful, perhaps crucial, question for critics to ask at this point is, "So what?"

There are several fundamental ways to answer such a question. First, a move by any major actor to become scene is worth examining because such a move frequently facilitates a kind of power that could lead to dangerous possibilities: leaders of movements mistake

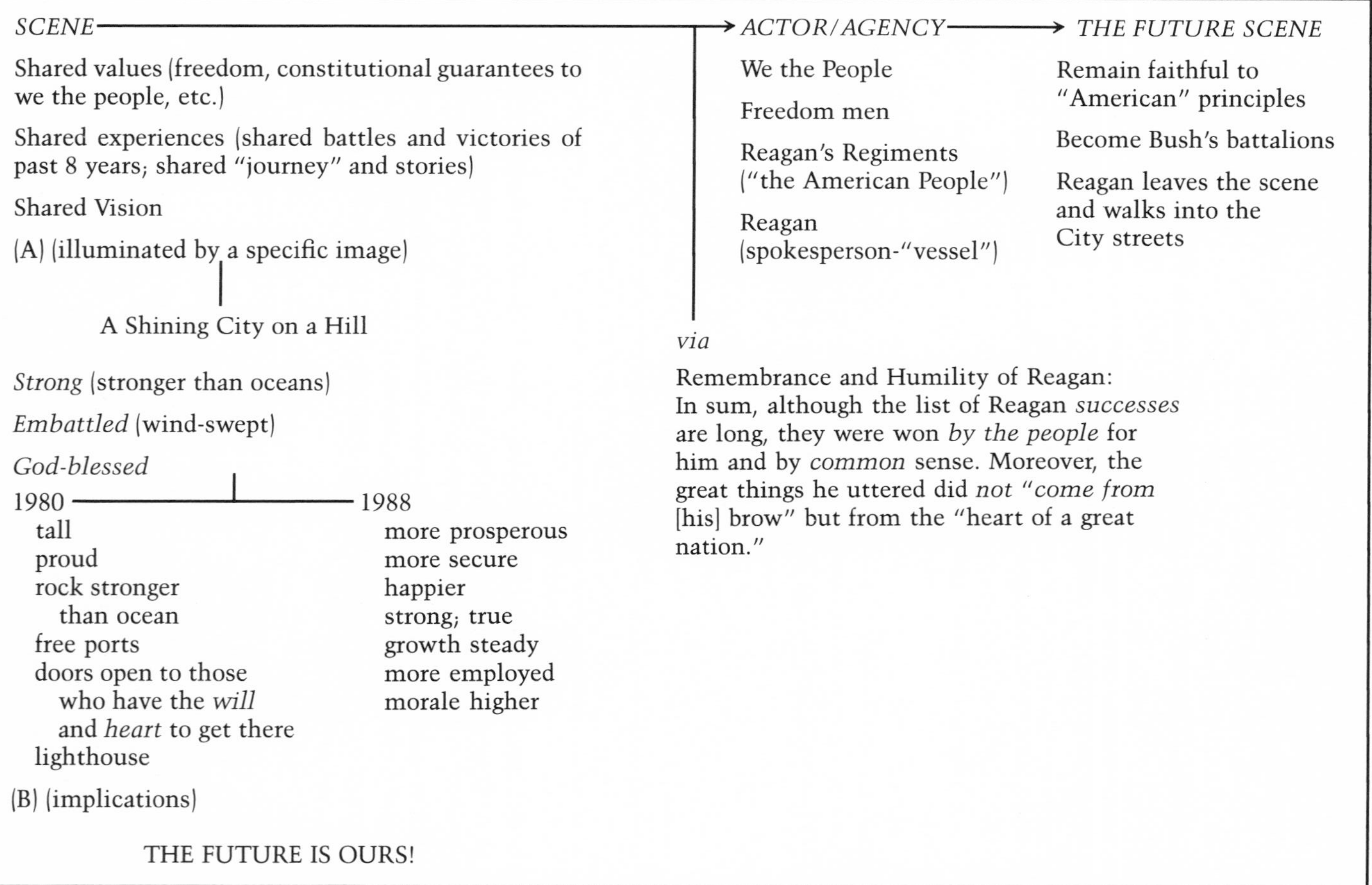

Figure 5. The Final Downward Transformation from Scene to Actor/Agency

themselves or can be mistaken for the movements; heads of nations can mistake themselves or can be mistaken for the nations they preside over temporarily.

Most typically in representative democracies the president may be perceived as either actor or agency; that is, as an actor (a president) or as an agency (a temporary office holder who performs a set of functions accruing to the presidency). In the United States suspicion of "imperial presidents," who would step beyond themselves and their temporary natures in an office, is the norm rather than the exception, from the misgivings of George Washington about the office he would come to hold, to underlying notions of a tripartite government able to provide a system of checks and balances against excessive power, to continuing variations on the belief that scenic power resides in the people; that is, that "we the people," rather than any one person, make up the fundamental visage of the American scene (dream).

Second, during the Reagan presidency in particular, the lines between appearance and reality, between style (strategy) and content (substance) became blurred. This is so both because of the constraints and focus of contemporary American politics and the particular conceptions of those within the Reagan administration.

Most Americans born into a television age get most of their political information from television news. Although there are exceptions (C-SPAN I and II, the PBS network, certain CNN coverage, and an occasional treatment by the three major networks), much of television news, driven by constraints of time, ratings, budget, and predominance of the visual, is relegated to "photo-ops" orchestrated by candidates and office holders, and to sound bites, now reduced from approximately one minute to twelve seconds. Moreover, news organizations rely largely on administration sources and a limited pool of experts for information and commentary, thereby facilitating control by a powerful elite of the stories that construct our notions of governing and campaigning.

Coupled with television's need for highly visual minidramas to capture the attention of an audience that has become accustomed to them, the Reagan administration, its media consultants, pollsters, speech writers, and political operatives by internal decisions and directives worked explicitly, as stated in Richard Darman's memo, to merge actor with scene, in this case to foster the equation "a vote for Ronald Reagan equals a vote for a mythic America."[29]

Since it seems unlikely that campaigning and governing will rely less on television or that ambition and conviction will yield few persons who thirst for power, critics might usefully focus some of their efforts on the inextricable relationships between strategy and substance. We have suggested here that a development of Kenneth Burke's dramatistic pentad may usefully provide one a way of looking at those relationships, particularly when they involve the transformation from actor to scene.

Part III

Foreign Policy Case Studies

7

A Rhetorical Ambush at Reykjavik

A Case Study of the Transformation of Discourse

W. Barnett Pearce
Deborah K. Johnson
Robert J. Branham

If the cold war can be said to have ended at a particular time and place, it was in Iceland on 11–12 October 1986. The ratification of treaties in Paris in November 1990 that formalized a nonbelligerent relation between East and West was something of an anticlimax, overshadowed by the new "crisis" in the Persian Gulf. In 1986, however, the arms race seemed both unwinnable and unstoppable. In this bleak context, the world was dazzled during two days in Reykjavik: the leaders of the two superpowers *almost* agreed to eliminate all nuclear weapons.

Although no agreements about megatons or missiles were reached at Reykjavik, something important happened. The discourse in which the American president experienced East-West relations was subverted; he found himself talking in a new manner, which meshed with that of his Soviet counterpart. Before Reykjavik, Reagan's foreign policy discourse had been fashioned primarily in the context of domestic politics. One result was a "zero-sum" approach to the United States–USSR balance of power: anything that weakened the USSR was seen as strengthening the United States, and vice versa. At Reykjavik, Reagan was engaged in face-to-face discussions with a Soviet leader who maneuvered him into publicly rejecting an elimination of all nuclear weapons. The result was to reverse public opinion around the world as to who was in fact the advocate of peace and to delegitimate the confrontational rhetoric in which Reagan had argued, simultaneously, that the United States occupied the moral high ground and that the United States was morally compelled to continue escalating the arms race.

We believe that this transformation was no accident. Mikhail Gorbachev orchestrated the meeting in such a way that Ronald Reagan was rhetorically ambushed.

The Meeting at Reykjavik

President Ronald Reagan and General Secretary Mikhail Gorbachev met on short notice less than a year after the Reagan administration's first summit meeting (November 1985, in Geneva). The trip to Iceland was somewhat casually announced by Secretary of State George Schultz (30 September 1986) only two weeks before the meeting began, in a report about the near-expulsion of a Soviet spy and the Soviets' decision to allow a dissident to emigrate to the United States.[1]

Reagan's staff worked to reduce expectations, acknowledging that Gorbachev had initiated the idea, and characterizing it not as a summit but an interim meeting between summits. Asked about the agenda, Reagan said, "All we've agreed upon is that we're going to have a meeting."[2] Just before his departure, the president explained:

> This will be essentially a private meeting between the two of us. We will not have large staffs with us nor is it planned that we sign substantive agreements. We will, rather, review the subjects that we intend to pursue, with redoubled effort, afterward, looking toward a possible full-scale summit. We'll be talking frankly about the differences between our countries on the major issues on the East-West agenda: arms reduction, human rights, regional conflicts, and bilateral contacts. We'll be talking about how we can—while recognizing those differences—still take steps further to make progress on those items and to make the world safer and keep the peace.[3]

We now know that Gorbachev's agenda was more specific than indicated in his letter proposing the meeting and that it was prompted in part by his disappointment with the Geneva summit.[4] In a report to the Supreme Soviet on 27 November 1985, the general secretary summarized the first summit meeting in Geneva as "important" and "significant," in part because the very fact of the meeting represented a change in the bellicose nature of the Reagan administration's foreign policy.[5] However, he expressed deep dissatisfaction with post-Geneva events:

> A strident campaign aimed at instigating anti-Soviet passions was started for the umpteenth time. Attempts were made again and again to portray the

Soviet Union as some kind of bugbear, to increase fears in order to get the latest military budget through Congress. The "evil empire" epithet has been trotted out. The President has again confirmed that he is not going to scrap this term.

All this could be put down to rhetoric, but, as I have already said, hostile rhetoric also ruins relations. It has a snowballing effect. Things are now far more serious . . . because everything in United States real politics remained as it had been.[6]

Particularly stung by the "malice and malevolence" in the United States' response to the Chernobyl accident in early 1986, he became convinced that the chances for a productive summit as scheduled in 1987 "were rapidly waning. Going to a new summit just to shake hands and maintain friendly relations would have been frivolous and senseless."[7] In this context, Gorbachev conceived the idea of "an interim Soviet-American summit in order to give a really powerful impetus to the cause of nuclear disarmament, to overcome the dangerous tendencies and to swing events in the right direction."[8]

We do not know what counsel Gorbachev received or what preparations he made for his rhetorical ambush. The ambush was sprung however, when Reagan found himself confronted with an adversary who was well prepared to propose and agree on specific issues at what Reagan thought was to be an agendaless nonsummit. The meeting was arranged in such a way as to deprive Reagan of the rhetorical resources on which he customarily drew in discussing international relations, and the content of Gorbachev's proposals toppled the particular structure of Reagan's rhetoric.

Gorbachev reported that he "planned thoroughly for the Reykjavik summit" and was prepared to agree "in the long run on the complete elimination of nuclear weapons." He brought with him "a set of drastic measures in draft form" consisting of "four points expressed in a page and a half" which could be understood by "the broad public." This latter feature was important. To break through the "old pattern," Gorbachev deliberately attempted "to make the world public a kind of party to our talks."[9]

Gorbachev succeeded in surprising Reagan with his agenda, but a successful ambush must be more than just a surprise. It must be executed in such a manner that the "victim" cannot respond effectively. Gorbachev accomplished this by creating a situation in which the discourse in which Reagan customarily conducted foreign policy was inappropriate and ineffective. Deprived of his usual rhetorical resources, Reagan was required to speak but unable to use the rhetoric in which he was accustomed to speak about the Soviet Union and nuclear arms.[10]

It is possible that there were two ambushes. The first consisted of the whole event and was designed to transform the discourse in which arms negotiations were being conducted. The second depended on the success of the first: after creating a different discourse, one which treated as "thinkable" the elimination of whole categories of weapons, Gorbachev then linked all of these proposals to a condition that SDI research would be confined to laboratories. Whether this was a deliberate trick or a logical extension of earlier discussions is unclear. The result, however, enabled Gorbachev to outflank Reagan as peacemaker. By linking elimination of nuclear weapons to the ABM treaty and thus to the Strategic Defense Initiative (SDI), Gorbachev proposed an alternative to Reagan's vision of a protective umbrella that would make nuclear weapons obsolete. Gorbachev's was a vision of trust and cooperation in international affairs, and Reagan did not accept it.

As a dramatic event, the ambush succeeded in surprising the world audience. Before Reykjavik Reagan portrayed himself as the reluctant but determined participant in the cold war, forced to act by the perfidy of the Soviets. After Reykjavik public opinion in the United Kingdom and Germany believed that the USSR was making a greater effort than was the United States toward nuclear arms agreement and that Reagan was to blame for "not accomplishing much" in their meeting.[11]

Although they assessed what had happened differently, all the participants hailed it as an important event. Secretary of State Shultz described the meeting as "a watershed in United States–Soviet relations."[12] Secretary General Gorbachev characterized it as a "landmark" which "signified completion of one stage in the disarmament effort and the beginning of another. We broke down the old pattern of talks and brought the Soviet-American dialogue out of what, I would say, was political fog and demagogy."[13] In a nationally broadcast address, President Reagan reported that "the implications of these talks are enormous and only just beginning to be understood."[14]

Although the meeting at Reykjavik is interesting from many perspectives, we were primarily struck by the claim that it improved the quality of communication between the United States and the USSR.

Moral Conflict About Arms Control in the Cold War

A major part of the communication between the United States and the USSR has focused on the potential for catastrophic military confrontation using "strategic" weapons. The sequence of accusations, denials, proposals, negotiations, and so forth may be looked at as an extended conversation in which the agents (the United States and the USSR) have remained the same, although each speaks through multiple and changing sets of actors. Nuclear weapons have been one of the primary reasons why the superpowers have continued to communicate with each other.

While talk about strategic weapons is preferable to war with those weapons, the quality of that talk may still be critiqued. The conversation between Reagan and Gorbachev before the Reykjavik meeting can usefully be understood as a "moral conflict" between incommensurate social realities. That is, their disagreements went beyond contrasting beliefs; although locked in conversation with each other by the specter of nuclear war, they had very different understandings of who they were, what each meant by what they said and did, what an acceptable outcome would be, and what counts as "good reasons" for taking or opposing particular actions.[15]

The structure of moral conflicts makes them often interminable. The best-intended efforts to *end* the conflict are understood by the other as strategic moves *within* the conflict; so interpreted, attempts at peacemaking often have the pernicious effect of intensifying the conflict. In addition to being interminable, communication in moral conflict has at least three other characteristics: nonsummativity, contextual misrepresentation, and moral attenuation.

Nonsummativity

All conversations have the potential to be the nonsummative product of the intentions and actions of the participants. The structure of conversation creates the possibility for a "logic of interaction" to develop in which the emerging pattern is something other than what the participants wanted or expected.[16] The outcome is "nonsummative" because it is historically contingent, not a simple fusion of two positions.

Without using the terms, Reagan and Gorbachev described the negotiations about arms control as nonsummative. Both claimed that they wanted peace;[17] both acknowledged that the other side *said* that they wanted to reduce or eliminate nuclear arms;[18] both frankly acknowledged that the negotiations in Geneva had yet to produce a workable agreement;[19] and both warned that the expensive, dangerous "race" to produce bigger and better nuclear weapons continued.[20]

Both Reagan and Gorbachev expressed frustration in their attempts to alter the direction of the logical force of the conversation. Shortly after Gorbachev assumed the duties of general secretary, the Soviet Union announced a unilateral moratorium on tests of nuclear weapons and publicly called for the United States to reciprocate.[21] In addition, Gorbachev made specific reference to the "zero option" in a major speech after the Geneva summit. At the very least, this could be seen as an attempt to set an agenda, and Gorbachev was very disappointed when nothing came of this initiative.[22]

President Reagan repeatedly called for the reduction of nuclear weapons and expressed disappointment (whether or not he was actually surprised) when these proposals were not accepted by the Soviets.[23] He conceived of a bold new strategy to replace that of "deterrence" through Mutually Assured Destruction (MAD): SDI, a high-tech impenetrable defensive shield that would render Soviet intercontinental missiles impotent.[24]

How do we account for the fact of nuclear escalation in the presence of so much talk about peace and disarmament? Both leaders employed the same rhetorical stratagem: they deplored the discrepancy between the pacific words and bellicose acts of the other.[25] The sequential interaction of these strategies produced a nonsummative pattern in which each participant's expressed intentions are frustrated.

Contextual Misrepresentation

Every utterance (or act) occurs in a context and derives its meaning from it. However, the relationship between text and context is often imperfect. Each utterance may be read as representing the context in which it occurs; that is, it contains lexical and other features which index the frame in which it is to be interpreted.[26] Such frames include "punctuations" of the sequence of events in which they occur, references to the relevant social systems or settings, clues to the perspective from which events are to be viewed, the infrastructure of evaluative judgments, and so forth.

Because they juxtapose incommensurate moral orders, the utterances made by participants in moral conflict normally misrepresent the contexts in which they occur. There were two contrasting misrepresentations in the United States–USSR dialogue before the meeting in Reykjavik.

Both Gorbachev and Reagan's statements to and about each other *underrepresented* the complicated connections between their countries.[27] Short, linear punctuations that portrayed their own actions as the calm, deliberate, but sadly necessary response to the other's provocation disabled them from the creativity and flexibility necessary to cope with moral conflict. Needing the greatest possible conceptual mobility to move comfortably within their moral order and to transcend its boundaries, they were tied by the logic of the interaction to impoverished representations of the conversation in which they participated.

On the other hand, the professional negotiators participating in the disarmament talks in Geneva used a much more complicated discourse which *overrepresented* the connections among participation in talks, regional conflicts, human rights, and economic policies. Their Byzantine discourse precluded the possibility of reaching agreements by requiring everything to be settled before anything could be, and by empowering the use of technically obtuse issues as stratagems to keep the discussion going. Cronen and Pearce suggested that there are two forms of confusion: one involves "simply" not knowing which of several well-defined alternatives is the correct one, while the other is a muddled mess in which the overall structure cannot be determined.[28] The disarmament talks were of the latter type. While there was a great deal of rigor in the definition of some of the particular components of the issue (for example, the meaning of "intermediate-" and "long-range" missiles; "strategic" and "tactical" weapons, and so forth), these elements existed within an amorphous structure of the talks themselves and thus could be used in an endless game of redefining the context in which no substantive agreements were needed or were likely to be produced.

Moral Attenuation

Moral attenuation occurs when each party in a dispute defines its own position as the embodiment of moral virtue, which is perceived to be rejected or ignored by the other. Both Reagan and Gorbachev located their respective countries' history in a sacred story that showed them as acting with high moral standards to accomplish historically important purposes in ways that deserve respect. However, each also portrayed the other as not deserving respect. The other's actions were taken as sufficient "proof" of the perfidies of the other's social and political system and history. As a result, each found himself living in a moral order that afforded himself the opportunity for a life of dignity, honor, even valor, but found that this moral quality was disqualified in conversation with the other.

For Gorbachev the sacred story consisted of the USSR's rise from a backward, underdeveloped, oppressive, uneducated Tsarist state through a heroic struggle in World War II to being recognized as a superpower in less than a century. This story permitted points of comparison between the United States and the USSR. Gorbachev noted similarities of interests between the countries (as well as profound differences), acknowledged mistakes and continuing problems in his own country, refrained from criticizing the Founding Fathers of the United States, and expounded a vision of international cooperation if the United States could curb the influence of its militarists. However, the contrast between the peaceful aims of the Soviets and the aggressiveness of the United States militarist-industrialist complex was the focal point. Gorbachev explicitly interpreted American actions in the Reagan era as a "one-sided approach clearly prompted by a striving for military superiority of the United States and NATO as a whole."[29]

The contrasts between the United States and the USSR in Reagan's sacred story were drawn much more sharply. The United States was a "shining city on a hill" whose people were the custodians of moral rectitude. The greatness of the United States was repeatedly expressed in specific comparison to other countries and was shown in the superiority of its military might, in its position of leadership in the world, and, most important, in the morality of its public and private conduct. The Soviet Union, on the other hand, was an "evil empire," founded by godless, evil men (Lenin, for example), bereft of divine leadership and authority, whose underlying motivation was world domination and international aggression. Reagan portrayed a future in which it was better to be dead than red.[30]

In the conversation produced by these discourses, there was little opportunity for a life of dignity and honor, much less a relationship that would permit disarmament. As Reagan repeatedly commented, "Nations do not distrust each other because they're armed; they arm themselves because they distrust each other."[31]

Gorbachev wondered just who Reagan thought he and the Soviet people were that they should meekly accept such characterizations and pointedly charged that the contemporary world is too complex to be dealt with in terms introduced in 1947 by President Harry Truman and Prime Minister Winston Churchill. He accused the United States of talking disarmament while deliberately pursuing the arms race as a means of overextending the Soviet Union's economic system, in the hope that the Soviet Union would collapse and a new era of American hegemony would occur.[32]

For his part, Reagan loudly proclaimed that he distrusted the Soviets, that they had violated every previous agreement, and that they were intent on subverting democratic governments around the world.[33]

Reagan's Rhetorical Stance

Reagan's rhetorical construction of foreign policy seems a fairly orthodox rendering of what has been called the "national security doctrine" in the United States.[34] Reagan's use of this rhetoric began at least as early as "The Speech," which he was hired to give in a fund-raising campaign sponsored by General Electric to support anticommunism during the postwar era. In the 1980 presidential campaign and during his first term, Reagan's political speeches remained remarkably consistent with his statements during this period. The anticommunist theme and tough-guy vigilante persona were immensely effective with domestic audiences but rendered him vulnerable to the ambush at Reykjavik.

In his first major foreign policy address as president, Reagan delivered his "zero-option" proposal, which countered calls for a nuclear freeze with the announcement of START as a vehicle for *reductions* of nuclear weapons.[35] This speech became an important reference for Reagan. For example, in the Eureka College address (his second major foreign policy speech, delivered 9 May 1982), Reagan said, "Last November, I committed the United States to seek significant reductions of nuclear and conventional forces."[36] The "last November" phrase became a stock expression for Reagan in this period, in which he constructed the act of delivering the "zero-option" speech as a demonstration that he had offered the Soviets a chance for arms reduction and they had rebuffed it.

Reagan's rhetorical strategy took a significant turn in 1984 and 1985 when the zero option began to mean the ideal of eliminating (not just reducing) all nuclear weapons.[37] Even Reagan noted the inconsistency between this utopian language and the call to arms against the demonic Soviet foe that he sounded during the 1980 campaign and until March 1983,[38] and insisted, "Let's not fool ourselves."[39] The rhetorical stance constructed through Reagan's inconsistent vision of utopian disarmament and persistent warnings about the "Evil Empire" enabled him to buy guns while looking for butter. He actually engaged in a build-up of military strength, while longing for negotiations as a ploy to coopt public sentiment.[40]

Reagan's utopian pronouncements did not signal a change in United States–USSR relations. Rather, they were a strategy for the 1984 elections in which Reagan sought to defuse the Democrats' criticism of his foreign policy by moderating his anti-Soviet discourse, expressing greater willingness to talk, and endorsing an ideal of weapons elimination.[41] By pairing negotiated reductions with the impenetrable defense, Reagan was able to envision "a world free from the threat of nuclear war. . . . Standing on the moral ground built by the Freeze movement, Reagan proposed the psychological fix as well as the technical fix. 'Star Wars' is an escape fantasy, presented by a reassuring authority figure."[42] The editors of *The Nation* translated Reagan's utopian statements in his 1985 United Nations address as a "trick of offering the impossible as a means of continuing the actual."[43] Reagan's pronouncements were also seen in this light by hard-line supporters. They were not threatened by Reagan's peace talk precisely because they saw it as empty.[44]

From the perspective of hard-liners, the more sweeping the vision the better, because it maximizes expression of the ideal (peace, freedom from nuclear fear) while minimizing the chance that the bluff might be called. Yet one might ask, if both Reagan's political opponents and supporters saw through Reagan's "deft" rhetoric—preaching ideals while still pursuing the American position in the cold war—then why would this strategy remain invisible to the Soviets? Why would Reagan remain blind to the possible "ambush" to come to Reykjavik?

Reagan's Antifreeze Strategy

Reagan finessed a series of moves aimed at squelching the freeze movement in the United States. The first step was to smear the freeze as a Soviet-inspired front. Commenting about freeze demonstrations in Ohio, he said: "They were demonstrating in behalf of a movement that has swept across our country that I think is inspired by, not the sincere, honest people who want peace, but by some who want the weakening of America, and so are manipulating many honest and sincere people."[45]

The second step of this strategy attempted to coopt the peace movement's language and expressions. In a radio address on 17 April 1982, Reagan said: "So, to those who protest against nuclear war, I can only say, 'I'm with you.' Like my predecessors, it is now my responsibility to do my utmost to prevent such a war. No one feels more than I the need for peace."

This statement seems incompatible at first glance with step one, but the apparent conflict is resolved through the use of the third step: "peace through strength." In this third stage, the Soviets were portrayed as ahead and intransigent in the arms race. Reagan insists that we can bring them to the bargaining table only through the weight of superior arms. As in his Central American policies, Reagan believed that talks are effective only if you are simultaneously fighting.

Step four in the strategy against the domestic movement consisted of citing the irrelevant opportunity costs of a freeze.[46] Criticizing a freeze as "not good enough" and as perpetuating a situation in which the Soviets have a superior position, Reagan claimed that what we need are reductions.

Opportunity costs refers to the benefits that have to be foregone when one policy option is chosen over another. Irrelevant opportunity costs are those that would be foregone regardless of the choice at hand. In this instance, Reagan identified the zero option and nuclear parity as the opportunity costs of accepting a "freeze." That these were irrelevant opportunity costs—that is, that Reagan had no intention of pursuing this option—was made clear by the next, and final, step of the strategy.

The construction of the Soviets in American public discourse as the Evil Empire, the fifth step, was pursued most vigorously during 1982 and 1983. The escalating rhetorical war against the Soviets warranted a military build-up. Announced exactly fifteen days after the "Evil Empire" speech, the Strategic Defense Initiative (SDI) was the trump card and the final stage of the strategy.

This strategy played on the same fears addressed by the freeze movement. Hertsgaard's reading of the public relations "apparatus" suggested that

> the two-track strategy . . . was based on the manipulation of fear. Recognizing that growing numbers of citizens were uneasy about the continuance of the arms race, the apparatus sought to quiet that fear by having the Great Communicator talk fervently about wanting radical cuts in nuclear arsenals. Promising reductions not only helped defuse the President's cowboy image, it diverted attention from the immense military buildup the administration was pursuing. At the same time, Reagan made sure to encourage a second fear, the fear of the Soviet Union.[47]

This strategy against the domestic peace movement may be expressed as building a discursive formation in the shape of an inverted triangle. The base of the triangle is fear: fear of nuclear weapons, distress at the escalation of the arms race, and enmity toward the Soviet Union. Two apparently contradictory but in fact mutually supporting idyllic visions oppose each other, but each rests on this base. One is the zero option, an image of a nuclear-free world achieved by rational, mature negotiation. The other is SDI, an image of a unilateral, impenetrable defense in a dangerous, nuclear weapons–ridden world. The rhetorical glue that binds these dissimilar elements together is captured in the slogan Peace Through Strength and Reagan's firm belief that success in bargaining and in deterring aggression come only from a position of overwhelming power. In this logic, anything that weakens the Soviet Union or strengthens the United States is a contribution to peace; anything that does the opposite is an irresponsible or malevolent act of war.

Reagan's Foreign Policy Discourse

The pattern of Reagan's public pronouncements during his administration was part of a broader discourse that enjoyed great popularity in the early 1980s. Jerry Falwell, president of the Moral Majority, might have been Reagan's speechwriter, so closely did the president's discourse resemble that of the New Christian Right.[48]

In this discourse foreign policy occurs in the context of morality, which is the highest of all contexts for public life. Morality is sharply divided into "good" and "bad," but these concepts are global, not further subdivided. For example, "good" is not defined but located within the Judao-Christian heritage as incarnated by the actions of the Founding Fathers of the United States. It now exists as a rhetorical resource on which right-thinking Americans can draw and is equated with a cluster of particular issues and values, such as the primacy of personal freedoms and individual liberties over those of social classes or governments. Similarly, "evil" is less defined than located as the opposition to all of these aspects of "good."

The global nature of these concepts accounts for the ease with which those using this discourse can generalize the attributes of those they support and those they oppose. For example, the Soviet Union's explicit atheism is a sufficient reason—in this discourse—to attribute to it all manner of negative characteristics. It is the veritable embodiment of evil, the oppressor of liberties (including freedom of religion), and the champion of hostility to the United States ("God's New Israel").

This vision of morality removes it from the sphere of "essentially contested concepts."[49] The paradigmatic moral question is whether one has sufficient resolve and courage to *do* what is right. This structure of the discourse facilitates a hidden circularity in the assessment of good and evil. The things which "good" people, institutions, and nations do is, by that fact, deemed as "good"; that which "evil" empires, institutions, and people do is *ipso facto* "evil."

The discourse in which Reagan conducted relations with the Soviet Union included a visualization of the "episode" of the interaction between them. In this vision, a righteous, mighty people inhabiting a "shining city on a hill" that is the inspiration for and envy of all other nations would, by its aggressive but purely defensive military posture, prevent any foreign encroachment or limitation of its actions, and these actions, by virtue of their moral rectitude, would . . . Here the vision becomes a bit blurry. At times it seems that the goal is to spread the American Way throughout the world in a form that far exceeds *Pax Americana* or economic hegemony. The frequently reiterated goal of taking the initiative or providing leadership and purpose in international relations (for example, in "Project Democracy") seemed designed to make little Americas of all the nations in the world. (This would be expressed, of course, as spreading the blessings of individual liberty and free enterprise among all nations.)

The image is blurred, however, because the Soviet Union itself is not included in this picture except as a villain that must be kept at bay. In fact, the image seems to *need* an implacable foe to sustain its dramatic narrative, and the Soviet Union is pressed into the role. The image becomes even more hazy when it comes to envisioning the pattern of interaction between the United States and the USSR. To put it in theatrical terms, the Soviet Union's role is crucial to the plot, but its character is underdeveloped and it is given only a few poorly written lines.

The discourse in which Reagan conducted international relations is profoundly Augustinian. After his conversion to Christianity, confronted by a world in which civilization was clearly about to be destroyed, Augustine wrestled with the problem of the relation between the pagan rhetorics of Rome and Greece (part of the civilization that he wanted to preserve) and Christianity (the doctrine that he believed to be true). His solution: there is no conflict, but there is an intransitive hierarchical relationship between them. Christianity contains Truth revealed by God, but the pagan rhetorics show how Christian ministers can present that truth effectively. Therefore, it is fully appropriate for a Christian to use Cicero or Aristotle as guides in writing sermons that will convince people to revere Jesus of Nazareth. Augustine's solution found a way to preserve rhetoric in "the City of God." It did so, however, by removing truth from the rhetorical domain.

In the same way, Reagan's discourse does not put truth or morality or the details of the sacred story within the arena in which rhetoric is epistemic. Public speaking, debate, conversation, and the like are not the means by which the validity of ideas is tested or the potential relationships among ideas discovered. Rather, this discourse envisions rhetoric as a means of persuading others of truth already known, of stirring the passions of the crowd, and, perhaps, as a surrogate for military actions, of bashing one's enemies. Reagan's foreign policy discourse envisions communication as a monologue, punctuated perhaps by the applause of the crowd, but not as an interactive sequence of messages in which each becomes the context for the others.

Gorbachev's Reconstruction of the Rhetorical Context

Since he burst on the international scene in 1985, Gorbachev's policies have been shocking but not improvised.[50] Both domestic and foreign policies have grown from a well-developed analysis of the contemporary situation in the Soviet Union, a vision to be pursued, and a set of strategies. Further, Gorbachev took unprecedented steps to publicize that analysis, vision, and strategy. In *Perestroika* he described his efforts as continuous with the tradition of Lenin. Taking what might be called a neo-orthodox[51] stance to Lenin, he challenged the misinterpretations of unnamed foreigners (surely he means Reagan) and those in the Soviet Union who read Lenin too prosaically.[52]

Perestroika presents the "sacred story" of the Soviet Union. As Gorbachev reconstructed that history, the past seventy years have seen unprecedented progress despite the stresses of two invasions and an economic depression. Because it followed the teachings of Lenin, the USSR has changed from a backward, poorly educated, agrarian state to a superpower, the international leader among socialist countries, and a modern, well-educated, technically advanced society.[53] Mistakes were made during this time, largely because of the problems posed by external enemies, but the leadership has learned from those mistakes and taken steps to correct them.[54] This progress has created a new situation, both permitting and requiring "new thinking" and a restructuring of Soviet society. One part of this new thinking is best summarized, Gorbachev admitted, by a non-Russian word: *dialogue*.[55]

Gorbachev's readiness to engage in international dialogue allowed him to avoid the conversational patterns of the cold war. Continued escalation of military expenditures, he argued, were economically disastrous and strategically unnecessary. But how might the pattern be changed? Virtually anything he said would be dismissed as empty propaganda, and even dramatic actions such as the unilateral moratorium on tests of nuclear weapons were regularly interpreted as simply a more devious stratagem within the same context.

As described by Gorbachev, the existing context of nuclear arms negotiations provided very little opportunity for progress. Three salient features precluded success. First, the state of the public discourse between the nuclear superpowers was poor. A lack of trust was explicitly affirmed; each side accused the other of violating past agreements and blamed the other for current problems. Second, the United States persisted in adding to its military capability. For example, the highly trumpeted SDI introduced a new element in the arms equations, the United States continued testing nuclear weapons after the USSR ceased, and the United States deployed the 131st strategic bomber equipped with cruise missiles, which violated the SALT II provisions.[56] Third, the continuing negotiations were making little progress due to the sheer complexity of the issues. Gorbachev described the negotiators as "choking on the endless discussions of dead issues. There were some fifty to a hundred alternatives in the air, the combination of which prevented any movement toward disarmament."[57]

Nevertheless, Gorbachev was convinced that something useful had happened during the Geneva summit. The face-to-face discussions for a short while opened up alternatives in the conversation between him and Reagan. These encouraging effects were undone, however, when the president reentered the spheres of influence of the militarists and anti-Soviet lobbies in the United States.[58] Gorbachev set himself to perform a rhetorical act that would transform the pattern of the existing conversation. Specifically, by creating the kind of discourse he had glimpsed in his private, informal meetings with Reagan in Geneva, he sought to reach agreements that would withstand the attempts of militarists on both sides to continue the cold war. Apparently, such a transformation could not be brought about by proposals ensconced in their normal rhetorics (both he and Reagan had offered bold proposals before) or by unilateral actions.[59]

The power of Gorbachev's rhetorical act at Reykjavik derived from five features. First, the suddenness of the invitation and the quickness of the subsequent meeting were strikingly unusual in the history of summitry. Further, its status as a "summit" was always equivocal. Usually summit meetings are carefully constructed to avoid surprises; this one was a surprise from the start, and the rules about what each was to do and expect from the other were always unclear. Whatever else this schedule reflected or produced, it conduced to an unusual degree of spontaneity in the face-to-face meetings. In particular, the American president arrived far less "scripted" than usual and less dependent upon a staff of technical experts.

Second, Gorbachev and his advisors had prepared a series of specific proposals. Although the content of these proposals was not new, the fact that they were presented in a sufficiently simple, direct, and unambiguous manner prevented them from being trapped in the complex rhetorics characteristic of the Geneva negotiations. As such, they shifted the discussions from the usual ceremonial confirmations of agreements previously reached by staffers to substantive negotiations between the leaders themselves.[60]

Third, as they had in Geneva, Reagan and Gorbachev spoke extensively face to face. This exploited the greatest weakness in Reagan's rhetorical approach to foreign policy. Designed as an arsenal for speeches *about* the Soviet Union, it was an exceptionally poor resource for conversations *with* a personable, articulate, and apparently sincere representative of the Soviet Union. Salient cold war touchstones for Reagan's domestic constituency are largely unavailable or unhelpful when one is speaking *to* Gorbachev or for a world (European) public in a joint press conference or communiqué. By meeting face to face with Reagan, Gorbachev effectively stripped him of his familiar, effective rhetorical resources. For good or ill, Reagan had to extemporize or find an alternative discourse in which to conduct business in these face-to-face situations.

Fourth, by accident or design, Gorbachev dissolved the rhetorical glue that held together Reagan's rhetoric about nuclear weapons. The twin visions of peace (the zero option and the impermeable defensive shield) were both grounded on an image of the implacable Soviet threat and linked to it by an argument that either image could be achieved only through a convincing counterthreat (Peace Through Strength). At Reykjavik, however, Gorbachev presented a compelling image of the Soviets as sincerely interested in peace. Not only did he arrive with a series of specific proposals, in the course of the talks he endorsed the most radical version of the zero option. Reagan found himself having to respond publicly to a proposal that he had himself advanced (in a very different context, to be sure) years earlier and which he had used to castigate the Soviet leadership.

By calling Reagan's bluff on the zero option, Gorbachev rendered the force of Reagan's rhetorical style not only irrelevant but counterproductive. Accounts differ about which leader introduced the notion of a complete bilateral elimination of nuclear weapons in their talks at Reykjavik, but it is unimportant who said it first. The effect was a radical change in the logic connecting the various points within Reagan's rhetorical stance. Before Reykjavik, Reagan portrayed the relationship between Gorbachev and himself as complementary, with Reagan as the peace-loving, innovative leader of the Free World and Gorbachev (like other Soviet leaders) the warlike, reactive, stumbling block to peace and prosperity. Now Gorbachev had not only claimed title to all the terms coveted by Reagan but evidenced his claim by endorsing the same vision of peace that Reagan had affirmed as the ideal, blocked only by Soviet intransigence. This forced Reagan to choose among undesirable options: he must either redefine the relationship as a symmetrical one in which both leaders strive for peace, or maintain the complementary relationship by presenting himself as warlike and reactive, the stumbling block to peace and prosperity.

Whatever else this dilemma produced, it does appear to have forced Reagan out of his customary rhetoric. His attempt to evade the dilemma took the form of positively connoting the underlying logic in his rhetoric, Peace Through Strength. He offered a linear punctuation of Gorbachev's willingness to bargain about nuclear weapons. The USSR's sudden amiability, he argued, did not evidence their character; rather, it was the result of "our" toughness, specifically, Reagan's build-up of American military forces.[61] This interpretation was immediately challenged by Gorbachev,[62] and the point quietly, obliquely conceded by the Reagan administration.[63]

Gorbachev noted that Reagan's response to the zero option was confused. First professing himself "reassured" and talking of the points on which they agreed, Reagan and his aides quickly began referring to all of the "divergences and disagreements" between them and produced "the familiar old sound that we heard at the Geneva negotiations."[64]

Reagan's attempt to reclaim the force of his rhetoric might have worked except that Gorbachev also outflanked Reagan as peacemaker. He proposed an alternative and, for many, a more compelling vision of peace. Whereas Reagan alluded to the zero option as a higher vision unfortunately unattainable because of the evil Soviets and turned his attention to SDI, Gorbachev showed himself as grasping the zero option as a realistic objective in the context of international cooperation, with the ABM treaty as a working example. In this frame, SDI is a flagrant provocation and threat to peace.

The rationale for SDI depends on the existence of an enemy equipped with nuclear weapons and poised for a "first strike." Gorbachev proposed a world in which he, at least, was no such enemy, thereby rendering unconvincing the rhetoric in which Reagan had justified SDI. In effect, Gorbachev had used on Reagan the same outflanking strategy that Reagan had earlier used on the domestic freeze movement.

Lessons from Reykjavik

It would be a mistake to draw simple conclusions from this analysis. Knowledge about rhetorical resources efficacious in transforming discourses is likely to be expressed as a compendium of examples and commentaries, the combination of which affords a kind of worldly wisdom facilitating the art of finding, as an old Greek once said, in any given situation, all the available means of improving the quality of the discourse in which conflicts are experienced.

With this caveat, we conclude with four observations. First, discourses *can* be transformed, even in such unpromising settings as international arms negotiations. The news from Reykjavik raised, and then frustrated, the expectations of people all over the world. For perhaps the first time in the wearying rhetoric of Mutually Assured Destruction came the sounds of a different voice, and the response was an outpouring of hope.

Further, the Soviets learned a new strategy: in subsequent arms negotiations, they continued to employ the tactic of offering back to the United States the proposals that the United States had previously made. For example, the United States had often made on-site inspections a precondition for weapons reductions. Strategically consistent with Gorbachev's public commitment to dialogue, and as a clever tactical ploy, the Soviets agreed, then extended the idea to include, for both sides, the right to inspect "any place, any time." Concerned about protecting their high-tech advantage, the Americans were tremendously disconcerted. More than any "no" or specification of limited accessibility, the Soviet strategy created a space in which the INF treaty was ultimately completed.[65]

Second, the robustness of "poor" patterns in the enactment of conflict should be noted. Because patterns of violence, hatred, and mistrust are powerful, it is difficult to talk about the discourses in which they occur. Even acts specifically designed to describe or change the discourse are customarily reinterpreted as moves *within* the discourse.

The power of the existing context was seen in the effects of Gorbachev's insistence that SDI research be limited to laboratory (rather than field) tests. This issue prevented an immediate agreement to eliminate all nuclear weapons.

Why so major a decision on such a relatively minor point? Here are four suggestions. (1) Gorbachev was under a strict brief from the Politburo not to concede on this point. He may have reached the limits of the constraints under which he was working.[66]

(2) This was an emotional stance for Reagan, to which his rhetoric had committed him so strongly that he could not back down; which identified the limits of his brief from his supporters and staff, beyond which he could not proceed; or which was based on a lack of technical expertise such that he misunderstood what was at issue.[67]

(3) Both found themselves politically overextended, unable to envision how they would explain themselves to their constituen-

cies. So radical a set of changes achieved so rapidly and so unexpectedly left both confronting a formidable rhetorical situation the day after Reykjavik. The furor among the United States' allies in Europe attested to the validity of this concern. Allied leaders judged the what-might-have-been as constituting a near abandonment of the American commitment to defend Europe. The same reaction occurred at home: "The Joint Chiefs, and key congressional personalities were equally shaken. A nuclear-free world was hard to envisage. Being able to verify any such millennial development was harder still to envisage."[68]

> The then national security adviser, Vice Admiral John M. Poindexter, warned Reagan that "we've got to clear up this business about you agreeing to get rid of all nuclear weapons."
> "But, John," replied the President, "I did agree to that."
> "No," persisted Poindexter, "You couldn't have."
> "John," said the President, "I was there, and I did."[69]

(4) They had come so far so fast that both Reagan and Gorbachev found themselves in uncharted territory, not trusting the meanings of the words they used or heard the other using, and perhaps they were frightened of the consequences of making a mistake.

These explanations for how such a minor matter prevented such a major event are not mutually exclusive, of course. They serve to remind us how powerfully the contexts of moral conflict reproduce themselves. However, the last explanation seems particularly consonant with our further observations about the transformation of discourse.

Third, the process of transforming discourses is not fully controllable by the rhetor. Because conversations are interactive, even the process of transforming discourse is cocreated. We suspect that self-styled revolutionaries are often surprised by the results of their activities; nonsummativity characterizes interventions into moral conflict just as it does the patterns of conversations in moral conflict.

The zero-option proposal emerged in the course of the conversation at Reykjavik; it was not the first proposal by either Reagan or Gorbachev, and there is some disagreement about who first introduced it. As Newhouse reconstructed the sequence of events, Gorbachev began with a proposal that would have cut strategic nuclear weapons in half and eliminated midrange weapons from Europe within ten years. The Americans countered with a proposal that would have eliminated all (but only) ballistic missiles in seven years. Noting that this favored the Americans, who have the advantage in bombers and cruise missiles, Gorbachev countered by suggesting that all nuclear weapons be eliminated within ten years. Reagan, by most accounts, responded favorably.

> "All nuclear weapons? Well, Mikhail, that's exactly what I've been talking about all along . . . get rid of all nuclear weapons. That's always been my goal."
> "Then why don't we agree on it?" Gorbachev asked.
> "We should," Reagan said, "That's what I've been trying to tell you."[70]

The logic of the interaction itself led both Reagan and Gorbachev to a point where they had not expected to be, for which they and their staffs were unprepared, and the consequences of which they were unable to envision fully. Given the stakes, who would blame them for getting scared?

This leads to our final observation, that the successful transformation of discourse is scary. It severs all those involved from the resources previously used to make the world coherent and to guide patterns of coordination within it.

> The unique species of historical event we call a revolution occurs when everything changes at once, not excluding the very categories used for gauging and shaping change. . . . It is in the nature of revolutions that no one can be an experienced citizen of the new order they bring into being. Those who fought for change, as well as those who resisted it, are confronted with the postlapsarian mandate to live their lives without a usable past.[71]

What happens if we transform, as Jean Bethke Elshtain insists we must, the discourse that sees war and peace as mutually exclusive opposites?[72] What happens if we set aside the discourses in which the "free world" is equated with "noncommunist" or "anticommunist" and opposed to the "Communist" world? What happens if the polarities between "us" and the various "others" are subsumed in a discourse that makes finer distinctions, and makes distinctions without implying oppositions? Whatever might be gained, something is lost: we are alienated from the familiar comforts and discomforts of a known discursive space and thrust, probably irretrievably, into an unknown world in which anything can happen.

With his ambush at Reykjavik, Gorbachev exposed the vulnerabilities of Reagan's foreign policy rhetoric and of the discourse in which it occurred better than any of Reagan's domestic critics before or since. He also opened a window into another discursive space in which arms negotiations can take place. The subtitle of Gorbachev's new book claims that it explains "how the two superpowers set the world on a course for peace." Did they? Well, to the extent that Gorbachev claims that "we"—not "I"—did, perhaps we are sailing a few points closer to the wind.

8

The Paranoid Style in Foreign Policy

Ronald Reagan's Control of the Situation in Nicaragua

Jeff D. Bass

Traditional political wisdom has conceived the function of foreign policy discourse to be that of creating and sustaining national consensus regarding the desirability and correctness of a leader's actions in this area. Along with effectiveness, political observers have often based their assessments of a president's performance in foreign affairs on his ability to gain popular and congressional support for his policies. The lack of such support, so the thinking goes, would severely hamper a president's capacity to initiate and carry out his foreign policy. Investigations of foreign policy discourse by rhetorical critics have largely proceeded on an uncritical acceptance of these assumptions and have consequently approached such discourse from the perspective of the consensus-gaining rationale.

If the arousal of political support is to be the primary criterion whereby a president's foreign policy discourse is to be evaluated (and so far it has), then Ronald Reagan's speeches on Nicaragua pose a significant problem for rhetorical critics. In the first place, public support for the Nicaraguan Contras never exceeded 41 percent, despite what has been termed "some of the most hair raising anti-communist portraits of the consequences of military inaction in the history of the cold war."[1] Furthermore, Reagan's repeated attempts to amplify the danger posed by a Communist Nicaragua provoked virulent criticism from several quarters. Political commentators, the liberal press, and Roman Catholic leaders denounced what they perceived to be Reagan's flagrant distortion of the Nicaraguan threat.[2] Congressional opponents of Reagan's Nicaraguan policy consistently alleged that the president's case was based on hyperbole, exaggeration, and unsubstantiated claims.[3]

But despite such criticism and the lack of a public mandate for action in Central America, Reagan was never deterred from seeking to destabilize and overthrow the Sandinista government. In fact, from 1983 through 1986 he annually succeeded in pressuring a reluctant Congress to approve grants in humanitarian and military aid for the Contras up to $100 million. In this particular case, the failure of a president's rhetoric to achieve consensus seemed to have had no effect on his ability to implement his policy.

This apparent contradiction of the consensus-gaining rationale of foreign policy discourse has caused analysts to attribute Reagan's success to a variety of extrarhetorical factors. Daniel Ortega's ill-timed trip to Moscow in 1985 and the inability of Democrats in both 1985 and 1986 to formulate viable alternatives to Reagan's policy are the factors most commonly identified by Washington experts.[4] By and large, the possibility that the president's speeches contributed in any way to his ability to carry out his policy has largely been discounted.

But such a dismissal of Reagan's discourse on Nicaragua overlooks the possibility that he may have intended his speeches to perform functions in addition to seeking consensus. A close examination of the nature of Reagan's rhetorical problem in justifying his policy of seeking the overthrow of the Sandinistas reveals the need for just such an additional function. I would argue that this was the need to control the agenda of the debate over his Nicaraguan policy and thereby nullify those situational factors that contribute to the effectiveness of opposition rhetoric. His ability to achieve this additional rhetorical purpose enabled him to overcome the lack of popular support for his policy.

The primary way in which Reagan accomplished this purpose has been suggested by Walter LaFeber. In discussing Reagan's reaction to criticisms of his policy toward El Salvador, LaFeber notes: "The president's words closely resembled the 'paranoid style,' as historian Richard Hofstadter had noted during the sixties, that frequently afflicted North Americans."[5] If Reagan's rhetoric toward El Salvador approached the "paranoid style," his speeches on Nicaragua are even clearer examples. In this essay I will follow LaFeber's lead in applying Hofstadter's conception of the paranoid style to Reagan's speeches.

In taking this approach I do not mean to imply that Reagan was either the first or only twentieth-century president to employ the paranoid style in foreign policy discourse. In varying degrees, most presidents of the cold war era have employed elements of the paranoid style in order to justify American interventionist actions. Dwight Eisenhower and Lyndon Johnson both utilized a conspiracy theme to rationalize their respective invasions of Lebanon and the Dominican Republic. All presidents since Harry Truman have sought to "demonize" the leaders of Communist states around the world. While Reagan's use of the paranoid style was more obvious than that employed by his predecessors, strategic similarities between his Nicaraguan speeches and the foreign policy rhetoric of other U.S. presidents are easy to detect.

But Reagan's ability to debate the question of Nicaragua on his own terms did not entirely result from his intensification of the paranoid elements present in cold war rhetoric. His use of the paranoid style must be examined in conjunction with the Contra offensive against the Sandinistas. There is evidence to indicate that these two factors worked in conjunction to enable the president to control the agenda of the debate. In the first place, the paranoid style allowed him to shift the focus of the debate from a question of the viability of his policy to the question of the nature of the Sandinista government and its aims. By utilizing the Contras, Reagan was able to provoke the Sandinistas into an increasing militarization of their country and a more aggressive posture. The president then employed such data to validate his original charges of Nicaraguan tyranny and aggression. He was thus able to refute opposition claims that he was distorting the situation in Nicaragua.

In this way Reagan was able to create advantages for himself that his predecessors lacked. Namely, he employed low-intensity warfare to complement and validate his rhetorical depiction of a foreign situation. Thus his interpretation of the situation in Nicaragua was never subject to the type of falsification that, say, Johnson's and Nixon's interpretations of Vietnam suffered. Unlike other presidents in the cold war era, Reagan was able to minimize the negative effects of a lengthy foreign venture.

In order to understand Reagan's motives for adopting the paranoid style, we must first examine the impact of the so-called "Vietnam syndrome" on the Reagan administration's vision of the U.S. role in Central America and the world. We must then note the extent to which this same syndrome impaired Reagan's ability to secure the support of the American public for his Nicaraguan policy. Failing to gain support, Reagan could only pursue his policy of "rolling back" communism in the Western hemisphere by subverting the influence of informed criticism. Consequently, he used the paranoid style in conjunction with the Contra offensive to negate those extra-rhetorical constraints favorable to the development of an effective informed opposition.

The Problems of the Vietnam Syndrome and the Informed Opposition

In many ways Reagan's policy toward Nicaragua seemed to defy the conventional rules governing political action. Although most presidents have sought to avoid making foreign policy a political liability, Reagan gave every indication of indifference to the political risks entailed in his posture toward Nicaragua. In his determination to overthrow the Sandinista government, Reagan disregarded public opinion, international law, and the criticisms of America's allies in Europe and the rest of the Third World.

Reagan's intransigence in this regard is most commonly attributed to his political background and beliefs. According to LaFeber, "The president and his closest White House advisers were inexperienced and ignorant of foreign policy. Their background and ideology led them to believe sincerely that the Soviets caused most of the world's problems, even in Central America."[6] But even the belief that the situation in Central America was serious enough to warrant approaching it in terms of a superpower confrontation does not adequately explain Reagan's continuing "obsession" with Nicaragua. As Richard Ullman noted in 1983, to behave as if Nicaragua "poses a dire threat to the stability of the Hemisphere and to the security of the United States, and on that basis to launch an unacknowledged but deadly war against [the Sandinistas], evinces a frame of mind that future historians are likely to discuss more in terms of pathology than in those of logic."[7]

Political beliefs and personality flaws aside, Reagan's reaction to the Sandinistas can also be attributed to the longstanding policies of the United States toward Central America. As Smith notes, concerns for national security have traditionally prompted the United States to take a heavy-handed approach to relations with states in the region. Before the Second World War, Washington's aggressive behavior toward Central America was predicated on the fear that "the failure of political institutions in the region to adjudicate civil conflicts or to discharge satisfactorily international obligations freely incurred [would make] these states standing targets for the expansionist designs of Washington's rivals."[8] During the era of the cold war, the fear of Communist expansion in the area served to intensify this traditional concern for national security on the part of U.S. officials.

Consequently, the United States has sought to exert a significant degree of "informal" control over the states of Central America through the medium of indigenous ruling classes. U.S. leaders have regarded any move toward revolution or radical reform within these states as a threat to the stability of the Western hemisphere. And, with the notable exception of Cuba, the United States has always acted swiftly to eliminate threats to the status quo.[9]

As can be seen, the revolution in Nicaragua represented a serious deviation from the norm in Central America. The replacement of the Somoza dictatorship (a regime solidly supported by the United States) by a leftist government independent of American influence would have been a cause for alarm for any president. Compounding the problem was the timing of the Sandinistas' takeover. The fall of the Somoza regime in 1979 happened to occur at a time when, in the wake of Vietnam, American hegemony throughout the world appeared to be crumbling. Perhaps inevitably, Nicaragua became identified as symptomatic of the Vietnam syndrome.

Here we must distinguish between the more commonly understood meaning of the syndrome and its meaning for Washington officialdom. The former is adequately described as "widespread public opposition to U.S. military involvement in Third World conflicts."[10] The latter meaning is more complex. Depending upon the context in which it is employed, the label can refer to several related attitudes, beliefs, and perceptions, the common bond between them being the idea of decline.

For the official mind of Washington in the early 1980s, the Vietnam syndrome manifested itself in the form of anxiety over a perceived erosion of U.S. power and influence throughout the world. Proponents of this view pointed not only to the American defeat in Indochina, but also to the foreign policy debacles of the Carter administration—the Iranian hostage crisis, the Soviet invasion of Afghanistan, the leftist insurgency in El Salvador, and the overthrow of Somoza in Nicaragua. These events seemed to signal a general U.S. retreat from the world and an inability to advance and protect its interests.

From this perspective the threat posed by Nicaragua was a dual one. In the first place, Nicaragua epitomized the twenty-year decline of American influence. While Nicaragua was neither a haven for U.S. investors nor large enough to threaten seriously a world superpower, it symbolized for the United States yet another loss of power in the international arena. Second, American strategists argued that the Sandinista victory established a dangerous precedent in Central America. In other words, the Sandinistas were "models for the Salvadorans and for revolutionaries in other Central American countries—living proof that movements such as theirs can triumph over a government that, at least until its last months, enjoyed substantial support from the United States."[11] It was argued that this trend, if allowed to continue, would eventually threaten the stability of other American-backed regimes in the region and create opportunities for hostile foreign influence on America's "back doorstep."

At the same time, however, many policy makers recognized that Nicaragua presented an excellent opportunity for America to reverse its "decline" and reassert its dominance throughout the hemisphere. This belief was wholly embraced by the Reagan administration. According to Kenworthy, Reagan entered office determined to prove that Vietnam was an "aberration." The accomplishment of this objective would entail a demonstration of his willingness "to use force in Third World settings." Central America and the Caribbean seemed an "ideal venue" for such a demonstration: "Minimal interference from other world powers coupled with a proximity to the United States that should make an appeal to the public on the basis of defense convincing."[12]

The only problem with this scenario was the American public. The opposition of Americans to another possible long-term military involvement in a Third World country proved too strong for Reagan to overcome. Although Americans were concerned about the spread of communism in the Western hemisphere, the idea that Nicaragua could seriously threaten U.S. security was difficult to accept.

Reagan's failure to secure a public mandate for a "war of liberation" against the Sandinistas created a significant rhetorical problem for him. In the absence of mass public support, Reagan's only means of rolling back the spread of communism in the Western hemisphere was a war by proxy. The Nicaraguan Contras rather than the American military were to carry the "banner of freedom" in Central America. With the Sandinistas strongly entrenched in power, however, Reagan's policy entailed a considerable investment in terms of time. And long-term foreign ventures risk the transformation of informed opinion into informed opposition.

Political scientists have rightly noted that presidential foreign policy discourse is directed primarily toward the American mass public. But they have also posited the existence of an "attentive" or "informed" public. The most distinguishing characteristic of the individuals who make up this segment of the total audience is that they are "aware of the major issues and well informed with respect to them."[13] Moreover, their level of interest in foreign policy matters is greater than that of mass audience members.

While the actual composition of the informed public varies somewhat according to the foreign event in question, this public's levels of awareness and interest can become problematic for presidents. Unlike the mass public, the informed audience does not respond as readily to rhetorical melodrama. Their experiences and knowledge better enable them to critique and evaluate presidential discourse. In short, the informed public is in a much better position than the majority of Americans to frustrate a president's attempts to construct and sustain national consensus on foreign policy issues. In the event of a questionable foreign venture, this public can coalesce with a president's other political opponents to form an informed opposition.

Much of the time, this opposition's potential impact on foreign policy is not realized. Most of American foreign policy crises are of brief duration with a minimal loss of American lives. Lyndon Johnson's 1965 invasion of the Dominican Republic and Reagan's 1983 assault on Grenada are noteworthy examples. In both cases, the actions in question were completed before any sort of opposition could solidify and become a factor in public opinion. Such was not the case, of course, with the Vietnam conflict. Here, the length of American involvement permitted the formation of a vocal and determined minority opposition that, in conjunction with a variety of other factors, prevented the maintenance of a national consensus. In other words, an informed opposition is more likely to organize and become a political factor if a crisis situation persists over a period of time.

It would be a mistake to overestimate the power of the informed opposition. As noted above, this audience segment is but one of many factors that can function to threaten consensus. But if Reagan's rhetorical strategy toward Nicaragua is any indication, I would argue that he did not ignore this group in constructing his appeals. By this I do not mean to imply that Reagan's target audience was meant to be those individuals and groups with a high level of knowledge and interest in Central American affairs. But I would suggest that Reagan's rhetorical purpose, at least in part, was to minimize the effects of such opposition.

If a "war by proxy" created Reagan's rhetorical problem, it also afforded the means to its solution. In other words, the Contra offensive against the Sandinistas presented Reagan with the opportunity to create the same sort of foreign situation in actuality that he was seeking to depict rhetorically. Reagan's ability to accomplish this freed him from the need to structure his appeals in conformity with a reality against which opponents could measure the veracity of his claims.

Reagan thus subverted the assumption that foreign events exist apart from the discourse employed to describe and interpret them. This assumption has always influenced the construction of a president's foreign policy rhetoric. Even though a foreign event is not directly accessible for personal inspection by the majority of Americans, its very facticity demands that leaders avoid deliberate falsifications in their rhetorical interpretations of it. This is not to say that presidents do not attempt to guide the public's perception of the event via the selective presentation of information, simplification, and dramatization. But to purposely mislead the public about foreign affairs entails the politically disastrous risk of having one's claims "unmasked" by opponents. Consequently, researchers have commonly attributed discrepancies between presidential discourse and the event or events it describes to misinformation or initial misunderstanding.[14]

Of course, to deny the existence of most foreign events would be ludicrous. But Reagan's discourse on Nicaragua forces us to consider the possibility that a leader can create a situation that validates his or her discourse. In other words, Reagan was not constrained by the facticity of the Nicaraguan situation; his use of the Contras and economic sanctions created a Nicaraguan "reality" congruent with his rhetorical interpretation of it.

Early observers of Reagan's policy toward Nicaragua do not seem to have considered the possibility that the president's support of the Contra war was an integral part of his rhetorical assault on the Sandinistas. More recent analysts, however, have argued that provoking the Sandinistas into an aggressive militarism was one of the objectives of U.S. policy as early as 1981, when the Contras were being organized by the CIA.

Peter Kornbluh takes this position in his analysis of Reagan's policy. He argues that the Reagan administration's strategy "was to force the Sandinistas to become in reality what administration officials called them rhetorically: Aggressive abroad, repressive at home, and hostile to the United States."[15] In support of this contention, Kornbluh cites two sources. The first is David MacMichael, a CIA intelligence analyst from 1981 to 1983, who claims that officials expected that the Contra attacks would force the Sandinistas to "clamp down on civil liberties within Nicaragua itself, arresting its opposition, demonstrating its allegedly inherent totalitarian nature and thus increase domestic dissent within the country."[16] The second source is that of an unidentified U.S. diplomat who claims that the purpose of the Contra war was to force the Sandinistas to react in "one of two ways. Either they'd liberalize and stop exporting revolution . . . or they'd tighten up, alienate their own people, their international support and their backers in the United States, in the long run making themselves more vulnerable. In a way, that one was even better—or so the idea went."[17]

Kornbluh's interpretation is intriguing. More important, his analysis allows us to identify the two primary rhetorical objectives of Reagan's Nicaraguan speeches. First and foremost, Reagan still hoped to attract the support of the mass public. The creation of a national consensus for more aggressive action against Nicaragua was always Reagan's fervent desire. But failing this, the president needed to be in a position to carry on his "war by proxy" without hindrance from opponents. In order to accomplish this, Reagan had to be able to set and control the agenda of any debate over his Nicaraguan policy and nullify any adverse effects of opposition arguments. He thus focused on three arguments (all characteristic of the paranoid style) about the Sandinistas in his speeches: (1) they were part of a larger Communist conspiracy; (2) they were demonic agents who brutalized and oppressed their own people; (3) their exportation of Communist revolution throughout the hemisphere made them a threat to the American way of life. Because the Contra offensive helped to validate each of these charges, the president's opponents were unable to refute his interpretation of the Nicaraguan situation and propose an alternative course of action. In the analysis that follows, I will examine each of these arguments as they were employed in Reagan's speeches from 1982 through 1986 and demonstrate their effect on the Nicaraguan debate.

The Application of the Paranoid Style to Nicaragua

In his essay entitled "The Paranoid Style in American Politics," Richard Hofstadter defends his use of the term *paranoid* to describe a certain rhetorical style "simply because no other word adequately evokes the qualities of heated exaggeration, suspiciousness, and conspiratorial fantasy."[18] Such a style, he argues, is "systematized in grandiose theories of conspiracy." Almost invariably, the paranoid rhetor finds the conspiracy in question to be "directed against a nation, a culture, a way of life whose fate affects not himself alone but millions of others."[19]

For the paranoid rhetor, the idea of conspiracy and powerful enemies behind it are the central elements in his or her vision of reality. Hofstadter ascribes the potency of the conspiracy idea to its status in the paranoid's mind as "*the motive force* in historical events. History *is* a conspiracy, set in motion by demonic forces of almost transcendent power, and what is felt to be needed to defeat it is not the usual methods of political give-and-take, but an all-out crusade."[20]

Reagan consistently employed the idea of conspiracy throughout his speeches on Nicaragua. He argued that in Nicaragua the agents of the international Communist conspiracy had succeeded once again in duping the world. In his speech of 16 March 1986, he advanced the following interpretation of the Nicaraguan revolution:

> In 1979, the people of Nicaragua rose up and overthrew a corrupt dictatorship. At first the revolutionary leaders promised free elections and respect for human rights. But among them was an organization called the Sandinistas. Theirs was a Communist organization, and their support of the revolutionary goals was sheer deceit.
>
> Two months after the revolution, the Sandinista leadership met in secret, and, in what became known as the "72-hour document," described themselves as the "vanguard" of a revolution that would sweep Central America, Latin America and finally the world. The true enemy they declared: the United States.
>
> Rather than make this document public, they followed the advice of Fidel Castro, who told them to put on a façade of democracy. While Castro viewed the democratic elements in Nicaragua with contempt, he urged his Nicaraguan friends to keep some of them in their coalition—in minor posts—as window dressing to deceive the West. That way, Castro said, you can have your revolution, and Americans will pay for it.[21]

Reagan concluded this scenario by noting that the Americans did pay for the revolution: "More aid flowed to Nicaragua from the United States in the first eighteen months under the Sandinistas than from any other country." When the real nature of the Sandinista government was discovered, U.S. aid ceased.

The above passage is also illustrative of another aspect of the paranoid style not directly discussed by Hofstadter. This is the posture or stance that a rhetor assumes in relation to the audience.[22] For the paranoid rhetor, this relationship is one of revelation. The audience is presumed to be ignorant or deceived by an apparently harmless situation or state of affairs. The rhetor, consequently, takes the responsibility of revealing to the audience the dangers inherent in the situation. Possessed of superior knowledge, insight, or information unavailable to anyone else, the rhetor perceives his or her duty to be that of alerting the unsuspecting masses to the fate that awaits them.

Such a posture was important to Reagan. In revealing the presence of a sinister reality in a world of innocuous appearances, Reagan utilized the posture of revelation to set the agenda of the debate on Nicaragua. Since what a president says about foreign situations is presumed to be true, if not always complete, Reagan used the inherent credibility of the presidency to ensure that the question of Nicaragua would rest on the affirmation or refutation of two claims.[23]

The first of these claims had to do with the nature of the Sandinistas themselves. For Reagan, the leaders of Nicaragua were demonic agents. His characterization of them encompassed two of the qualities that paranoids typically attribute to their enemies.

The first of these qualities is the wholly condemnatory nature of their actions. In paranoid discourse, the heinousness of the enemy justifies behaving in the same manner toward him or her. According to Hofstadter, "This enemy seems to be on many counts a projection of the self: both the ideal and unacceptable aspects of the self are attributed to him. A fundamental paradox of the paranoid style is the imitation of the enemy."[24]

To some extent, this paranoid element was employed in Reagan's speeches to justify U.S. support of the Contras. To be sure, Reagan never explicitly stated that the Contras were fighting a "dirty" war against the Sandinistas. He compared them to the French Resistance in World War II and stated that "the freedom fighters of Nicaragua have pinned down the Sandinista Army and bought the people of Central America precious time. We Americans owe them a debt of gratitude. In helping to thwart the Sandinistas . . . the resistance has contributed directly to the security of the United States."

Despite such effusive praise, however, the president had to realize that his audience would be aware of various acts of savagery on the part of the Contras and their CIA mentors. One such act was the 1984 mining of the Nicaraguan harbors of Corinto, Puerto Sandino, and El Bluff. Another was the secret publication of a Spanish-language CIA manual entitled *Psychological Operations in Guerrilla Warfare.* Among other things, the booklet advocated targeting judges, police, and government officials for assassination and even recommended "arranging the death of a contra supporter to create a 'martyr' for the cause."[25] In both instances, public and congressional reaction was largely negative. Consequently, Reagan tried throughout his speeches to make the Sandinistas appear much worse than his own forces.

But Reagan's main purpose in defaming the Sandinistas was to challenge his critics to refute his claims. In this respect, a second quality of the enemy is most salient: the "sins" of the enemy are those the audience would find most repugnant. Hofstadter notes that paranoid discourse is carefully based on "moral commitments that can be justified to many non-paranoids."[26] In his catalogue of Sandinista atrocities, the president was very careful to focus his audience's attention on those examples of Sandinista oppression that most Americans would find objectionable.

In his speech of 1983, for example, Reagan cited the following list of Sandinista crimes:

> The Government of Nicaragua has imposed a new dictatorship: it has refused to hold the elections it promised; it has seized control of most media and subjects all media to heavy prior censorship; it denied the bishops and priests of the Roman Catholic Church the right to say mass on radio during Holy Week; it insulted and mocked the Pope; it has driven the Miskito Indians from their homelands—burning their villages, destroying their crops, and forcing them into involuntary internment camps far from home; it has moved against the private sector and free labor unions; it condoned mob action against Nicaragua's independent human rights commission and drove the director of that commission into exile.

In essence, Nicaragua had become a typical dictatorship in which freedom of the press and religion were suppressed, private enterprise discouraged, and human rights violated.

In his 1986 address Reagan claimed that "there seems to be no crime to which the Sandinistas will not stoop—this is an outlaw regime." In addition to the crimes of 1983, Reagan added complicity in the drug trade: "The Sandinistas have been involved themselves in the international drug trade. . . . top Nicaraguan Government officials are deeply involved in drug trafficking."

In most of his speeches Reagan directed much of his invective against the Sandinistas' policy of religious persecution. In his address to the nation of 9 May 1984, Reagan described this persecution in the following manner:

> The Sandinistas engaged in anti-Semitic acts against the Jewish community, and they persecuted the Catholic Church and publicly humiliated individual priests. When Pope John Paul II visited Nicaragua last year, the Sandinistas organized public demonstrations, hurling insults at him and his message of peace. On this last Good Friday, some 100,000 Catholic faithfuls staged a demonstration of defiance. . . . Nicaraguan Bishop Pablo Antonio Vega recently said, "We are living with a totalitarian ideology that no one wants in this country"—this country being Nicaragua.

In his 1986 speech Reagan charged that Protestants had also suffered the most extreme forms of persecution and abuse. In support of this claim he related the following gory narrative:

Evangelical pastor Prudencio Baltodano found out he was on the Sandinista hit list, when an army patrol asked his name: "You don't know what we do to the evangelical pastors. We don't believe in God," they told him. Pastor Baltodano was tied to a tree, struck in the forehead with a rifle butt, stabbed in the neck with a bayonet—finally his ears were cut off, and he was left for dead. "See if your God will save you," they mocked. Well, God did have other plans for Pastor Baltodano. He lived to tell the world his story—to tell it, among other places right here in the White House.

Despite the highly emotive quality of passages such as the above, one cannot conclude that Reagan's discourse is devoid of logic. But the logic employed is that of the paranoid style. As noted above, paranoid discourse is generally based upon "moral commitments" and "defensible assumptions" whose claims to validity are not in question. On the basis of these assumptions, the paranoid rhetor carefully selects "facts" to "prove" that his or her vision of reality is correct. To this extent, paranoid rhetoric "is, if not wholly rational, at least intensely rationalistic."[27]

Such can be said for Reagan's discourse. He sought to "prove" that the Sandinistas were Communist agents who had tyrannized their own country. In order to demonstrate the validity of this claim, the president linked the Sandinista regime with the repression of those institutions in Nicaragua that Americans would regard as indispensable to political democracy: freedom of religion, freedom of speech and the press, and respect for minority rights. By means of illustrations and the testimony of disaffected Nicaraguans, Reagan adduced the evidence necessary to support his original contention.

The president's demonization of the Sandinistas was not without effect. His detractors were placed in the position of having to prove that Nicaragua was not the expansionistic Communist dictatorship of Reagan's portrayal. And this they could not do. While they asserted that Reagan had overreacted to the Nicaraguan situation, they were forced to admit that the country was far from being an open type of political system.

Typical of such an admission was that of Richard Ullman. A critic of the administration's Nicaraguan policy, Ullman visited Nicaragua in 1983 and found various Sandinista policies disconcerting. He noted that the Nicaraguan government had undermined its pretensions of nonalignment by consistently voting with the Soviet bloc on major issues before the United Nations. Cuban advisers continued to play a guiding role in the Nicaraguan military and in the intelligence and internal security apparatus. Early supporters of the revolution had gone into exiled opposition.

He also reported that the country under the Sandinistas was becoming increasingly intolerant of potential dissenting groups. While the Sandinistas had instituted programs to increase literacy among the peasants, these same programs were used to indoctrinate the peasantry in Marxist ideology. Roman Catholic medical workers were sometimes harassed by Sandinista youth groups. Miskito Indians had been driven from their homelands and forcibly resettled in other parts of the country.

Finally, freedom of expression was extremely limited in Nicaragua under the Sandinistas. After a state of emergency was declared in 1982, the Sandinistas adopted a policy of prepublication censorship and occasional shutdowns of the press. In addition, they suspended the right to liberty and habeas corpus, the right to freedom of travel, and the rights of association and peaceful assembly.[28]

But despite these problems, Ullman denied that Nicaragua was the repressive despotism of Reagan's discourse. He noted that "North Americans and Western Europeans long resident [in Nicaragua] find simply bewildering the Administration's characterizations and the blatantly dual standard that it employs when it compares Nicaragua with other states in the region."[29]

Ullman suggested that the restrictions on personal and political freedom were not the result of any inherently evil tendencies of the Sandinistas themselves. Rather, these restrictions stemmed from the fact that attacks by the CIA and the Contras had the effect of placing the country under a state of siege. He expressed bewilderment over the reasons why the president continued to pursue a policy that was ultimately self-defeating. Reagan's alleged purpose in supporting the freedom fighters, Ullman noted, was to force the Sandinistas to institute democratic reforms. And yet the Contra attacks had produced just the opposite effect; they had made Nicaragua "more militarized, more monolithic, and more repressive . . . and in its foreign policy more stridently anti-American, more dependent on the Soviet Union and on Cuba, and therefore more willing to do their bidding."[30]

Ullman's interpretation of the reasons for Nicaragua's growing militarization was one that Reagan flatly rejected. In his address to the nation on 9 May 1984, he asserted:

> Nicaragua's own military forces have grown enormously. Since 1979 their trained forces have increased from 10,000 to over 100,000. Why does Nicaragua need all this power? Why did this country of only 2.8 million build this large military force?
>
> They claim the buildup is the result of the anti-Sandinista forces. That's a lie. The Sandinista military buildup began 2½ years before the anti-

Sandinista freedom fighters had taken up arms. They claim the buildup is because they're threatened by their neighbors. Well, that, too, is a lie. Nicaragua's next door neighbor Costa Rica doesn't even have an army. Another neighbor, Honduras, has armed forces of only 16,000.

The Sandinistas claim the buildup is in response to American aggression. And that is the most cynical lie of all. The truth is they announced at their first anniversary, in July of 1980, that their revolution was going to spread beyond their own borders.

In essence, Reagan made the question one of whether the Sandinistas' acts were a response to American-backed aggression or a natural consequence of their political ideology and program. As can be seen, his critics had to admit the facts of Nicaraguan military buildup and domestic repression. Their claims regarding the underlying reasons for these acts could not erase the fact that the acts were committed. Reagan, on the other hand, had merely to point to the acts as "proof" of his claim that the Sandinistas were Marxist agents of aggression. The resolution of this impasse could not be definitively achieved. In this case, the presumption rested with the president. The facts on which his paranoid style was based could be explained but not refuted.

Reagan's defamation of the Sandinistas supported his third claim about them. This claim related to the nature of the threat posed by Nicaragua to the Western hemisphere. Reagan's assertions on this point were clear and unmistakable. In a 1982 speech to the Organization of American States, he noted "the tightening grip of the totalitarian left in . . . Nicaragua" and stated that the country "has served as a platform for covert military actions. Through Nicaragua, arms are being smuggled to guerrillas in El Salvador and Guatemala."

In his 1983 speech to Congress, Reagan expanded on his perception of the Nicaraguan threat. He explicitly linked Nicaragua to the leaders of international communism by asserting that the Sandinistas were "supported by weapons and military resources provided by the Communist bloc." Furthermore, the ability of Nicaragua to wreak havoc in Central America had dramatically increased: "Nicaragua's new army numbers 25,000 men supported by a militia of 50,000. It is the largest army in Central America supplemented by 2,000 Cuban military and security advisers." The Soviet Union, East Germany, Libya, and the PLO had also supplied advisers to the Sandinista forces. According to the president, the goal of Nicaragua was "as simple as it is sinister—to destabilize the entire region from the Panama Canal to Mexico." After noting the proximity of Nicaragua to Texas, Reagan argued that U.S. inaction would cause the "safety of our homeland [to] be put in jeopardy."

Perhaps his 1986 speech contained the most graphic indictment of the Nicaraguan threat. In his opening sentence, Reagan stated that his purpose in speaking was to inform Americans about "a mounting danger in Central America that threatens the security of the United States. This danger will not go away; it will grow worse, much worse, if we fail to take action now." After alluding again to the nearness of Nicaragua to America, Reagan argued that the danger of Nicaragua must now be seen in international terms:

> Using Nicaragua as a base, the Soviets and Cubans can become the dominant power in the crucial corridor between North and South America. Established there they will be in a position to threaten the Panama Canal, interdict our vital Caribbean sea lanes, and ultimately, move against Mexico. Should that happen, desperate Latin peoples by the millions would begin fleeing north into the cities of the southern United States, or to wherever some hope of freedom remained.

In addition to the Communist threat to geopolitical and economic security, Reagan claimed that Nicaragua had become a haven for "all the elements of international terror—from the P.L.O. to Italy's Red Brigades." He quoted Libya's Muammar Qaddafi on the reason for the presence of these elements: "Nicaragua means a great thing, it means fighting America near its border. Fighting America at its doorstep." Later in the speech, Reagan urged Americans to "ask yourselves, what in the world are Soviets, East Germans, Bulgarians, North Koreans, Cubans and terrorists from the P.L.O. and the Red Brigades doing in our hemisphere, camped on our doorstep? Is that for peace?"

Reagan's depiction of the Nicaraguan menace in his speeches was a veritable litany of the contemporary world's most apocalyptic fears; war, terrorism, refugees. Furthermore, the authors of this coming nightmare of anarchy and disorder were the Sandinistas.

If Reagan's defamation of the Sandinistas was difficult to refute, his assessment of the threat posed by them was equally hard to disprove. The effect of this claim was most clearly evident in the congressional debates of 1986 over continued funding for the Contras. Here administration opponents were acutely aware of the president's advantage on the question of the Sandinistas' despotic actions in Nicaragua. For the most part, those opposing Contra aid did not dispute the president's characterization of Nicaragua. In fact, his critics joined Reagan in roundly denouncing the Sandinistas' totalitarian leanings. But this agreement made it difficult to argue that Reagan's predictions of the expansion of communism in Central America were incorrect.

Reagan's opponents in Congress were thus placed in a difficult argumentative position. Their primary objection to the Nicaraguan policy was a "slippery slope" argument that the continued support of the Contras would eventuate in another Third World entanglement of the Vietnam type. The administration countered this claim by declaring that the actions of Sandinistas indicated a determination to destabilize other states in the region. In this case, the United States would have to confront them eventually. Supporting the Contras, however, would enable the United States to avoid such an outcome. The debate was thus stalemated. The Sandinistas' aggressive posture made it impossible for opponents to prove conclusively that the president's predictions of Communist expansion in the region would not occur. As Muravchik observes: "The President's estimate of the stakes was either right or wrong. It could be challenged on empirical or analytical grounds, but it could not logically be challenged on the ground of its consequences."[31]

Compounding the problems of Reagan's opponents was their inability to discern the will of their constituents. To be sure, the American public was still opposed to aid to the Contras. On the other hand, the president's discourse did have the effect of arousing concern about the presence and possible spread of communism in Central America. Muravchik offers the following explanation for this apparent contradiction: "The same electorate that invariably tells pollsters that it favors increases in government services and decreases in taxation now was telling them that it was anxious to stop communism in our hemisphere but reluctant to go to much trouble or accept many risks in order to do so."[32] But the president's opponents could provide no alternative to the president's policy. This made defeat of Reagan's request politically risky. According to Muravchik, "Moderate Democrats in Congress grew fearful that defeating the President's program—unpopular as it seemed—without offering any genuine alternative would leave them open to blame for whatever troubling developments might unfold in the region."[33]

In this manner, Reagan was ultimately able to fund his "war by proxy" against the Sandinistas without the backing of the American mass public. By adroitly utilizing the Sandinistas' aggressive posture to validate his paranoid style, he controlled the questions to be debated regarding Nicaragua and frustrated his detractors' efforts to modify his policy.

My use of the paranoid style as an approach to Reagan's discourse should not be misinterpreted. I have not been concerned with advancing the claim that Reagan himself was an authentic paranoid in Hofstadter's sense of the term. Instead, my intention here has been to explore his use of paranoid elements and strategies as part of an overall strategy to set and control the agenda of the Nicaraguan debate. But even in advancing this somewhat modest claim, I do not mean to imply that Reagan's use of the paranoid style was unique to his approach to foreign policy. As I noted at the beginning of this essay, every president since World War II has employed rhetorical strategies in foreign policy discourse that resemble the paranoid style. In the final analysis, then, why are Reagan's speeches on Nicaragua worthy of study? Their worth, I would argue, lies in the fact that they force us to take a more sophisticated approach to the functions that foreign policy discourse can perform. This claim will be developed at length in the final section of this essay.

The Functions of Presidential Foreign Policy Discourse

At the beginning of this essay, I noted that critics and political observers have largely viewed the sole purpose of a president's foreign policy discourse to be that of securing public and congressional support for his actions. This assumption seems obvious when one examines examples of presidential foreign policy rhetoric produced during the cold war era. These speeches, for the most part, have seemed less concerned with dispassionate analyses of policy formulation than with the mobilization of popular support for a predetermined course of action.[34]

As I have tried to demonstrate above, however, Reagan's Nicaraguan speeches suggest that the ultimate goal of foreign policy discourse is not the gaining of consensus. Neither is it, as Wander has argued, the stifling of public debate.[35] These purposes are but complementary *means* of gaining for presidents as much freedom as possible to act as they see fit in the arena of world politics. In other words, presidents have typically sought to free themselves from the constraints of public opinion and political criticism via the rhetorical creation of consensus.

Because foreign events exist apart from the discourse used to interpret them, the creation and maintenance of consensus has always depended on a president's ability to monopolize the interpretation the public receives of an event. Consequently, those foreign policy situations in which leaders have typically enjoyed the most success in this regard have been interventions of relatively brief duration. Lengthy foreign ventures jeopardize a president's monopoly of information and make possible the formation of informed opposition groups.

Herein lies the significance of Reagan's paranoid strategy. By using the Contras to create a situation congruent with his interpretation of it, Reagan nullified the problem of time. Even though his "war by proxy" lasted for almost seven years and cost taxpayers millions of dollars, he continued to enjoy argumentative presumption. To be sure, his more extreme inferences and exaggerations were ridiculed by critics. But his ability to control the situation in such a way as to make the Sandinistas conform to his characterization of them confounded his opponents and prevented the mobilization of an ambivalent mass public. In other words, Reagan was able to force his critics into the position of having to disprove his interpretation of the Nicaraguan situation. Because of the Contras, they were never able to do this decisively.

This leads us to consider the question of what effect (if any) Reagan's paranoid strategy will have on the future of democratic participation in foreign policy. At present, the potential for any future impact of this strategy appears to be minimal. In the first place, the Bush administration seems to have little use for protracted proinsurgent policies. As was demonstrated by the U.S. invasion of Panama, Bush appears to regard the military removal of troublesome foreign leaders as a cheaper and more politically expedient course of action.

Another factor currently mitigating against the revival of Reagan's strategy is the atmosphere of cooperation between the major powers of the East and West. Like Wander's argument of "prophetic dualism" (a closely related variant of the paranoid style),[36] the success of proinsurgency policies and their rhetorical justifications depend upon at least a minimal audience acceptance of the facticity of a hostile state capable and desirous of threatening U.S. security and interests. With Mikhail Gorbachev's commitment to the economic and political reform of the Soviet Union, no such external enemy now exists. While certain Arab leaders such as Muammar Qaddafi and Saddam Hussein have taken the place of the Soviet menace in the rhetoric of George Bush, it is doubtful they will be seriously perceived as the architects of a worldwide conspiracy whose goal is the conquest of the Western industrial democracies.

But it should be noted that proinsurgency is a situation-specific policy. If another Nicaragua-like situation presents itself, there is every reason to suppose that Reagan's paranoid strategy would be resurrected. Such is the position taken by those analysts who predict that the use of low-intensity warfare by the United States against hostile Third World governments will increase in the future. The fact that such "client" states are no longer subject to the restraints of powerful "patrons" like the Soviet Union may actually increase the likelihood of this type of scenario.[37]

If this in fact becomes the case, then Reagan has bequeathed to his successors an effective rhetorical strategy for the justification of a policy of proinsurgency. In essence, he has demonstrated that in some areas of foreign policy the support of the mass public is less important than the nullification of factors favorable to the development of informed opposition groups. Should a president succeed in achieving this sort of nullification, the possibility of reasoned policy debate is seriously diminished. For only in the clash of competing interpretations of foreign reality can an electorate become apprised of all the options available to the state for the solution of a foreign problem. Only in this manner can the public exercise its will in the conduct of foreign policy.

9

When the Shoe Is on the Other Foot

The Reagan Administration's Treatment of the Shootdown of Iran Air 655

Marilyn J. Young

Ronald Reagan must have been the most studied sitting president in the history of our discipline. Yet the impression remains that no one has truly penetrated the veil of success that enabled Reagan—one of the least communicative of presidents—to retire with the label the Great Communicator. If he has a lasting impact on nothing else, rhetoric and the criticism of rhetoric may never be the same. Traditionally, we have considered the artifacts of rhetorical criticism to be individual speeches, the rhetoric of an entire movement, or the rhetorical collection of a single individual. The presidency of Ronald Reagan argues for a conception of the rhetorical transaction as a complex episode: a conception wherein the entire constellation of rhetoric surrounding a specific event is treated as the rhetorical text, with each individual transaction regarded as part of the whole. The rhetoric of crisis situations illustrates the significance of this approach.

Two such crises occurred during the Reagan years: when the Soviets shot down Korean Air Lines Flight 007 in 1983 and when the United States shot down Iran Air Flight 655 in 1988. The purpose of this essay is to compare the rhetoric surrounding the destruction of KAL 007 with that following the destruction of Iran Air 655. There are two points of comparison: the Reagan administration's rhetoric in 1988 with that of 1983; and the Reagan administration's position in 1988 with the Soviet argument in 1983.

These two tragedies are instructive for a number of reasons. They provide impetus to the notion of rhetoric as a series of apparently discrete but related discursive events. Further, they illustrate the ways in which the text of an individual component of the rhetorical episode interacts with its context, the ways in which context, or the discursive environment, constrains discourse options, and so on in an infinitely reflexive progression. In each of these instances, one cannot gain a complete understanding of the situation unless one examines the entire constellation of administration rhetoric surrounding the shootdown. Indeed, I would argue that it is necessary to revisit the Reagan administration's rhetoric after KAL in order to fully comprehend the rhetorical antecedents and constraints operating in the Airbus situation. These two events were inextricably linked, a circumstance recognized by Reagan and his advisers as they sought to forestall comparisons with KAL and to control public opinion at home and abroad. Tactics used by the administration include the episodic approach itself: multiple speakers (four in this instance), each with a different role; concomitant with this was the low profile of the president (in contrast to his KAL posture). The goal was management of the context through redefinition (recontextualization); the strategy was to salvage Reagan's gulf policy by refocusing the debate on the culpability of Iran for pursuing the war with Iraq. This study examines those efforts, concluding that justificatory rhetoric accompanied by attempts to shift blame are hallmarks of crisis rhetoric in general.

On the night of 31 August/1 September 1983, Korean Air Lines Flight 007, bound for Seoul from Anchorage, Alaska, strayed some three hundred miles off course, entering Soviet airspace over the Kamchatka Peninsula and again over Sakhalin Island. At 18:26 GMT (3:26 A.M. local time) a Soviet interceptor pilot fired two missiles at the airliner; all 269 persons aboard perished in the Sea of Japan. This episode represented a nadir in our relations with the Soviet Union: the president and his administration, spurred by the influence of their own anti-Soviet rhetoric and the horror of the event itself, adopted an essentially exhortative style designed to gain universal condemnation of the USSR from both a domestic and a foreign audience. U.S. leaders excoriated the Soviet Union in every available forum, seizing the opportunity presented by what they perceived as a clear moral imperative to make a political statement to the world. Administration officials hastened to take advantage of what appeared to them to be an act which revealed the "true nature" of the Soviet system, an act for which there was no provocation, no possible justification. American officials exhorted not only their domestic constituency but the entire world to join in condemnation of the Soviet barbarians.[1]

Examples of such administration oratory abound:

Ronald Reagan (1 September): . . . [an] appalling and wanton misdeed . . . inexplicable to civilized people everywhere.

Ambassador Charles Lichenstein (2 September): Let us call the crime for what clearly it is: wanton, calculated, deliberate murder.

Ambassador to the U.N. Jeane Kirkpatrick (6 September): The destruction of KAL 007 was . . . a deliberate stroke designed to intimidate—a brutal, decisive act meant to instill fear.

Assistant Secretary of State Lawrence Eagleburger (6 September): [The USSR] is not bound by the norms of international behavior and human decency to which virtually all other nations subscribe.

Secretary of State George Shultz (8 September): This brutal Soviet action has vividly displayed the Soviet Union's lack of concern for the human lives involved.[2]

Shultz, of course, set the tone in his 1 September announcement of the tragedy: "The United States reacts with revulsion to this attack. Loss of life appears to be heavy. We can see no excuse whatsoever for this appalling act."[3]

In particular, the language used by President Reagan reveals both his ideological stance and the political position the United States maintained vis-à-vis the tragedy. Word frequency counts indicate that the term most used to describe the downing was *massacre*, followed by *tragedy*, *attack*, *crime against humanity*, *murder*, *brutality*, and *atrocity*. Of the USSR he says, "They have totally failed to explain how or why"; they "have flunked the test"; "they continue to distort and deny the truth."[4]

In an extended analysis of the aftermath of KAL 007, a colleague and I viewed rhetorical context as a micro concept, exploring the ways in which public knowledge, the rhetorical situation, and rhetorical exigencies[5] both shape and are shaped by public discourse, particularly in a crisis atmosphere. Yet the larger context for this campaign is equally important.

Reagan came to the office of president with an anti-Soviet attitude, a promise to restore America to glory, and a determination to increase our military strength at home and abroad. Throughout the Reagan administration, the rhetoric of virtually all administration personnel resonated with these themes; September 1983 was no exception. It is significant, for example, that the early players in the crisis—William Casey, Richard Burt, Lawrence Eagleburger—could capture, in the initial Shultz announcement on 1 September, the tone of Reagan's rhetoric. Even though they had not talked with the president, they knew what to say and how to say it.[6] Surely, the rhetorical environment for U.S.-Soviet relations was established at least for the near term.

On 3 July 1988, almost five years later, the USS *Vincennes,* on patrol in the Persian Gulf, fired two radar-guided missiles at Iran Air Flight 655; all 290 persons aboard perished. Obviously, times had changed since the 1983 incident. Relations with the Soviet Union had reached an all-time high as Reagan and the American public courted Mikhail Gorbachev, the charismatic new Soviet premier. In contrast to the KAL 007 incident, when the Americans precipitated an international crisis atmosphere, the U.S. handling of the Airbus tragedy was measured, with the administration deliberately seeking to avert an atmosphere of crisis.[7]

A closer examination of the disparate rhetoric of 1983 and 1988 is revealing. In order to test empirically the notion that there were salient similarities between the rhetorical stance of the USSR in 1983 and that of the United States in 1988, as well as profound differences in U.S. rhetoric concerning the two tragedies, a colleague and I performed a content analysis of the primary U.S. documents for both incidents. These included all official statements and speeches by members of the Reagan administration.[8] The hypothesis for the pilot study was that the administration's rhetoric would differ markedly between the two bodies of text (that of the KAL tragedy and that of the Airbus tragedy); that the 1983 rhetoric would use accusatory and condemnatory language while the 1988 rhetoric would be primarily justificatory (as was that of the Soviet government in 1983). Coding frequencies appeared to support this hypothesis.

Some of the categories developed for the study yielded interesting results. For example, whenever a tragedy of this type occurs, one serious question on the minds of the public is that of intentionality. There are almost no references to either shootdown being *un*intentional. However, with respect to the KAL incident, there were twenty-eight instances of claims of intentionality. These fell into two types: the first (twelve codings) are all examples of narrative description of the act itself, designed to bring the audience into the action, such as "the Soviet SU15 interceptor deliberately circled back around behind the Korean 747, the better to aim his heat-seeking missile." The remaining sixteen codings are accusations of an intentional massacre, claiming that the Soviets, aware that the aircraft was a passenger plane after tracking it for two and a half hours, simply decided to shoot it down. In contrast, there were no references to the Airbus incident as an intentional shootdown, despite the fact that the United States admitted firing at the airliner after following it on radar and attempting voice communication.

A major theme of the U.S. campaign in 1983 was the depiction of the USSR as outside the pale of civilized nations. Eleven of the twenty-eight coded instances indirectly accused the Soviet Union of being uncivilized or barbaric; the remaining seventeen depicted the United States and its allies as civilized. This rhetorical device was designed to draw Soviet client states into condemnation of the shootdown. The key to civility in the Airbus incident is the prosecution of the war in the Persian Gulf; the United States accused Iran of being uncivilized in this context in five of the seven codings and took the opportunity to deplore the use of chemical weapons. In the remaining two instances, the United States used the distinction to shift the blame for the shootdown: Iran (not the United States) endangered innocent civilian lives because of its continued prosecution of the war—a theme redolent of Soviet accusations against the United States for not warning the Korean aircraft, for sending it on a spy mission, and so forth. Shifting the blame for the tragedy seems to be a hallmark of this type of justificatory rhetoric.

Twenty-three of the thirty-seven references to moral codes or human rights in 1983 accuse the Soviets of being immoral, of threatening norms of civilized behavior, and so on. Four times the administration referred to the USSR's duty to the families of the victims. Only three instances refer directly to the United States as moral, though the implication is clearly there as a result of the exclusion of the USSR. In contrast, references to moral codes in the Airbus incident are limited to the moral obligation to provide compensation to the families of the dead; both of these codings occur in Vice-President Bush's address to the United Nations and are linked to the value the United States places on human life (perhaps in contrast to Iran and the USSR?).

As noted above, the descriptors chosen to reference the act of the downing are indicative of ideological and political stance. Comparison of the language used in describing these two tragedies is illuminating in this regard. In 1983 the terms most frequently used to refer to the act itself were *tragic* (seventeen references), *criminal* (ten), *brutal* (five), *massacre* (four), *atrocity* (three), *wanton* (three), *heinous* (three), *appalling* (two), *inhuman* (two). Other words used to describe the event included *horrifying, barbaric, outrageous, shocking, unprovoked, flagrant, callous, unspeakable.*[9] Obviously, such expressions were used to increase the public's sense of outrage: at last, the administration believed, the Evil Empire had committed an act that conclusively demonstrated the bankruptcy of its cruel system. Neutral terms (*act, shootdown, incident*) occurred only twelve times in the entire sample.

In contrast, the most frequently occurring terms describing the destruction of Iran Air 655 were *tragic* (eleven references), *incident* (ten), *terrible human tragedy* (two), *accident* (two), *disaster, unfortunate*. While not entirely neutral, these terms carry a different affective load, more empathic, compassionate, almost apologetic. The recurring use of *incident* is especially noteworthy because of the distancing effect of such antiseptic terminology; not surprisingly, it was the most frequently used word in the 1983 *Soviet* rhetoric.

In addition to descriptions of the act itself, it is useful to consider the most frequently used words in the rhetoric of the Reagan administration, regardless of the referent. When one omits the obvious and unimportant terms, such as articles, the terms used most often, in order of appearance, are:

KAL: Soviet, aircraft, Union, international airliner, civil, world, civilian, security, law, human, military, incident, 269 [number killed], destruction, attack, facts, safety, families, civilized, norms, commercial, community, fact, passengers, responsibility, unarmed, force, shooting, rights, crime, destroyed, target, tragedy, compensation, missile, behavior, peace, shoot, evidence, innocent, tragic, rules, strayed, tracked, obligation, standards, demand, explanation, mission, prevent, violence

These are the terms that occurred at least ten times; for perspective, *international* occurred one hundred times; *civilian* occurred thirty-two times; *demand* occurred ten times.

AIRBUS: Gulf, Iran, war, States, United, policy, hostages, reparations, incident, situation, investigation, security, compensation, information, probes, review, safety, attack, defense, life, premature, report, *Stark*, tragedy, responsibility, advisories, diplomatic, friends, negotiations, shooting, facts, know, dialogue, interests, answer, appropriate, briefing, claims, plane, shipping, threats, action, airliner, cause, civilian, discussion, downing, informed, international, Iraq, normal, official, accident, Airbus, airplane, arms, danger

Because of the smaller corpus, I have reported terms that appeared at least five times; for perspective, the term *war* occurred forty-three times, *incident* occurred nineteen times, *tragedy* occurred nine times.

In both instances, one of the most frequently used terms is the name of the rhetorical adversary: *Soviet* in the case of KAL and *Iran* in the case of the Airbus. In each case, the named adversary was identified as the party responsible for the tragedy (massacre, disaster, and so forth). If simple repetition has persuasive impact, and we are told it does, such naming cannot be accidental.

Even a casual perusal of the two lists reinforces a sense of difference. Words used to describe the Airbus incident were in almost every instance more detached, more dispassionate than those used in 1983. Indeed, among the high-frequency words, only twelve occurred in both rhetorical episodes: *international*, *civilian*, *security*, *incident*, *attack*, *facts*, *safety*, *responsibility*, *shooting*, *tragedy*, *compensation*. Obviously, some of the omissions were strictly context-bound, such as *Soviet* and *Gulf*. But it is significant that, for example, the number of dead was mentioned fewer than five times in the U.S. version of the Airbus shootdown; by comparison, that was one of the most frequently used references in the KAL rhetoric. Similarly, the Soviets continually referred to the number of dead in its domestic coverage of the Airbus shootdown,[10] although they had mentioned the number of passengers and crew killed on KAL 007 only twice during the entire four-year period studied.[11]

Other terms that might have recurred in 1988, but do not, include *passengers*, *families*, *norms*, *unarmed*, *target*, *evidence*, *innocent*, and *tracked*. Similarly, there were terms used exclusively in 1988, including *policy*, *reparations*, *situation*, *investigation*, *information*, *review*, and *dialogue*.

Some of the terms that occurred in both instances did so in inverse order of frequency. For example, *civilian*, which is near the top of the list of words used in 1983, occurs near the bottom of the 1988 corpus. Clearly, the United States wanted to play up this aspect of the KAL plane and downplay it in the case of the Airbus. In fact, there were *no* words describing the aircraft among the most frequently used terms in the Reagan administration's Airbus rhetoric; indeed, the only term in the Airbus list that clearly characterizes the craft as civilian is *airliner*, a word that occurred fewer than nine times in the entire episode. By way of contrast, *international airliner* was among the most frequently occurring terms in the KAL list, appearing nearly one hundred times.

Such lexical choices were part of a larger strategy that included themes, claims, and choice of spokespersons—all intended to advance a specific coherent response to a rhetorical situation. Perhaps Reagan's advisers had learned a lesson from the chaotic days of September 1983, when poor coordination cost the United States much of its rhetorical advantage,[12] for the 1988 campaign was more carefully orchestrated. The number of spokespersons was limited, and the military took the lead in dealing with the press.

The most striking difference between the two episodes from the

U.S. perspective rests in the rhetorical behavior of the chief executive. During the KAL crisis Reagan maintained a very high profile, taking over from George Shultz on 1 September and leading the charge to excoriate the Evil Empire. Of course, in 1988 the United States was on the defensive rather than the offensive; nevertheless, Reagan himself had very little to say beyond his initial statement, a three-paragraph announcement that expressed regret and not much else.

In 1983 Reagan had addressed the American public—and the world—with audio tapes accompanied by graphics; after the Airbus was shot down, on the other hand, the president made only three other recorded statements: a letter sent to both houses of Congress,[13] a brief exchange with reporters in which he misstated the facts,[14] and another informal exchange with reporters on 11 July 1988, during which he discussed the question of compensation.[15] There were two other official pronouncements, each attributed to the assistant to the president for press relations: the first, dated 11 July 1988, referred to a presidential decision not to alter U.S. policy in the Persian Gulf and announced the offer of *ex gratia* compensation; the second, on 19 August announced the president's receipt of the report of the investigation and expressed support for the actions of Defense Secretary Carlucci.

There can be little doubt that a major factor in determining Reagan's low profile was his unwillingness to promote echoes of KAL; from the beginning, administration spokesmen fended off the invidious comparisons. In an early report, even before Admiral Crowe made the official announcement of the tragedy, Jim Miklaszewski of NBC News anticipated many of the developments to come: "Some folks here at the Pentagon are beginning to try to fashion somewhat of a damage control scenario. They're afraid that the obvious comparisons will be made . . . between this incident and the KAL 007 shootdown."[16] Not surprisingly, someone at Crowe's press conference inquired about the comparison; Crowe was prepared: "The fundamental differences are, of course, that [Sakhalin] was not a war zone, there was not combat in progress, there was not combat there normally; and, secondly, the KAL 007 was not warned in any way, form, or fashion."[17]

Nevertheless, as Eric Engberg of CBS News pointed out, "US officials [were] bracing for an onslaught of criticism from foreign adversaries comparing this shootdown to the Soviet attack on the KAL plane." Engberg correctly predicted that the U.S. defense would be that the Iranian aircraft had been warned and was shot down only because of erroneous information.[18] In a later report, Engberg noted that U.S. officials were stressing (as an argument against the comparison) the belief that the Iranian plane was taking hostile action.[19]

On 4 July Reagan spoke briefly with reporters awaiting him at Andrews Air Force Base; he referred to the shootdown as an "understandable accident" and rejected any comparison to the Soviet downing of KAL: "With regard to the Soviets comparing this to the KAL shootdown, there was a great difference. Our shots were fired as a result of a radar screen of a plane approaching it at quite a distance. Remember, the KAL [*sic*], a group of Soviet fighter planes went up, identified the plane for what it was, and then proceeded to shoot it down. There's no comparison."[20] There can be little doubt, however, that the Reagan administration was acutely aware of the potential for comparison for its own behavior in this instance with the actions of the Soviets in 1983. Indeed, sensitivity to this possibility drove much of the response to the tragedy of Iran Air 655.[21] The constraints derived from the earlier incident first manifested themselves in the administration's attempt to avoid a sense of crisis: President Reagan remained at Camp David until his scheduled return to Washington on 4 July.[22] In addition, early comments stressed the defensive nature of the *Vincennes*'s action as Reagan's advisers focused attention on possible retaliation by Iran.[23]

The Reagan administration should be given credit for quickly acknowledging the U.S. role in the loss of the airliner; but no one should be surprised that this was done with an eye to forestalling criticism and the comparisons with KAL: "One official said that 'everyone was mindful' that the Soviet Union had suffered in world public opinion after shooting down a Korean Air Lines 747 in 1983 because of initial Soviet refusal to acknowledge what it had done or express regret for the incident."[24] One Reagan assistant noted that the president was "determined to avoid the mistake made by the Soviet Union." Instead, he "went right out and tried to deal with it. And that put us in a pretty good position on the world scale as well."[25] White House aides considered being open and straightforward the key to damping criticism and thus retaining administration policy in the Persian Gulf.[26] Regrets would be expressed by the White House, but the announcement of the tragedy would be handled by the military, thus distancing the president from the unfortunate actions of the military forces under his command.[27]

Reagan's advisers recognized that the audience for any information released about the incident was twofold: in addition to domestic concerns, consideration had to be given to the possibility of condemnation from abroad. In fact, though this was foreshadowed early, it never really materialized: foreign reaction—notably Soviet reaction—was generally subdued, particularly when compared to the outcry following the KAL shootdown—an outcry largely orchestrated by the United States.[28]

Not surprisingly, the event received considerable play on the two U.S.-sponsored international radios: Voice of America and Radio Free Europe/Radio Liberty. What is interesting for our purposes in this essay is the treatment given to the inevitable comparisons with KAL. Voice of America made no mention of the earlier incident in its rather extensive coverage of the Airbus shootdown, except in reporting the statement of President Reagan disputing the parallels (4 July 1988).[29] However, VOA offered three "background reports," which are provided to the foreign-language broadcast services for optional use. The first of these merely restates Admiral Crowe's response to the question of comparison. A later background report went into more detail on that parallel and mentioned other instances in which passenger craft were brought down by military fire. This dispatch also cited the comments of Thomas Foley, House majority leader: "I think there are great differences. I think the candor and speed with which we accepted responsibility for the accident, tragic as it was, is important—and I think we should continue to provide to all of our citizens and to the world the fullest information that the investigations disclose."[30]

Interestingly, the reporter included the following analysis after the quotation from Foley: "Representative Foley was alluding to the fact that in 1983 it was nearly a week before the Soviets admitted their responsibility for the disappearance of the South Korean Airliner. The Soviets charged the plane was on a spy mission and they never expressed any regret over the incident. In Sunday's incident in the Gulf, President Reagan said the United States deeply regrets the loss of life aboard the Iranian Airliner." The piece concluded by repeating the comments from Admiral Crowe.

While there is no clear indication how extensively these reports were used by the services, the material did appear in a (news) special "Focus" program and a condensed version called a "Close-up." The "Close-up" reviewed the events of 3 July, revisited U.S. policy in the Persian Gulf, including some negative comment on that policy in light of the shootdown, and concluded with the statement from Representative Foley and the observation that President Reagan had ordered a full investigation.[31] Another background report prepared 4 July 1988, summarized worldwide reaction by referring to comments published in the newspapers of various countries. No comparisons to KAL were mentioned, and the reported commentary was generally subdued.

Meanwhile, Radio Liberty was taking a very different tack. Although considered by its personnel as an "alternate news source," most of the discussion of comparisons to the 1983 incident occurred in this forum.[32] On 4 July, for example RL broadcast a twenty-minute program that brought together a wide range of information and commentary including a statement by a member of the International Institute for Strategic Studies in London, who said, in part, that "the US version of the downing is trustworthy and excluded any comparison between this incident and the KAL incident."[33] The next day, RL broadcast a "roundtable" discussion that focused on the incomparability of the two tragedies. Another program, "Events and People," featured the views of two French papers contrasting the 1983 and 1988 incidents.[34] A program on Thursday, 7 July, featured the news that the United States was planning to compensate the families of those who died on Iran Air 655; the comparison to the Soviets' refusal to compensate the families of the KAL victims was unspoken but obvious.[35] The significance of these broadcasts is twofold: clearly they were designed for a foreign audience, since RL and VOA material cannot be disseminated in the United States. One can only conclude that they were part of the larger damage-control scenario—to damp the fires of worldwide reaction. Second, these particular broadcasts were beamed directly into the Soviet Union, an obvious effort to draw a clear contrast between the U.S. action and that of the Soviet government five years earlier.

Most observers felt the strategy was successful both at home and abroad. Certainly, if the length of time a story remains in the news is any indication, the efforts of the president's advisers were enormously effective. KAL was front-page news for nearly a month; by contrast, Iran Air was old news by 6 July, reappearing a week later at the time of the United Nations debate and again briefly in early August, when the report of the investigation was released.

That the strategy achieved its rhetorical and political goals may be ascribed to the fact that the Reagan administration was acutely aware of the context in which the Airbus tragedy had occurred. Reagan himself was conscious that the charges leveled by the United States in 1983 could be turned around, with this country now placed under the glare of harsh international attention; accordingly, the president was constrained by the U.S. role in the earlier episode. Exemplifying what McKerrow has termed "the irreversibility of rhetorical postures once choices have been made,"[36] the experience of the KAL shootdown established an unavoidable context within which the Airbus disaster would be viewed by the American public and by the world at large. As such, it could not be ignored by the president.

Thus, conscious of the comparisons to KAL and anxious to avoid widespread condemnation, the Reagan administration promised full disclosure of its own investigation and cooperation with that of the ICAO—the United Nations body that had also investigated the Korean Air Lines episode.[37]

Nevertheless, such comparisons endured—as noted, the earliest occurred during an NBC news brief at 1:22 P.M. on 3 July, even before any official announcement had been made. CBS brought in Seymour Hersh, author of *The Target Is Destroyed*, as part of a short piece on the comparisons.[38] In a later broadcast that same day, NBC noted the long memories people have about such incidents and the assurances made about the capabilities of U.S. equipment in similar circumstances.[39] On 4 July the *New York Times* printed an extended comparison, which went into the details of the earlier tragedy. As noted above, the administration spent considerable time attempting to forestall such efforts, pointing out that the Iran Air shootdown had occurred in a war zone during combat while the aircraft was flying at low altitude.[40]

Such a response overlooks the points of comparison between the two tragedies, however. In both instances, for example, there were real questions raised about identification of the aircraft as civilian. In 1983 formal investigations concluded that the Soviets did not know they were shooting at a civilian airliner because the fighter pilot had made no attempt to identify the plane before firing his missiles; all Soviet claims of poor visibility due to the nighttime conditions fell on deaf ears. Similarly, Iran expressed surprise that the crew of the *Vincennes* could mistake an A-300 Airbus for an F-14; U.S. spokespersons pointed out in response that the airliner was nine miles away under conditions of limited visibility (two to five miles).[41] Despite all this, there existed in both instances the possibility of radar confusion between two aircraft: KAL with an RC-135 reconnaissance plane patrolling the Soviet coast and Iran Air with a C-135 on the ground at Bandar Abbas.

Both the United States and the Soviet Union claimed that the two downed aircraft had engaged in hostile behavior before the launch of missiles. The Soviets argued that KAL had flown with its lights off,[42] performed evasive maneuvers, and failed to respond to radio signals or warning shots. In like manner, the U.S. government charged that the Airbus had flown toward the navy ship in a hostile manner, had failed to respond to numerous signals to identify itself, and ultimately had failed to alter its course; at one point, the United States even claimed that the aircraft had been descending and increasing speed—an attack mode.[43] Both governments claimed the right of self-defense: the United States because of hostilities in the area and the Soviet Union because KAL had flown over sensitive military installations. Tom Wicker effectively summed up the true similarities in these two tragic encounters: "And what about all those statements that the destruction of Flight 655 was entirely different from the Soviet shooting down of KAL 007 in 1983? That disaster took place in highly sensitive Soviet airspace on the night of a planned missile test, when an overly excited Soviet air defense, acting on inadequate information, fired too hastily. That's just about what the Navy is expected to report that the *Vincennes* crewmen did in the Persian Gulf on July 3."[44]

Still more compelling are the similarities in the way the incidents were handled by the two governments responsible—the Soviet Union and the United States—a similarity that, ultimately, is enlightening. In both instances, the government allowed the military to anchor the public relations effort, diffusing it of any parochial political overtones. In the case of KAL, Marshal Nikolai Ogarkov, commander of the Soviet air defense forces, made a stunning presentation of the Soviet case to the world press, submitting to questions from both foreign and domestic reporters. Likewise, Admiral William Crowe, chairman of the Joint Chiefs of Staff, made the formal announcement of the Iran Air shootdown and submitted to questions from the press. Perhaps the major difference between these two statements was the timing: Admiral Crowe spoke on 3 July, the same day the incident occurred, while Ogarkov's press conference did not take place until nine days after the shootdown. It is also important to consider the content of the statements, for while Crowe announced the tragedy to the world, the initial Soviet statement (not made until 2 September, two days after the event) was terse and brief and said nothing about their military shooting down anything.

Admiral Crowe: After receiving further data and evaluating information available from the Persian Gulf, we believe that the cruiser *USS Vincennes,* while actively engaged with threatening Iranian surface units and protecting itself from what was concluded to be a hostile aircraft, shot down an Iranian airliner over the Strait of Hormuz.

The US Government deeply regrets this incident.

A full investigation will be conducted, but it is our judgment that based on the information currently available, the local commanders have sufficient reasons to believe their units were in jeopardy and they fired in self-defense.[45]

TASS: An unidentified plane entered the airspace of the Soviet Union over the Kamchatka Peninsula from the direction of the Pacific Ocean and then for the second time violated the airspace of the USSR over Sakhalin Island on the night from August 31 to September 1. The plane did not have navigation lights, did not respond to queries and did not enter into contact with the dispatcher service.

Fighters of the anti-aircraft defense, which were sent aloft toward the intruder plane, tried to give it assistance in directing it to the nearest airfield. But the intruder plane did not react to the signals and warning from the Soviet fighters and continued its flight in the direction of the Sea of Japan.[46]

Despite some important differences, there is similarity between the final paragraph of the Crowe statement and the TASS release. Both presage an effort to distance the government from the tragedy and to shift blame from the military to the hapless airplane and "those who sent it to its fate."

It should not be surprising that the chief executives of both countries maintained a low profile. Andropov's was so low as to be nearly invisible: not until 28 September 1983—fully four weeks after the tragedy—did the Soviet chief executive offer any public comment, and then he buried it in a more comprehensive, written policy statement:

The sophisticated provocation masterminded by the United States special services with the use of a South Korean plane is an example of extreme adventurism in politics. We have elucidated the factual aspect of the action thoroughly and authentically. The guilt of its organizers, no matter how hard they may dodge and what false versions they may put forward, has been proved.

The Soviet leadership expressed regret over the loss of human life resulting from that unprecedented, criminal subversion. It is on the conscience of those who . . . masterminded and carried out the provocation, who literally on the following day hastily pushed through Congress colossal military spending and are now rubbing their hands with pleasure.[47]

In 1988 Reagan, following Andropov's earlier example, issued only a single brief statement. This announcement bears striking resemblance to the terse comment issued by TASS on 3 September 1983.[48] Reagan's statement is quoted here in its entirety.

I am saddened to report that it appears that in a proper defensive action by the *USS Vincennes* this morning in the Persian Gulf an Iranian airliner was shot down over the Strait of Hormuz. This is a terrible human tragedy. Our sympathy and condolences go out to the passengers, crew, and their families. The Defense Department will conduct a full investigation.

We deeply regret any loss of life. The course of the Iranian civilian airliner was such that it was headed directly for the *USS Vincennes*, which was at the time engaged with five Iranian . . . boats that had attacked our forces. When the aircraft failed to heed repeated warnings, the *Vincennes* followed standing orders and widely publicized procedures, firing to protect itself against possible attack.

The only US interest in the Persian Gulf is peace, and this tragedy reinforces the need to achieve that goal with all possible speed.[49]

Both statements focus on the shifting of blame for the shootdown. While Andropov summarizes the Soviet argument, Reagan foreshadows the position of his administration: the responsibility for the tragedy should be laid at the feet of Iran for its prosecution of the Persian Gulf war.

Transferring the guilt for the event away from the aggressor is not the only result of such an approach; a secondary effect is the distancing of the political arm of the government from the ramifications of the act. It is this distancing process that allows the transformation of a factual event into an interpretive one—a strategy at which the Soviets are past masters and the Reagan administration their contemporary equals.[50] The ultimate failure of Soviet strategy in the aftermath of the KAL shootdown was the government's inability to effect this conversion.[51] The United States was more successful: the Airbus became a symbol of the Persian Gulf war. The ultimate fact of the shootdown, and the errors that caused it, were lost in the alchemy. What this means is that the true similarity in the pronouncements of Ogarkov and Crowe, as well as Andropov and Reagan, lies in the use of what Hugo refers to as "cant"—a pious verbal subterfuge.[52] In his perspicacious essay on the subject, Hugo further observes that "when governments have to choose between different courses of action, they are accustomed to argue either that a particular decision was intrinsically meritorious or else that it was forced upon them by the objectionable act of another government. . . . Such arguments are often untrue, usually misleading and always irrelevant. The significance of an action depends more on its concrete results than on its antecedents or on the merits of the motives that prompted it."[53]

To some extent, of course, the noted similarities between Soviet rhetorical behavior in 1983 and U.S. rhetorical behavior in 1988 existed because the rhetorical exigencies of the two situations were similar. In both instances, after all, the two countries were potentially on the defensive. However, I would argue that the United States—unlike the Soviets in 1983—never really assumed a defensive posture; it is true that U.S spokespersons justified the attack on the airplane by invoking the wartime environment, but American rhetoric was aggressive and active in shifting the blame and in managing public opinion. It is possible that the similarity of rhetorical exigencies is a true yet incomplete explanation; I would suggest that these tactics signify a strategy of avoiding responsibility.

Crowe's press conference and Reagan's statement are two of four primary rhetorical artifacts which supply the framework—and therefore the context—for viewing this event; the third is Vice-President Bush's address to the United Nations Security Council on 14 July and the fourth is Ambassador Vernon Walters's follow-up address to that same body on 20 July.[54] Over the three weeks following the shootdown, culminating in Walters's speech, four primary themes were developed.

1. THE ACTION TAKEN BY THE *VINCENNES* WAS A PROPER, DEFENSIVE MOVE. Immediately upon receiving word of the shootdown—before anyone was aware that the aircraft involved was civilian—the United States went on the offensive in establishing the defensive nature of the attack on the airliner. The earliest reports on 3 July, as recorded by NBC News, referred to the downed plane as an F-14—a fighter plane—engaged in hostile action against the cruiser.[55] Later bulletins described the airliner as in an attack mode, descending and increasing speed. Virtually all reports referred to the hostilities that had been taking place immediately prior to the shootdown.

2. THE UNITED STATES WOULD ACCEPT RESPONSIBILITY BUT NOT BLAME. The claim for the defensive nature of the *Vincennes*'s action lay the groundwork for this position, of course. This was a necessary step to lead the audience to the conclusion found in number 4 below. It also left open the question of payment of *ex gratia* compensation, which is the driving force behind the dialogue that follows. The statements of Admiral Crowe and President Reagan carefully avoid the question of blame, an approach which is graphically illustrated in this exchange from a State Department press briefing:

Q: You mentioned that the Iraqis, right up to the attack on the *Stark*, they admitted responsibility. Are we about to do something similar? Or are we still saying that the captain was justified in shooting down the plane?
Phyllis Oakley: We said, clearly, we regretted the loss of life. It was a terrible accident. We have admitted obviously, the airplane was destroyed by a missile from our ship. But beyond that I just have no further comment.
Q: To put things clearly, we are not accepting responsibility as of now?
Oakley: Well, I think that would depend upon how "responsibility" is defined. . . . Clearly, the Iranian Airbus was destroyed by a missile from the US ship, the *Vincennes*. That's very clear. We have expressed our regret over this incident, the loss of life. We are looking at all the facts and trying to

obtain them before we get into the exact definition of, I think, what you would mean by "responsibility."[56]

3. THE INCIDENT WAS A HUMAN TRAGEDY THAT COULD BE SEPARATED FROM THE QUESTION OF ITS APPROPRIATENESS UNDER THE CIRCUMSTANCES. This subterfuge allowed the U.S. government to accept responsibility without accepting blame; it also enabled the administration to magnanimously offer compensation to the families of the victims, also without accepting ultimate responsibility for the tragedy—in other words, *ex gratia*. It is, nevertheless, a step that was not taken by the Soviets in 1983, an important consideration in Reagan's efforts to avoid the KAL comparison.

4. THE REAL CAUSE OF THE TRAGEDY WAS IRAN'S REFUSAL TO END THE WAR WITH IRAQ. This message, tantamount to arguing that the cold war was the proximate cause of the destruction of KAL 007, was introduced by Reagan and culminated in Bush's presentation to the United Nations Security Council. Bush made three major points in this address: the importance of the Security Council and the gravity of the questions it considers; the intransigence of Iran in prolonging the war and refusing to accept U.N. Resolution 598; and the recklessness of the Iranian government in allowing a civilian airliner to proceed over a combat zone.

This last theme deserves special mention, for it represents an old tactic that appears to be endemic in postmodernist rhetoric: keep mentioning it and people won't notice it isn't true. From the beginning of this episode, spokespersons for the Reagan administration referred to the lack of concern shown by the Iranians in allowing the aircraft to fly through hostile territory. Yet the airplane was well within the approved commercial corridor and was a regular flight whose number appeared on the printed schedule available to U.S. combat vessels, including the *Vincennes*. Nevertheless, the drumbeat of U.S. rhetoric ignored the practical realities of commercial air traffic control and hammered home the notion that Iran was somehow to blame. Indeed, it is interesting to trace the evolution of Iran's alleged role in the shootdown: from the belief that the downed aircraft was a fighter plane, to the argument that it was approaching the *Vincennes* in an attack mode, to the suggestion that perhaps the Iranians had deliberately sent the airliner to its death, to a more generalized responsibility placed on the shoulders of air traffic controllers at Bandar Abbas.[57]

The completion of the strategy came some seventeen days after the event—20 July 1988—when Vernon Walters, U.S. Ambassador to the United Nations, spoke to the Security Council following its vote on the draft resolution on the shootdown, the proverbial "slap on the wrist" administered to the United States. In the final four paragraphs of his address, Walters effectively sums up the Reagan administration's rhetorical position:

> We intend to stick with our effective Gulf policy. As the Vice-President stated in this chamber last Thursday, once tensions decrease and the threat to Western interests dissipates in the area, then the level of our naval presence will naturally be reduced. We reject any suggestion that the current Western naval presence in the Persian Gulf is somehow an intrusion; it is not. It is a force for peace. It is there in support of regional states whose interests and those of the West face a very real threat. The legitimacy of the Western naval presence in the Gulf is simply not subject to question.
>
> The Iran Air 655 incident was a tragic accident. The United States has expressed its deep regret over the loss of life and has conveyed sincere condolences to relatives of the victims. As the Council is aware, the United States has offered to pay *ex gratia* compensation to the families of the victims—not as an act of charity, not on the basis of any legal liability but, rather, as a sincere humanitarian gesture. We do so without apology for the action of the *Vincennes*, which was taken in justifiable self-defence in the context of unprovoked attacks from Iranian forces, which bear a substantial measure of responsibility for the incident.
>
> Having initiated our own military investigation, the Unitd States joins in endorsing the actions taken by the Council of the International Civil Aviation Organization (ICAO) to commence its investigation of the Iran Air incident. We look forward to cooperating with ICAO in that investigation and in the efforts that the President of the ICAO Council and the ICAO Secretary-General will be undertaking to improve civil aviation safety in the Gulf and to study possible improvements to ICAO Standards and Recommended Practices generally to prevent a recurrence of tragic incidents of this kind.
>
> It is in this context that the United States lends its support to the resolution just adopted by the Council today. We believe this resolution puts the unfortunate events of 3 July in proper perspective. It is our hope that this action of the Security Council will serve as an urgent reminder to the international community that we cannot permit this senseless conflict in the gulf to continue. The risks are too great and the price in human suffering and material destruction is too high. We members of the Security Council bear a special responsibility to provide leadership in this regard. Let us all rededicate ourselves to this vital task and do what we can to encourage the belligerents to walk through the door of opportunity now open before them into an era of lasting peace.[58]

The conclusion of Walters's speech represented the apotheosis of the U.S. strategy. The presence of U.S. warships in the gulf was no longer even the proximate cause of the tragedy; responsibility had been shifted from the offender to the offended. In short, the United States was able to control the terms of the debate, to redefine the situation so that it encompassed the totality of the Persian Gulf war rather than focusing on the act of the shootdown of an Iranian plane by a U.S. warship. In this way, both Bush's and Walters's U.N. addresses served a normalizing function similar to that of Andropov's policy statement in 1983: summing up the incident, placing it in perspective, and filing it away in the storehouse of public knowledge.[59]

Ironically, while the United States worked hard to prevent the invidious comparison to the 1983 incident, they nevertheless exploited existing worldwide public knowledge about Soviet behavior in its aftermath. The actions of the United States stood in clear contrast to those of the Soviet Union five years earlier: the United States had admitted shooting down the airliner, offered *ex gratia* compensation, initiated an internal investigation, and cooperated in the ICAO's investigation; the unspoken comparison was also unmistakable.

But the most remarkable aspect of Walter's speech is the level of cant, particularly in the closing paragraphs, quoted above. Here, and throughout its post-Airbus campaign, the United States adopted both types of justificatory rhetoric described by Hugo: the intrinsic merit of its own position (the legitimacy of the U.S. presence in the Persian Gulf) and that the action was forced on them by the actions of the Iranian government, first in allowing the plane to traverse hostile airspace and then by engaging in the war at all. Finally, the destruction of the Airbus becomes an icon for the cost of the gulf war and a talisman for condemnation of the Iranian government. That the Iranians ultimately shouldered most of the blame is evidenced by the inability of that government to secure enough votes to effect a U.N. condemnation of the American *action.*

As noted, the U.S. campaign to deflect attention from the destruction of the Iranian airliner was more successful than the Soviet effort had been in 1983. Public opinion domestically and worldwide was more accepting of the American action. In concrete terms, both the U.N. Security Council and the ICAO took action that was much less condemnatory than that taken by those same bodies in 1983.[60] Certainly, the Reagan administration was aided by the environment in which the incident occurred: most observers recognized the right of any vessel to protect itself from attack (this point undoubtedly gained credibility in the aftermath of the *Stark* tragedy);[61] the airliner *was* flying in a war zone, although in an approved commercial corridor; there *had* been prior hostilities involving the *Vincennes* and Iranian gunboats; and, perhaps most important, Iran was perceived by many as dragging its feet in negotiating a peace agreement with Iraq. In addition, Italian ships in the area reported monitoring the *Vincennes*'s attempts to contact and warn the Iranian plane. Thus, the United States escaped much of the condemnation that had been visited upon the USSR five years earlier.[62]

This last, of course, represents one of the most frustrating aspects of the Reagan presidency. Reagan's ability to shed responsibility for the outcomes and impacts of his policies (read: rhetorical positions) is legendary. The Airbus incident is no exception, and it is this aspect of the tragedy that is instructive for our purposes here. For it illustrates the manner in which institutional structures may be used to expand or inhibit the range of responses available in an emergency.

Reagan's disengagement from the day-to-day operations of his administration is well documented. In this situation, as in others, that tendency served him well, for it was not seen as unusual for him to remain aloof, to allow the military or members of his cabinet and staff to face the public and the press. Thus, his decision not to make an early return to Washington was seen as normal and defused the situation of any sense of urgency.

Sending Vice-President Bush to the United Nations was a brilliant move that served many purposes. It was generally perceived as a boost to Bush's presidential ambitions, but there is no reason to assume that Reagan would have obliged the vice-president had it not suited his own purposes to do so; indeed, the decision was taken carefully. The prestige of Bush's office allowed the United States to set the terms of the debate, to focus attention on the issue behind the shootdown.[63] Bush provided, in effect, a surrogate target, and substituting him for the regular ambassador emphasized the gravity of the event without giving it urgency.

Bush's speech served as the capstone of the rhetorical campaign, a campaign marked by executive disengagement and a diffusion of roles among several spokespersons, with the military taking the lead and U.N. Ambassador Walters occupying the mop-up position. It resembled closely the Soviet campaign five years earlier, when the military also took the lead, and the thrust of the effort was the shifting of responsibility from the Soviet government and its policies to the United States and its intrusive surveillance. Unlike the Soviets, however, the United States was able first to extricate itself and its policy from the ramifications of this incident, and then successfully to reframe the event in its own terms. This is dramatically different from the approach taken by the same administration in 1983, when the framing, or contextualizing, was established in the first hours following the tragedy; the struggle in that incident was to maintain the context. Nevertheless the differing roles of the United States in those two instances is borne out in the language used to describe them. Despite the desire of the United States to blame Iran for the shootdown of the Airbus, the need to dissociate U.S. policy from the tragedy dictated a different rhetorical style from that used in blaming the Soviets five years earlier. Interestingly, in both incidents, the United States claimed the mantle of civilized behavior for its own, forcing other nations to align with civilization or against it.

Bush's role in the Airbus campaign also illuminates the episodic nature of events such as this one and the way in which that attribute informs the protective quality of administrative discourse. The general style of the Reagan presidency—the dissociation of the chief executive from day-to-day operations—was well suited to a decentralized rhetorical approach. It was not accidental that the four primary rhetorical artifacts of the Airbus controversy were delivered by four different individuals. The structure of the executive branch not only shielded the president, it made his isolation seem unremarkable. Thus, the episodic style mirrors the management style that Reagan employed throughout his presidency, offering him the choice of being out front on an issue (as with KAL) or disengaging (as with the Airbus).

In those circumstances where protection of the executive is the paramount concern, as usually was the case with Reagan, the episodic nature of discourse does not depend on crises to manifest itself. Reagan's entire presidency was carefully orchestrated; his advisers were forever attempting to control the context in which his administration, and its rhetoric, was viewed.

But the harsh glare of a crisis exacerbates these effects, with its natural demand for information—information that is often in the hands of many different individuals, including reporters. In these cirucmstances, the public looks to its leaders to establish the context—or meaning—for a given event; the leadership can then elect to step forward or to leave the task in the hands of surrogates. In either case, understanding of cataclysmic events depends on examination of the entire constellation of rhetoric surrounding that event.

It is possible, of course, for the episodic approach to expand, rather than constrict, the opportunities for response to a rhetorical situation, as officials are called to account for their actions or to share the knowledge they have accumulated. In the case of the Airbus tragedy, as in that of KAL 007, however, the administration attempted to control the context in order to limit the response. In 1983 Reagan's advisers sought to fix responsibility; in 1988 many of the same players sought to avoid it. By recognizing and examining the episodic nature of this discursive environment, the critic can more fully elucidate the manner in which the Reagan presidency managed public understanding.

Part IV

Domestic Policy Case Studies

10

The Reagan Attack on Welfare

Michael Weiler

Why do citizens of a rich country like the United States tolerate substantial poverty in their midst? What accounts for their apparent "acceptance of [the] chronic inequality, deprivation, and daily indignities"[1] borne by a significant percentage of their fellow citizens?[2] Are Americans simply selfish and uncaring? Public opinion surveys suggest otherwise. For over three decades, when Americans have been asked if they favor more federal assistance to the poor than is provided presently, a majority have answered yes. Only a small minority have favored a reduced federal effort.[3] Yet, assessing public attitudes toward poverty is not so simple. For when the same question has been asked about federal "welfare" expenditures, a majority have always said no. Americans want to help the poor, but they hate welfare.[4]

There is no logical contradiction here, of course. People can object to a particular method of helping the poor without opposing help in principle. But if the assortment of existing antipoverty programs is unacceptable, then what sort of effort will the public support? Are there clear alternatives a majority would favor?

The answer is a qualified no. There appears to be no great support for innovative approaches aimed at helping the people now aided under welfare programs such as Aid to Families with Dependent Children, Food Stamps, and Medicaid. The proposal in the 1960s for a guaranteed annual income, for example, did not come close to the public support it would have needed to have a serious chance of congressional passage.[5] A majority of Americans, however, have been willing to support programs targeted not at the working age poor, but at the elderly, war veterans, or the disabled.[6] Indeed, these programs are not in most cases understood as "welfare" at all. Welfare has come to mean benefits provided to able-bodied, working-age adults, and most Americans cannot accept the notion that such people are appropriate recipients of government aid.[7] Nor does it seem to matter much that an increasing share of AFDC benefits go to single mothers with school-age or preschool children.[8] Despite the "feminization" of poverty, welfare programs continue to arouse hostility.

Public support for national antipoverty efforts rests, therefore, as much on to whom the help is given as on what kind of help it is. The success of poverty programs is not measured simply in terms of providing a minimally sufficient material life for those who are poor. Nor even are partial reductions in poverty enough.[9] In order to have any chance of widespread support, welfare must promote independence from welfare. Most Americans seem convinced that it does not and cannot.[10]

In this essay I want to investigate the ways in which the conservative attack on welfare and the "welfare state" have promoted these attitudes toward poverty and welfare. This attack, especially during the past two decades, has been increasingly successful. Social benefits targeted specifically at the poor have shrunk in real terms.[11] This trend has continued even through the 1980s, a period of uninterrupted economic growth.[12] Why?

I will suggest that an important part of the answer lies in the rhetoric of those who for the past twenty years have attacked the social welfare state. In the 1980s it would be hard to find a better example of this kind of rhetoric than Ronald Reagan's discourse on poverty and poverty policy. Though Reagan's was by no means the only voice, his views on welfare enjoyed wide currency, not only because the presidency offered a uniquely powerful platform for them but because Reagan's political philosophy gave prominent place to such attacks. He pressed the case against the welfare system both during the 1980 presidential campaign and in the first two years of his presidency. Thereafter, poverty issues received less emphasis but remained, nonetheless, an integral part of his domestic policy discourse.

Reagan's central political challenge was to cut poverty programs while at the same time avoiding the appearance of indifference to the needs of the poor. Meeting this challenge was essentially a rhetorical project. To pursue it successfully, Reagan needed, in Murray Edelman's words, those "special bits of language [that] convert immoral action into the promotion of the public interest. These phrases . . . can transform such actions as impoverishing large numbers of people . . . into evidence of effective leadership."[13]

The "welfare mess," the "truly needy," the "safety net": these were Reagan's special bits of language. Embedded in his arguments against poverty policy, they were the reductions that helped him justify policies that increased poverty in the short term and failed to address its fundamental long-term causes.

Reagan's case against the welfare state, though logically complete, was made considerably more powerful by his incorporation of two interrelated discourses: populism and American mythology. He used populist appeals to attack his liberal opponents and, by extension, their support of poverty programs. He used a mythologized version of American history to undermine the legitimacy of the poor's claims on common resources and of their advocates' case for government help.

I will investigate how Reagan's anti–welfare state rhetoric helped to promote public tolerance of poverty and acquiescence to policies restricting aid to the poor. Of course, he did not work alone. Other advocates contributed. But as the example par excellence of this species of political discourse, Reagan's attacks on the welfare state are worthy of particular attention. And because the "welfare mess" theme figured so prominently in his rhetoric not only during his presidency but throughout his political career, no treatment of Reagan's rhetorical legacy could be complete without attending to it.

Background

In its first term the Reagan administration proposed deep cuts in several social welfare programs and the complete elimination of others. Though congressional action left Reagan far short of his goals, overall social welfare spending by the federal government declined in this period by nearly 10 percent from what it would have been at prior policy levels.[14] Aid to Families with Dependent Children and Food Stamps were cut by about 14 percent, far less than requested, but substantially nonetheless, not only because welfare recipients must live close to the margin, but because the reductions occurred at the height of the 1982 recession.[15]

During Reagan's first term, the poverty rate increased by 3.3 percent, plunging an additional nine million Americans below the poverty line. Some estimates suggest that half of this increase can be attributed to cuts in federal social welfare spending.[16] And this does not include the effect of cuts in programs not usually listed in the welfare category such as subsidies for public housing. Federal spending on such housing declined from around thirty billion dollars to less than ten billion during Reagan's first term.[17]

Clearly then, the Reagan administration, especially its first term, coincided with a significant reduction in the level of social welfare benefits provided to the poor,[18] and with a significant increase in poverty. But these developments should be understood not as an historical aberration, but as the culmination of a decade-long retrenchment of federal antipoverty efforts. Indeed, the cuts of the early 1980s, while significant in themselves, were painful primarily because they followed an eight-year period in which nominal AFDC benefit levels did not increase at all and real levels declined by 30–50 percent across the nation. Nor was this pattern reversed during 1983–89, years of continuous economic growth.[19]

Like his policy proposals, Reagan's anti–welfare state rhetoric also was rooted in an earlier period. Attacks on the social welfare state have been a staple of conservative rhetoric (Republican and Democrat) since the inception of Franklin Roosevelt's New Deal. Their potency is reflected both in the limited scope of Roosevelt's initial efforts to create a national social welfare system and in the attenuated version of that system that finally emerged at the end of World War II.[20] Indeed, throughout its history, American political ideology has been hostile to a national welfare policy.[21]

Ronald Reagan's first forays into politics date from the end of World War II, when he began making speeches on national issues.[22] Though his movie career did not end finally until 1964 *(The Killers)*, throughout the 1950s Reagan turned increasingly to politics. His tenure as president of the Screen Actors Guild from 1947 to 1952[23] placed him in the middle of congressional investigations of communism in Hollywood, and he responded by turning the SAG into an instrument for opposing the Communist menace.[24] Beginning in 1954, in parallel with his job as host of the popular "General Electric Theatre," Reagan toured GE's plants nationwide, extolling the virtues of the company as the embodiment of the entrepreneurial spirit. During his eight-year contract with GE, Reagan gave hundreds of speeches. As time went on, he shifted increasingly from company cheerleader to America's jeremiah, talking of the dangers to the nation of communism abroad and welfare statism at home.[25]

In 1964 Reagan entered the national electoral arena for the first time. In a paid televised speech supporting Barry Goldwater, he claimed to hear from those backing Goldwater's opponent Lyndon Johnson, "another voice [that] says that the profit motive has been outmoded; it must be replaced by the incentives of the welfare state."[26] To his general indictments of the welfare system, Reagan appended the following illustration, a foretaste of his rhetoric in presidential (and gubernatorial) campaigns to come: "Not too long ago, a judge called me from Los Angeles. He told me of a young woman who had come before him for a divorce. She had six children, was pregnant with her seventh. Under his questioning, she revealed her husband was a laborer earning $250 a month. She wanted a divorce so that she could get an $80 raise. She is eligible for $330 a month in the aid to dependent children program. She got the idea from two women in her neighborhood who had already done that very thing."[27]

Reagan's attacks on welfare spending in 1964 were motivated in part by the Johnson administration's efforts, before the election, to build support for the War on Poverty.[28] Johnson's initiative was, for the first time in American history, to make the long-term poor a primary emphasis of social welfare policy. Though the Social Security Act of 1935 had included relief for the poor, the assumption of most of its sponsors was that such relief would not be more than a temporary expedient necessitated by the Depression.[29] The War on Poverty, by contrast, though sold as a way to help people escape poverty,[30] in effect placed the long-term, working-age poor (and their children) at the center of the public's image of the social welfare system.

The bifurcation of social *security* and social *welfare*, inherent in the New Deal's embarrassed confrontation with American poverty, was vivified and calcified in the rhetoric of those advocating and attacking the War on Poverty. The social security system provided for a more comfortable retirement after years of productive work, rewarded military service, and helped the temporarily unemployed. Its beneficiaries were annuitants, compensables, and workers. The social welfare system, however, was targeted at "recipients," people who received without first having given.

Thus, for Americans who are not poor, "welfare" stands for a system that aids "them" rather than insures and compensates "us." Broad-coverage programs such as Social Security, Unemployment Compensation, and Medicare can be seen as existing in a different category from poor-centered programs such as AFDC, Food Stamps, and Medicaid. The former group does not distinguish (for the most part) between rich and nonpoor recipients; the latter does.

Ironically, since the 1960s, by far the most rapid expenditure growth has occurred in Social Security and Medicare, both of which provide relatively equal benefits to poor and nonpoor elderly people alike.[31] Yet, critics of increasing social welfare expenditures have focused not on these programs, but on those aiding the poor exclusively.[32]

Reagan attempted to capture the Republican presidential nomination both in 1968 and 1976, and his anti–welfare state rhetoric in both campaigns reflected what had become, thanks to a withering conservative attack and the deficiencies of existing antipoverty programs, the widespread perception that the War on Poverty was an expensive failure. Summarized neatly as "the welfare mess," the War on Poverty became a persistent albatross around Democratic necks, one whose implications they are still trying to escape.

In 1976 Reagan was still talking of "welfare queens" cheating the system. The relatively minor case of abuse detailed in his 1964 speech gave way to a Chicago woman with "eighty names, thirty addresses, and fifteen telephone numbers," collecting, in addition to AFDC checks, "veterans' benefits on four dead husbands who never lived—for a total estimated income of a hundred and fifty thousand dollars."[33]

In the 1976 campaign Reagan's attacks on federal policy were muted somewhat by his having to run against a Republican incumbent. But by 1980, with a Democrat in the White House, the way was open for an all-out attack on welfare. During the 1970s most Americans' incomes had risen little if at all. The oil price shocks of 1973–74 and 1979 had brought home all too vividly the reality of an era of limited consumption. Federal tax increases (including FICA payroll taxes) were squeezing many taxpayers. As the 1980 election neared, double-digit inflation and interest rates, and historically high unemployment, created an atmosphere of justified alarm.[34]

In these circumstances, anti–welfare state rhetoric found a particularly receptive audience. Resentful and frustrated middle-class Americans, having to face for the first time in their lives the possibility that neither they nor their children would be better off economically than their parents, wanted to know what and whom to blame, and what could be done to redress the situation. Reagan's answers were seductively simple. Social spending run amok had produced economic malaise. "Tax and spend" liberals were responsible. Social spending must be cut and the rascals thrown out.

Since social spending could be reduced to welfare, and welfare to welfare queens, attacks on the social welfare state had considerable appeal to middle-class Americans. The system benefited "them," not "us"; indeed, its lavish wastefulness and mistaken design hurt us without really helping them. Americans might in general want to help the less fortunate, but not at the expense of their children's future, nor in wasteful and counterproductive ways. And because programs aimed at the poor had come to loom so large in attacks on social spending, it was easy to believe that reductions in poverty spending alone would have a substantial impact on the economy as a whole, though such spending was in fact a relatively small part of the federal domestic budget.

Ronald Reagan neither invented nor substantially reshaped anti–welfare state rhetoric. His attacks on welfare spending did not differ from those of others pressing the same claims. Nor did Reagan's rhetoric during his presidency change greatly from what it had been during the late 1960s and the 1970s, as he made his circuitous but relentless way to the White House. Moreover, Reagan did not fabricate the material basis of his rhetoric's effectiveness. Inflation, unemployment, stagnant incomes, disappointed expectations: all of these were real conditions as the 1980s began. Without them, the case against social welfare spending could not have seemed so persuasive.

But if Reagan's case against welfare spending was unoriginal, he nonetheless pressed it with exceptional persistence and skill, interweaving his arguments about poverty and poverty programs with populist appeals and references to a mythologized version of American history. The result was a discourse that rendered intelligible and acceptable policies that, though failing to provide a long-term solution to the problem of poverty, did, in the near term, result in reducing such aid as the poor received. Here was an instance, in Murray Edelman's trenchant phrase, of "words that succeed and policies that fail."[35]

Reagan's Supporting Discourses

Perhaps better than any American president in recent memory, Ronald Reagan understood the importance of flattery to political success. People like you better if you continually say nice things about them, and Reagan said a lot of nice things. His appeals to populist sentiment and his stories of America's past made voters feel wise, politically significant, and part of a great historic crusade. There was a negative side, however, to the hymn of praise that Reagan sang so sweetly through most of the 1980s. For he understood also that one way people have of feeling better about themselves is to think ill of someone else. The people are not just wise but wiser than arrogant elites; the people are not just politically significant but significant despite the efforts of elites to repress them; the people are not only predestined to play an important role historically, but must do so against those forces that would thwart that destiny.

Populism

Populism is a discourse, a set of appeals supporting a single theme: the popular will as the source of legitimacy for the exercise of political power. Considered as a set, these appeals do not themselves constitute a political ideology or program; they lack the coherence of the former and the specificity of the latter. Rather, populist appeals are articulable to a wide range of ideologies and programs: from libertarian to fascist, from liberal to socialist.[36]

"Power to the people" is the essential populist cry. It suggests more than the exercise of power in the *interests* of the people; all governing regimes claim to act in their people's interests.[37] Populism implies that the people themselves should decide how power is to be exercised; the people's wisdom, not elite paternalism, should determine how the general welfare is to be served.

Populist discourse responds to situations where the people's power allegedly is threatened or has been usurped. Such discourse has both a deliberative and an epideictic dimension: populist appeals may support courses of action to maintain or restore the people's power, or they may celebrate wisdom of the people and condemn those who would stand in the way of applying it to problems of public policy.[38]

Populist discourse always advocates the power of the people against the power of elites. Who are "the people" whose empowerment it recommends? This is not always a simple question. For while populist rhetoric tends toward universalism, its purposes are often sectarian and undemocratic. This contradiction arises particularly in cases where the group seeking power is a minority with interests in conflict with other groups both powerful and weak.

To resolve this contradiction, it is necessary to define the popular will not by counting votes but by seeking essences. The members of the aggrieved group, because of their unique characteristics, become the "real," the "essential" people, the part that can stand synecdochically for the societal whole. If the agrarian life can be seen as typifying all that is best in American social traditions, then farmers' interests, though undeniably provincial, are more than theirs alone; they are in an essential way the interests of all Americans. If the class consciousness of wage workers, though obviously particular, is most closely connected to the material reality in which all must live, then working class ascendancy is justified. Thus, through an essentialist line of argument, populist discourse can be articulated to sectarian ideologies and agendas, and the particular/universal contradiction inherent in it can be resolved.

A second kind of contradiction also confronts populist discourse.

In modern mass societies, "power to the people" does not mean power to every farmer or wage worker; it means power to their leaders. It is the leaders who assert the right of the people to rule, and implement that rule should their assertions succeed.

Populist leadership often involves pressing a political program on behalf not of a minority recognized as such but of a vaguely defined majority. In the former instance, the links between leaders and group members are likely to be far more real and direct than in the latter. Indeed, when leaders claim to speak for all (or nearly all) the people, they are speaking for what does not exist, a homogeneous body politic whose members' interests can be comprehended by a single political program and represented by a single leader or ruling group.[39]

Explicitly, Ronald Reagan's brand of populism was clearly of this universalist type. His appeal was to all Americans minus the liberal elite (presumably a very small group) that oppressed them. He spoke for ordinary citizens against "the Government." Yet, implicitly, his rhetoric also contained a sectarian element. For by emphasizing the people *versus* the government polarity, he was able to imply a second conflict, one between the interests of the middle class and those of the poor. As clients of the government, poor people were, at least by implication, as much parties to the usurpation of popular (middle-class) power as were their bureaucratic overseers.[40]

The celebration of the common wisdom of the average citizen was one of Reagan's most often used and effective themes. To exploit it fully, we find him integrating it with the full range of devices in his considerable rhetorical arsenal. Sometimes he appealed to populist sentiment explicitly, as in the following passage from his first inaugural address: "We hear much of special interest groups. Well our concern must be for a special interest group that has been too long neglected. It knows no sectional boundaries or ethnic and racial divisions, and it crosses political party lines. It is made up of men and women who raise our food, patrol our streets, man our mines and factories, teach our children, keep our homes, and heal us when we're sick—professionals, industrialists, shopkeepers, clerks, cabbies, and truckdrivers. They are, in short, 'We the people,' this breed called Americans."[41]

Reagan's definition by enumeration of "we, the people," suggests his philosophy of social welfare spending. There are two sets of "people" here. The first uses functional categories; it identifies people according to what they do. The second is a list of types of occupations; it names people according to what they are. All but one of the items in either category is a paid function or occupation; "the people" are overwhelmingly employees. The one exception is those who "keep our homes." Though this category could refer to maids and housekeepers, the use of the word *home* suggests the unpaid labor of the housewife and, significantly, a good deal more. A home is a place where families dwell, and, traditionally, keeping a home encompasses not just housework but taking primary responsibility for raising children. The "kept home," therefore, amounts to a metonymic reduction of the traditional family unit: bread-winning father, loving wife and mother, their children. The home is not the world of single mothers, and certainly not of those on welfare.

This exclusion is made clearer in Reagan's opening reference to "special interest groups" of whom "we hear." Since "the people" stand apart from these groups, and the people are distinguished chiefly by their gainful employment, Reagan can only be referring to those groups who make claims on the rest of "us" *and* do not work. Poor people on welfare are the obvious culprits.

For Reagan, indeed for any political leader, the "we" in "we the people" must be plausible. It is not enough to communicate admiration of "the people." The point of the populist strategy is to convince an audience of a speaker's identification with them, of his membership in their group. This identification is aided by proof of a speaker's awareness of the people's concerns and aspirations, but far better actually to have shared them. Consider this excerpt from a speech Reagan gave to the United Brotherhood of Carpenters and Joiners in 1981:

> You know, a speaker always hopes that he can identify in some way with his audience. Well, my first summer job, when I was 14 years old, was with an outfit remodeling homes for resale. And before the summer ended, I'd laid hardwood floors, shingled roof, painted ceilings, and dug foundations. . . . I remember one hot morning. I'd been swinging a pick for about four hours. I heard the noon whistle blow. I'd been waiting for that sound. I had the pick up over my shoulder ready for the next blow, and when I heard that whistle, I just let go and walked out underneath it, let it fall behind me. And I heard a loud scream and then came some very strong, profane language, and I turned around. And the boss was standing right behind me, and that pick was embedded in the ground right between his feet. Two inches either way and I'd have nailed him.[42]

In just a few lines, this artful narrative at once endows Reagan with a union man's experience, aligns him with working stiffs against their boss, and effaces through acknowledgment its obviously strategic character. The story's conflict between worker and boss reminds us that populism is a matter not simply of being for the people, but also of being against what or whom the people are against. This negative element is crucial and in Reagan's discourse never fails to emerge. The enemy is the government and that "ill-assorted mix of elitists and special-interest groups who see government as the principal vehicle of social change, who believe that the only thing we have to fear is the people, who must be watched and regulated and superintended from Washington."[43]

Populism is about common people against elites, the ruled against their rulers, the powerless against the powerful. By and large, Reagan's political opponents were not able to come fully to grips with his success at speaking to the economic and social grievances and resentments of many American voters in the 1980s. By focusing negative feelings on an at times ill-defined but symbolically potent concept of big government, Reagan provided a broad foundation for his domestic policy positions and for his personal popularity. His populism proved to be a sturdy platform from which to launch attacks on the social welfare state.

American Mythology

For a people to be constructed symbolically, it must be endowed with a past, one that makes members of an audience feel good about themselves in the present and reassured about the future. A people's past should also provide historical evidence for the correctness of a political leader's policy prescriptions. Any reconstruction of the past for present political purposes is the stuff of myth in that to be effective, it must present a coherent, and for that reason selective, narrative that can explain contemporary events, resolve contradictions, and point people in a clear and optimistic direction. I am labeling Reagan's version of U.S. history "American mythology" because it falls well within the mainstream of how most American citizens have been encouraged most often to define their past: as a record of economic individualism and social volunteerism. Though Reagan's history of America is unoriginal, his considerable skills as a speaker and his avuncular ethos in general made its presentation during his presidency particulary affecting.

Economic individualism is more than simply a form of economic organization. Indeed, to emphasize only its macroeconomic aspects would be to undermine its ideological force. Often summarized by the term *entrepreneurialism*, economic individualism is a matter of the personal spirit.[44] Because it goes to the very core of what is viewed as human nature, it is an influence that transcends the economic realm. In a 1983 radio address Reagan observed that "entrepreneurs have always been leaders in America. They led the rebellion against excessive taxation and regulation. They and their offspring pushed back the frontier, transforming the wilderness into a land of plenty. Their knowledge and contributions have sustained us in wartime, brought us out of recessions, carried our astronauts to the Moon, and led American industry to new frontiers of high technology."[45] In other words, the entrepreneurial spirit has pervaded American history from first to last, making possible what is to be understood as a record of continuous economic and technological progress.

But Americans are not economic actors exclusively. The vigorous independence of the entrepreneur is complemented in Reagan's story of America by the cooperative spirit of Americans as social actors. This means emphasizing instances of voluntary association for the purpose of solving social problems. For Reagan, these instances add up to an inherent characteristic of the American personality, and because the problems he mentions are generally local, this de Toquevillian spin to the history of the United States amounts as well to a powerful argument in favor of nongovernmental, local solutions to contemporary social problems.

Michael Rogin has commented on Ronald Reagan's tendency to look to American movies, especially of the 1930s and 1940s, as guides to what America was really like, and by extension, as narratives depicting the better political nature that its people should seek to recover.[46] One need not accept Rogin's claim that Reagan really believed in the literal reality of these depictions to recognize their prominence in his discourse. In a 1981 speech to the National Alliance of Business, for example, Reagan offered a long quotation from a speech by Gary Cooper in *Mr. Deeds Goes to Washington* to answer the question, What exactly is volunteerism? To Cooper's "best" description, Reagan added, "Over our history, Americans have always extended their hands in gestures of assistance. They helped build a neighbor's barn when it burned down, and then formed a volunteer fire department so it wouldn't burn down again. They harvested the next fellow's crop when he was injured or ill, and they raised school funds at quilting bees and church socials. They took for granted that neighbor would care for neighbor."[47]

Voluntary, nongovernmental solutions are to be preferred to governmental initiatives because they are consistent with America's historical traditions, and by extension, with the better natures of individual Americans. Voluntary initiatives are more likely to work because they avoid the inefficiencies and insensitivities to specific local conditions typical of centralized government. Finally, government solutions are to be avoided because they preempt and discourage voluntarism generally.[48]

These are familiar arguments, and conservatives have used them for a long time. In Reagan's hands, however, they, along with impassioned paeans to the entrepreneurial spirit, and in combination with Reagan's potent brand of populism, formed the foundation for an effective attack on the social welfare state, particularly on programs serving the needs of poor Americans.

Reagan's Attack on the Welfare State

Reagan's discourse about poverty embodied both a moral and a practical dimension. He argued for a narrowed conception of government's *responsibility* to help the poor and for increased skepticism regarding its *capability* to do so. His arguments were primarily definitional and causal. His analysis of the nature of the welfare system, and of its sources, pointed toward a sharply reduced federal role.

Reagan's attack on the social welfare state was most intense in the early years of his presidency. The 1981 and 1982 congressional sessions were the high-water marks of his success in achieving substantial cuts in welfare programs.[49] After 1982 perhaps he felt that few cuts were politically possible. Or, perhaps Nicaragua, Grenada, Star Wars, and nuclear summitry simply crowded out other concerns. In any event, the incidence in his rhetoric of references to welfare waste and abuse declined. Accordingly, I have confined myself in this section primarily to those early years of his presidency when his campaign against social welfare spending was most vociferous and successful.

A number of critics of Reagan's rhetoric have argued that viewing it as narrative is the best way to understand its power.[50] This point of view implies that the most important instances of Reagan's discourse actually appear in narative *form*. However true this may be on other issues and themes, in his statements about welfare Reagan employed a relatively straightforward problem-cause-solution argumentative sequence. He defined the problem of poverty policy, specified its causes, and justified his solution. Historical narratives played a supporting evidentiary role. But in his attack on the welfare state, Reagan, consistently and explicitly, made arguments.

In the 1960s Great Society advocates identified poverty as the problem to be addressed. Their analysis of its causes as primarily social and economic implied the array of programs that came to be united under the War on Poverty. For Reagan in the 1980s, the problem was not poverty at all, but wasteful and counterproductive poverty policy. Its causes were liberal softheartedness (and softheadedness) and bureaucratic arrogance; the solution was to restrict welfare programs to a smaller group than benefited from them now: the "truly needy."

Reagan's redefinition of the problem, however, did not mean he could escape the acknowledging of poverty itself. It existed, and something had to be done (at least rhetorically) about it. Reagan met this challenge by interweaving with his analysis of poverty policy a complementary analysis of the nature of poverty and the poor. Thus he was able to claim that his solution to the welfare mess not only would save money but would be more beneficial to the poor than the programs he attacked.

Problem

The alleged problems with social welfare spending are familiar. Much of it is ineffective, failing to achieve its intended objectives and thus wasting a lot of taxpayers' money better spent elsewhere, and it is counterproductive, encouraging long-term dependence on government aid for recipients who could support themselves. In other words, welfare hurts everybody *and* welfare hurts the poor. This second argument is important because if it were simply a matter of welfare helping the poor at the expense of everyone else, then to attack welfare spending would be to admit that the interests of the poor were relatively unimportant. Reagan made clear that he recognized some of the implications of this position in a 1982 speech, when he asked rhetorically:

> How can limited government and fiscal restraint be equated with a lack of compassion for the poor? How can a tax break that puts a little more money in the weekly paychecks of working people be seen as an attack on the needy? Since when do we in America believe that our society is made up of two diametrically opposed classes—one rich, one poor—both in a permanent state of conflict and neither able to get ahead except at the expense of the other? Since when do we in America accept this alien and discredited theory of social and class warfare? Since when do we in America endorse the politics of envy and division?[51]

Here, Reagan appeals explicitly to an important theme of American historical mythology, the absence in the United States of social and economic class conflict, and the relatively stable political consensus this absence has made possible. To associate his position on the welfare state with this theme, he must avoid the charge that he represents the interests of those who are relatively well-off against those who are not. Making the poor into victims of welfare instead of its beneficiaries is therefore a crucial strategic step.

Yet, at the same time that his arguments explicitly deny class conflict in America, his supporting evidence implies a related but more rhetorically potent class configuration, the poor *versus* the middle class. Notice that the trade-off in his example is the working person's paycheck *versus* the requirements of the needy. These are the two groups really in conflict. The tension between rich and poor becomes, in this sense, something of a straw man.

This raises the question of why Reagan feels it necessary to belabor the effects of social welfare spending on the economic well-being of the taxpaying public, and belabor them he does. Inflation, high taxes and interest rates, large deficits, even an impoverished national defense are all laid at the doorstep of excessive welfare expenditures. If welfare hurts the very people it is supposed to help, then why not make that fact the primary support for reductions in welfare spending? Certainly, it is a fully sufficient reason for such cuts. Why spend so much more time, as a cursory look at Reagan's discourse reveals that he does, on the macroeconomic effects of welfare on the populace as a whole?

Consider the following from a 1982 address: "Since 1960 Federal spending has increased nearly 700 percent. That's much faster than our ability to pay for it. This spending was excused in the name of fairness and compassion. But it turned out that fairness and compassion also meant local governments losing control of their communities; working people, small business, and pensioners being hit by record interest rates, inflation and taxation. And that golden era of growth that we once knew in this land gradually slipped away."[52] Fairness and compassion have much to answer for. Of course Reagan talks here of a sham compassion, one that actually hurt the poor not only by promoting dependence but by producing double-digit inflation, in effect a regressive tax that hurts the poorest most of all. There is, then, no logical contradiction in what he says. But in analyzing Reagan's discourse, it is important not just to render full-fledged his case against the welfare state, with each claim given the place it would logically occupy, but to take note of the varying degrees of emphasis different claims receive. In hundreds of instances, Reagan has described the harmful effects of welfare spending in the zero-sum terms exemplified above. It is working people, business people, and retired people whose interests are primary, and whose interests are hurt. Reagan's discourse trades heavily on latent middle-class hostility toward federal budget expenditures on the poor.

This hostility is by no means baseless. The 1970s were a decade in which middle-class incomes ceased to grow or even declined. Social welfare spending did increase rapidly. Many War on Poverty initiatives were wasteful failures. Inflation did take an increasing toll, especially as it propelled middle-class taxpayers into higher tax brackets without increasing their real purchasing power.[53] Reagan's discourse did not fabricate these material conditions. Anti–welfare state discourse was well developed and effective long before the 1980 election. Many middle-class voters were persuaded to see themselves as hurt by transfer payments to the poor, and in some ways indeed they were. Reagan found himself in a position to cultivate an already fertile rhetorical field.

Reagan's initial proposals to reduce domestic spending cut a wide swath. He tried in his first term to tighten eligibility for Social Security benefits by striking several hundred thousand disabled persons from the roles and forcing them to reestablish their eligibility in appeal hearings which were virtually impossible to arrange. Eventually, the courts forced him to restore these benefits.[54] He was more successful in efforts to reduce Medicare expenditures. The administration established for specific medical services and procedures allowable charges beyond which the government would not reimburse physicians and hospitals, and deductibles were increased.[55]

But Social Security and Medicare benefit the elderly, poor and nonpoor alike, and a relatively large percentage of this group are regular voters. Unsurprisingly, Reagan's attacks on these programs were muted. More often, he found himself on the defensive, reassuring elderly Americans that their benefits, especially Social Security, would not be cut. Stiff opposition in the Congress and from lobby groups representing the elderly forced him either to retreat from or dissemble about his efforts to reduce the size of these entitlement programs.

Not so with Reagan's assault on programs targeted primarily at the poor. The more aggressive strategy he used in these cases rested in large part on the potential for characterizing welfare recipients as unworthy of help, an approach unavailable in the case of the elderly as a group. David Zarefsky, Carol Miller-Tutzauer, and Frank E. Tutzauer have drawn attention to Reagan's division of the poor into two groups, the truly needy who deserve help, and the rest, presumably a substantial percentage, who do not.[56] The first group was to be protected by a "safety net" of social programs. The second included those obtaining benefits fraudulently. Their existence in allegedly substantial numbers was crucial to his argument that welfare budgets could be cut without hurting those who really needed benefits; indeed, "Reagan maneuvered his way around excessive criticism by subtle changes in focus—from safety net for the truly needy to waste, inefficiency, and fraud."[57] Virtually every time he discussed the subject, Reagan emphasized abuse of the system, the elimination of which could produce substantial savings. In his 1983 State of the Union message he asserted, "Our standard here will be fairness, ensuring that the taxpayers' hard-earned dollars go only to the truly needy; that none of them are turned away but that fraud and abuse are stamped out. And, I'm sorry to say, there's a lot of it out there. In the food stamp program alone, last year, we identified almost $1.1 billion in overpayments. The taxpayers aren't the only victims of this kind of abuse. The truly needy suffer as funds intended for them are taken not by the needy, but by the greedy. For everyone's sake, we must put an end to such waste and corruption."[58]

Whether or not Reagan actually intended savings from the control of fraud to be put toward increased benefits for the truly needy, one wonders why such increases would be necessary if, as he asserted constantly, those benefits were adequate already and would be kept so by his administration. Nor is it clear how food-stamp fraud reduced benefits to those who qualified for them legitimately. The reason this and other entitlement programs had grown so much in the 1970s was that recipients were viewed as being entitled to a certain level of benefit based on need, rather than to whatever was left of a finite benefit pool after fraud was deducted. Once the Reagan budget cuts began to be felt, however, the truly needy were indeed hurt as general budget reductions in programs such as food stamps resulted not in increased efforts to root out cheating but in benefit cuts across the board.

Reagan's arguments about welfare fraud, though not expressive of the harsh reality of the welfare policies during his first term, are consistent with his strategy of justifying cuts first on the basis of the general welfare—in this case, the interest of all taxpayers in seeing that their taxes not be used to enrich welfare cheaters—and second, on the basis of the welfare of the poor—in this case, by asserting a connection between welfare fraud and inadequate benefit levels for the truly needy.

Reagan often cited another kind of harm to the poor from welfare: its demoralizing impact on the initiative of recipients who could work. These people cannot be classified as truly needy, yet one is reluctant to make them simply coequals with welfare cheaters. The following excerpt from a 1981 speech suggests why a third category of recipient is needed: "You know that a job at $4.00 an hour is priceless in terms of the self-respect it can buy. Many people today are economically trapped in welfare. They'd like nothing better than to be out in the work-a-day world with the rest of us. Independence and self-sufficiency is what they want. They aren't lazy or unwilling to work, they just don't know how to free themselves from that welfare security blanket."[59]

What follows is Reagan's paraphrase of a letter he received while governor of California, from a welfare recipient so described, thanking him for liberating her from the welfare mentality. "When our reforms began," Reagan explains, "she just assumed that the time had come and that somehow she would be off welfare." She took her three children to Alaska, where she found a job and the self-respect of which welfare had deprived her. Of her new life she concludes, "It sure beats day-time television."[60]

Though, in Reagan's account, this woman had saved six hundred dollars from her "so-called poverty," she was not really a cheat. She occupied that all-important middle ground between the pure-in-heart and the purely avaricious. As with most of us, her sins were largely a function of her opportunities. When temptation was removed, she responded by discovering her virtuous side.

In sum, Reagan's discourse divides the poor into three types: the truly needy whose benefits must and will be protected, the cheater who defrauds the taxpayer, and the addict who can break the habit but only with appropriate treatment. This last group is worthy of special emphasis in Reagan's view, because here is a case where the government, by getting out of the welfare business, can be not just a punisher of miscreants but a savior of souls.

Though Reagan never indicated how many welfare recipients fell into each of the three categories, the fact that those in two of the three did not deserve help suggested that substantial cuts could be made in benefits without hurting the truly needy. And, as Zarefsky, Miller-Tutzauer, and Tutzauer note, "by focusing on waste and fraud, Reagan instilled a fundamental belief in millions of middle- and upper-middle income Americans . . . that if someone lost benefits then that person must not have been truly needy."[61]

Reagan's definition of the welfare problem, then, ranges across several types of harm that together form a compelling reason to initiate change. Social welfare spending transfers income from the middle class to the poor, hurting the middle class directly, and contributes to runaway federal spending generally, thus fueling inflation, high interest rates, and higher taxes. Welfare programs demoralize that segment of the poor who could and would find work if properly motivated. Welfare benefits often go to cheaters rather than to the truly needy.

Causes

Reagan's more or less obvious solution to the welfare mess is foreshadowed not only in his appraisal of the problem but particularly in his explication of its causes. His causal analysis helped him to take advantage of and indeed to reinforce two overlapping oppositions: ordinary citizens against governing elites, and the middle class against the poor.

Reagan's discourse offers three major reasons for the waste and ineffectiveness of current poverty policy: misguided compassion, bureaucratic inertia, and liberal elitism. In a 1982 radio address Reagan conceded that "many of the spending proposals are motivated by sincere compassion, compassion we all feel. But pretty soon," he added, "if we aren't careful, we find ourselves inventing miracle cures for which there are no known diseases. That's what led to the situation in 1980, when Federal spending was increasing by 17 percent and the inflation rate was 12.4 percent. Interest rates reached a 100-year high of 21 and one-half percent." This was a compassion that the country could ill afford, and, moreover, these programs, though "started with the best of intentions and the dedication to a worthwhile cause or purpose, . . . have not succeeded in their purpose."[62]

In place of softheartedness run wild, Reagan offered a soberer compassion, one that feels just as deeply but sees clearly. He charged himself with weeding out unsuccessful and counterproductive programs, and serving the interests not just of the poor but of everyone. At other times, however, Reagan could be less charitable toward proponents of social welfare spending. In a 1982 speech he brushed off suggestions that his proposed cuts would hurt the poor, and then referring to his opponents as "sob-sisters," described them as "the very people whose past policies, all done in the name of compassion, brought us the current recession. Some of their criticism is perfectly sincere. But let's also understand that some of their criticism comes from those who have a vested interest in a permanent welfare constituency and in government programs that reinforce the dependency of our people. Well, I would suggest that no one should have a vested interest in poverty or dependency, that these tragedies must never be looked to as a source of votes for politicians or paychecks for bureaucrats."[63]

Here is the dark side of liberal motivation. Vote buying, patronage, and bureaucratic self-interest are the charges. Add to this his frequent accusation of elitism and the distrust of "the people" it implies, and Reagan had constructed a compelling causal explanation of welfare waste and excess, one that complemented nicely his admission that some liberal support of welfare spending was "perfectly sincere." Whether due to naiveté, greed, or arrogance, liberals had foisted upon the American people a wasteful, unnecessary, and harmful welfare system, one largely responsible for the federal government's fiscal crisis and its consequences.

Murray Edelman has observed that "a particular explanation of a persisting problem is likely to strike a large part of the public as correct for a fairly long period if it reflects and reinforces the dominant ideology of that era."[64] American ideology emphasizes the existence of opportunities for all able-bodied, willing citizens to succeed economically if only they will apply themselves with effort and determination. The notion that a substantial group of American adults might need long-term economic assistance is one that goes down hard for many.

Reagan's description of the causes of the welfare mess could not have been plausible without this ideological foundation. Additionally, his causal analysis played to the traditional American fear of concentrations of political power and contempt for bureaucratic elites. By identifying social welfare programs with wasteful national bureaucracies, arrogant liberals, and idealism gone sour, Reagan was able to reinforce his identification with average Americans while at the same time supplying a necessary logical element of his case against the welfare state. Having defined the problem and having explained how it came about, he was in a position to offer the inevitable solution: cuts in welfare spending.

Solution

Reagan constantly reassured the public that the truly needy would not be disadvantaged by the cuts he proposed. In 1981 he adopted the term "safety net" to describe the social welfare benefits necessary to assure at least the bare survival of worthy recipients.[65] Certainly, these people were to receive no more than this minimum. "Some of these cuts will pinch," he warned in a 1981 speech, "which upsets those who believe the less fortunate deserve more than the basic subsistence which the governmental safety net programs provide."[66]

Which programs constituted the safety net was never quite clear. In fact, Reagan's chief domestic adviser Martin Anderson is said to have admitted that the "safety net" never had any definite content; it was a term used to categorize whichever programs at a particular time could not be cut without unacceptable political risk.[67] And though Reagan seemed to pledge not to tamper with benefits in the safety-net category, his representatives occasionally admitted that cuts were possible here as well.[68] These quibbles aside, however, the safety-net concept was central to Reagan's efforts to reassure the squeamish that the truly needy would emerge from his budgets unhurt.

The names of policies, as distinct from the policies themselves, can come to symbolize a leader's commitment to values and goals. The invocation of these names can reassure the public of this commitment even when the policies are inconsistent with it. As Murray Edelman observes, "a 'policy,' then, is a set of shifting, diverse, and contradictory responses to a spectrum of political interests. But its name is quite another phenomenon, with a different function, offering a ground for ignoring the inconsistencies to people inclined to do so. The name typically reassures, while a focus upon policy inconsistencies and differences might be disturbing."[69] Thus, the "safety net."

The safety net's capacity to reassure was enhanced by Reagan's support of it with frequent references to his social welfare record while governor of California. Often he would cite welfare policies there as proof that his program of federal reform could save money while increasing benefits for the needy. In a 1981 speech he concluded that "when we finished, we not only over a three-year period had saved the taxpayers of California $2 billion, we were able to give the welfare recipients the first cost-of-living increase they'd had since 1958—an average increase of 43 percent in their grants. And it was simply just the application of common sense and the fact that these are your neighbors there that you're trying to help, and you're better able to know what to do for them than Washington is 3,000 miles away."[70]

The historical example is a powerful kind of evidence. If it worked in California, then why not here, even if "here" is the entire country? Reagan's story reveals a second argumentative strategy. When confronted with the question of the adequacy of his social welfare budget, Reagan often would associate himself with whatever benefits actually were available, even if he had fought previously to have them substantially reduced. Used often with this strategy was the topic of quantitative degree, in this case, the expression of welfare benefits in dollars without regard to how these dollar amounts corresponded to levels of need. Thus we hear that even after the cuts in fiscal year 1982, "we're still subsidizing 95 million meals a day, providing $70 billion in health care to the elderly and poor, some 47 million people. And we're still providing scholarships for a million and a half students. Only here in this city of Oz [Washington] would a budget this big and this generous be characterized as a miserly attack on the poor."[71]

Though he avoided comparisons of welfare spending with welfare need, Reagan did favor another kind of comparison: social welfare expenditures in the 1980s *versus* expenditures in previous periods. These could be expressed as percentage increases of specific programs or as increases in the percentage of the total federal budget, or perhaps of the gross national product devoted to social welfare. In both cases, substantial growth could be documented.

A final form of proof for the humaneness of his welfare-reform proposals made direct connection with the American mythos on which his antiwelfare rhetoric was founded. An ever-present theme of Reagan's discourse, as indeed of conservative discourse generally, was the notion of a past world superior to the present, and therefore of a world to be regained: in other words, bring back the good old days. In a 1981 speech Reagan recalled that "before the idea got around that government was the principal vehicle of social change, it was understood that the real source of our progress was the private sector. The private sector still offers creative, less expensive, and more efficient alternatives to solving our social problems. Now, we're not advocating private initiatives and voluntary action as a half-hearted replacement for budget cuts. We advocate them because they're right in their own regard. They're a part of what we can proudly call 'the American personality.'"[72]

Here, we are to understand the private sector not just in economic but in social terms; we are to appreciate the historically proven potential of economic individualism *and* social volunteerism. Though no additional proof is presented, we are to assume that the needs of the poor were met adequately in this former time, whenever it might have been. Only in recent American history has it seemed necessary to erect a massive welfare state edifice to do what private resources had done earlier. Why did this happen?

Ronald Reagan's anti–welfare state discourse offers a coherent and appealing answer. The welfare state grew not because of the failure of private institutions to deal with poverty but because liberal politicians, believing in government's ability to solve all social problems and wanting to augment their own power, set in motion a welter of programs that, by destroying the incentives for the poor to improve their own lives, created ever more poverty, thus justifying ever-increasing government expenditures. The only way to arrest and eventually to reverse this trend is to limit welfare benefits only to those who are truly without the wherewithal to survive on their own. This will not only help the poor to lead independent, self-respecting lives but will ameliorate the disastrous economic consequences for all Americans of social spending run amok.

The appeal of this explanation rests on its claim that the same course of action, cuts in welfare spending, is the best policy for both the general population and the poor. Moreover, it is a solution grounded in the most cherished traditions of American history as they are commonly understood. Finally, it is a solution that provides an outlet for middle-class resentments of national political elites; that is, resentment of those people into whose neighborhoods minority children are never bused, and next to whose houses low-cost public housing is never built.

Conclusion

Ronald Reagan's discourse is an important example of how traditional conservative arguments and historical evidence based on American myths can be melded with populist appeals to produce a potent mix. Though his rhetoric was far from unique, Reagan's position and ethos allowed him to play a uniquely significant role in the attack on the social welfare state. The cuts in poverty spending during his first term came primarily in 1981, a time when his electoral mandate seemed most irresistible. As president he was willing to employ his considerable ethos on the side of reducing federal spending on poverty in the early 1980s, and keeping it reduced for the remainder of his tenure. Though his impact on the welfare debate cannot be quantified, it seems reasonable to suggest that the persistence and skill with which he pressed his case, albeit a case whose time had come, was not without significant effect.

While suggesting some link, unquantified and unquantifiable, between Reagan's antiwelfare rhetoric and welfare spending cuts, it is important to remember that the most profound impact on social welfare spending of the Reagan presidency must be measured not just directly as an attack on welfare programs, but indirectly as the preemption of social spending through the inducement of a long-term fiscal crisis. Whether by mistake or design,[73] the continuing budget deficit inherited from the Reagan years has "defunded" social welfare initiatives for years to come, and this, as much as the success of his early attacks on welfare programs, has locked in most of the reductions in social welfare outlays during the early 1980s.[74]

The capability of the federal government, at least in the near term, to fund programs to help the poor may indeed have become as limited as conservatives have long suggested. In such circumstances, arguments for why the poor need not or should not be helped may acquire increasing force. The political powerlessness of the poor as a group leaves them largely helpless to contest such trends. As Edelman has noted, "It is largely problems that are damaging to the groups with few resources for influence that are treated as fated, uncontrollable or invisible."[75]

The foregoing analysis of antiwelfare rhetoric in general, and of Ronald Reagan's antiwelfare rhetoric in particular, amounts to a partial explanation of how it is that poverty and the poor are "always with us" despite the professed desire of most Americans to eliminate it. As I have suggested, there are many other parts to this story, parts whose interconnections with rhetoric and with each other must be investigated for a full-fledged account to emerge. But if rhetoric is not a sufficient cause of public tolerance for poverty, it is certainly a necessary one, and it may be that to reverse the situation, rhetoric will be the place to start.

11

The City as Marketplace

A Rhetorical Analysis of the Urban Enterprise Zone Policy

DeLysa Burnier
David Descutner

Ronald Reagan did not capture the White House in 1980 by acting as a champion of cities. During the campaign he seldom addressed urban issues and made it clear that urban policy would not be a top priority if he was elected. To offset criticism that he cared too little for cities, Reagan did endorse one urban policy, the urban enterprise zone. The enterprise zone differed from established urban programs in that it required no intergovernmental aid and called for a reduced federal presence in cities. Zones instead offered the prospect of rebuilding the urban "marketplace" through business tax incentives.

The zone initiative is worth examining from a rhetorical standpoint for several reasons. First, it attempted to redefine the nation's understanding of urban problems and their solution. Second, it provided Reagan with a platform from which to criticize Lyndon Johnson's urban policies, thereby establishing ever more sharply the philosophical differences between himself and previous presidents. Third, the initiative defused the Democrats' charge that he had no urban policy. Finally, the proposal enabled the rhetoric of entrepreneurial capitalism to be injected into urban policy discourse.

Our essay adopts Kenneth Burke's account of the concept of substance as the organizing construct for our rhetorical analysis of the zone initiative. First, we explicate Burke's understanding of substance, concentrating on the concept's paradoxical character and its relation to ambiguity and rhetoric. His theoretical understanding is supplemented with the more applied work of Deborah Stone, a political scientist who addresses the rhetorical elements of public policy from a perspective consonant with Burke's. Second, we present a brief history of the federal enterprise zone policy. Third, we probe the substance of enterprise zone policy discourse to discover how the policy's "story" is articulated through a set of rhetorical practices.

Substance, Paradox, Ambiguity, and Rhetoric

Burke notes that the concept of substance lost "prestige" at the hands of the modern empiricists, who dismissed it as a metaphysical fiction without epistemological significance. The empiricists' recommendation that the term *substance* be discarded, however, moved Burke to warn that "in banishing the term, far from banishing its functions one merely conceals them."[1] Indeed, one of Burke's main aims has been to expose the concealed functions that the concept of substance serves in all forms of discourse. What makes substance such a powerful force in discourse, according to Burke, is precisely the paradoxical nature of the concept itself.

Burke's discussion of the "paradox of substance" turns on the following observation: "The word 'substance,' used to designate what a thing is, derives from a word designating something that a thing is not. That is, though used to designate something within the thing, intrinsic to it, the word etymologically refers to something outside the thing, extrinsic to it."[2] This paradox means the concept of substance embodies an "unresolvable ambiguity" insofar as defining the intrinsic nature of things, which is the traditional notion of substance, requires appealing to extrinsic, contextual factors. The paradox of substance in effect becomes a paradox of definition because of the ambiguity necessarily involved in defining things by reference to that which they are not. Defining something in terms of something else introduces ambiguity because "no two things or acts or situations are exactly alike."[3] That ambiguity allows the intrinsic and extrinsic to "change places," such that in definitions the extrinsic stands for the intrinsic. The paradox of substance, then, reveals that definition operates by way of negation and inevitably yields ambiguity because it relies simultaneously on intrinsic and extrinsic reference.

The resulting ambiguity represents a "major resource of rhetoric."[4] For example, the indeterminacy of definition allows one word to carry multiple and even conflicting meanings, all the while appearing univocal, which in turn generates the "considerable vagueness" of reference that Burke indicates can be used to rhetorical advantage.[5] Closely linked to the ambiguity of substance is identification. Identification occurs when persons become "consubstantial" with others, which means they become "substantially one" with each other while remaining "substantially distinct."[6] Also dependent on the ambiguities of substance are the four master tropes of metaphor, irony, metonymy, and synecdoche.

Burke contends that "metaphor is a device for seeing something in terms of something else," suggesting that within the substance of any metaphor are at least two substantially different thoughts "acting together."[7] The substance of irony is likewise ambiguous, for within any ironic statement are two substantially distinct significations, or what Burke calls "A and non-A."[8] Similarly, metonymy's "basic 'strategy' is . . . to convey some incorporeal or intangible state in terms of the corporeal or tangible."[9] That incorporeal states can be, in Burke's word, "reduced" to corporeal states owes to the paradoxical nature of substance itself. Finally, synecdoche requires understanding the part in terms of the whole, or vice versa. The part may be used, in Burke's word, to "represent" the whole only because of the ambiguity of substance.[10]

So it is precisely the ambiguity of substance that makes possible the linguistic transformations whereby, for example, two terms are merged into a third that stands as a metaphor. Burke maintains that "it is in the areas of ambiguity that transformations take place; in fact, without such areas, transformation would be impossible."[11] The unanswered question so far is how Burke's theory of the interrelationships among substance, paradox, ambiguity, and rhetoric can be used to inform critical analysis of public policy. Stone's *Policy Paradox and Political Reason*, which takes an approach in line with Burke's theory, answers that question with its study of the rhetorical elements of public policy.

The Rhetorical Elements of Public Policy

Stone contends that "because paradox is an essential feature of political life," policy scholars must begin to address the origins and implications of ambiguity resulting from paradox.[12] She agrees with Burke that ambiguity is intimately tied to paradox and represents "the most important feature of all symbols."[13] Moreover, her account of how ambiguity effects transformations in policy discourse accords with Burke's account: "Why is ambiguity so essential in politics? What role does it play? Ambiguity enables the transformation of individual intentions and actions into collective results and purposes. Without it cooperation and compromise would be far more difficult, if not impossible."[14] Even her claim that ambiguity facilitates "cooperation and compromise" concurs with Burke's thesis that rhetoric works "to induce cooperation in beings that by nature respond to symbols."[15]

Furthermore, Stone's view of metaphor and synecdoche also squares with Burke's view, and her insightful analysis of how each trope functions in policy discourse will guide our study. For example, she illustrates how synecdoche is used to simplify complex public problems and how metaphor is used strategically to define problems and their solutions. Equally important to our study is Stone's treatment of how stories function as "tools of strategy" in policy discourse. Stone notes: "Stories provide explanations of how the world works. These explanations are often unspoken, widely shared, and so much taken for granted that we are not even aware of them. They can hold a powerful grip on our imaginations and our psyches because they offer the promise of resolution for scary problems."[16] Her notion that policy stories "offer the promise of resolution" fits with Burke's proposition that all narrative forms involve the "creation of an appetite in the mind of the auditor, and the adequate satisfying of that appetite."[17] Indeed, policy stories create an appetite for action among the public by depicting "scary problems" and then satisfy that appetite by taking action through proposed policy solutions. For example, Ronald Reagan first told a story of urban policy failure and city decline before proposing the zone initiative as a solution. Having established the consonance between Burke and Stone, we will continue to draw from both as we clarify our critical method and then use that method to examine the rhetorical substance of enterprise zone policy discourse.

Studying the Rhetorical Substance of Public Policy

Burke contends that the concept of substance is "so fertile a source of error, that only by learning to recognize its nature from within could we hope to detect its many disguises from without."[18] Specifying "from within" how the "substance" of the enterprise zone policy's discourse operates rhetorically is the principal task of our study. For guidance we turn to Weldon Durham's investigation of Burke's concept of substance and various naming strategies. Durham states: "Phenomena become real only insofar as they are nameable and only insofar as they are perceptibly arranged. Thus, substantiation, consciousness of reality, occurs at the point of contact between the finite (vocabulary) and the infinite (the world around). Substance is consciousness as it is enacted in the way people use symbols."[19] The individual "naming strategies" that Durham interpolates correspond to Burke's "terms for substance." Naming accomplishes strategic transformations, and particular "naming strategies" that Durham interpolates correspond to Burke's terms for substance.

Durham's "geometric strategy of naming" follows from Burke's term of "geometric substance," and focuses on the rhetorical role of negation in definition. For example, Reagan defines the substance of the enterprise zone policy by saying it is *not* the same substance as that of the failed Great Society urban policies. Durham's "familial strategy" follows from Burke's "familial substance," and focuses on the role of "ancestral connotation" in definition. For example, Reagan also defines the enterprise zone policy by noting its philosophical consubstantiality with capitalism and the related ideals of entrepreneurship and individual freedom.

Each "naming strategy" thus defines the substance of enterprise zone policy discourse in different ways, and thereby furthers the overall cause of promoting identification with the policy's values and goals. The relative power of each "naming strategy" owes to the distinctive story it tells and the distinctive use it makes of metaphor and synecdoche. We attempt to show how specific policy stories, with their accompanying metaphors and synecdoches, effect the transformations necessary to the policy's rhetorical success. With our method and aims so clarified, we turn next to a brief history of the federal enterprise zone policy.

The Enterprise Zone Policy

From the beginning of his presidency, Ronald Reagan made it clear that the only federal urban policy the nation needed was a national economic policy devoted to "long-term, non-inflationary economic growth."[20] The *President's Urban Policy Report* stated plainly that the "foundation for the Administration's urban policy is the Economic Recovery Program."[21] The single exception to this "hands-off" approach was the urban enterprise zone initiative. The enterprise zone, by reducing government taxes and regulations, is designed to create a free-market environment in specially designated and geographically defined urban areas in need of revitalization. The underlying assumption is that reducing taxes and regulations will permit market forces to work more freely within the designated areas. This in turn will create more opportunities for private firms and entrepreneurs to create jobs—particularly for disadvantaged workers.[22]

The zone concept originated in England but was introduced to the United States by the American Legislative Exchange Council, an organization of largely conservative state legislators, and by Stuart Butler, a Heritage Foundation economist and policy analyst. Butler was particularly influential in circulating the concept throughout the Washington policy community, and he generated most of the initial analysis.[23] Republican Jack Kemp introduced the first zone legislation in May 1980, and Democrat Robert Garcia joined him as the bill's cosponsor in June.

The zone concept gained a larger audience when Ronald Reagan, then a presidential candidate, announced his support in a September 1980 speech delivered against the grim backdrop of New York City's South Bronx. The severely blighted South Bronx has long stood as a national symbol of urban policy failure. After criticizing older federal programs, Reagan promised to push for enterprise zone legislation. Overall, the program was expected to meet the "two somewhat contradictory expectations" of liberating "the nation's entrepreneurial spirit" and revitalizing declining urban areas.[24] Even more, the zone initiative served as a "philosophical illustration of a wider strategy for economic and social problem definition, relationship, and solution."[25] This "wider strategy" became the New Federalism as outlined in Reagan's 1982 State of the Union address, where he proposed that power should be returned to the states. Reagan claimed that the federal government had become "too intrusive" in domestic policy, and therefore it was time to make the states responsible for the majority of the country's domestic programs (for example, Food Stamps and welfare).[26]

With respect to urban problems, the "wider strategy" entailed working through the marketplace in order to solve significant public problems like urban blight and poverty. The strategy rests on three assumptions that follow from a strict, free-market philosophy. First, economic opportunity, jobs, and wealth are created through an unobstructed marketplace. Second, people have both the potential and the desire to be entrepreneurs. Third, marketplace freedom and individual entrepreneurial potential have been stifled by government taxes, regulations, and previous urban programs. In short, less government, not more, is needed, and a free working marketplace should be the government's goal. Once such a marketplace is established, individual initiative and hard work will take over, and the nation's problems gradually will be solved.

When the administration introduced its zone bill in March 1982,

President Reagan emphasized the "wider strategy" theme by claiming that zones "are based on an entirely fresh approach for promoting economic growth in the inner cities." Previous urban programs depended on "heavy government subsidy and central planning," whereas enterprise zones "remove government barriers freeing individuals to create, produce, and earn their own wages and profits."[27] In sum, market forces rather than government programs would lead to the revitalization of depressed urban areas. In Butler's terms, local economies would be rebuilt from the bottom up, instead of having "outsiders imposing a plan on the neighborhood."[28]

The bill proposed that a maximum of seventy-five enterprise zones be created over three years, with state and local governments making the actual nominations. Any area nominated would have to meet several distress criteria. Once designated, area businesses would become eligible for tax incentives and regulatory relief. In his zone message Reagan stressed that the designation process would be competitive since more than two thousand areas qualified. Therefore, "a key criterion in this competitive process [would] be the nature and strength of the State and local incentives to be contributed to the zones."[29]

For reasons including the newness of the concept, partisan politics, and estimates of lost tax revenue, the zone legislation failed to pass Congress until late 1987. The legislation that Congress passed, however, contained no tax incentives. Simply, it allowed the secretary of the Department of Housing and Urban Development to designate up to one hundred zones, of which a third must be in rural areas. Areas receiving zone designation would be eligible for HUD regulatory relief and would be given priority consideration for other HUD programs. Bush's HUD secretary, Jack Kemp, delayed the designation process until Congress passed tax-incentive legislation. With incentives in place, the secretary believed the program could help eliminate poverty by "unleashing the enormous potential of entrepreneurial capitalism."[30]

The Rhetorical Dimensions of Enterprise Zone Policy Discourse

Creating a favorable reception for a new policy requires placing it in a context that supplies background and establishes boundaries of meaning. Geometric naming strategies facilitate such a reception by defining policy substance through extrinsic reference, and their governing principles are negation and consistency. The principle of negation is used to establish a favorable policy identity by referring extrinsically to what the policy is not in order to define what it is. The principle of consistency is used to place a policy in an environment with which it is consistent, suggesting that the match between the two is timely and appropriate.

In his zone message, Reagan employs the principle of negation to identify, or place, the enterprise zone policy. He explains that it should not be confused with the failed urban programs of the 1960s. Unlike those programs, enterprise zones do not depend on "heavy government subsidies and central planning," nor do they impose "taxes, regulations, and other government burdens on economic activity."[31] Enterprise zones instead emphasize the marketplace and the removal of government burdens as the means by which urban areas will be revitalized and the 1960s legacy of zero economic growth reversed.

Animating Reagan's story of urban decline is his synecdochic use of a particular policy, the Model Cities program, to represent the whole of urban policy in the 1960s. He cites the Model Cities program as the "prime example" of the "old approach," and observes that it is the "direct opposite" of the enterprise zone policy. His choice of Model Cities is strategic because it was a controversial program that became a symbol of failure for Republican critics of the Great Society. With this synecdoche Reagan makes concrete simultaneously what the Model Cities program represented and what the enterprise zone initiative will *not* represent. The "old approach" failed to solve the problems of urban blight and unemployment, while the enterprise zone policy promises to be a "fresh approach" that promotes economic growth in the inner city. Typical of the negative contrast Reagan draws is his claim that the "old approach" designated areas for assistance in an "automatic or routine" manner, whereas the "fresh approach" will make designations through a "competitive process . . . consistent with the Enterprise Zone theme of creating an open market environment by removing government burdens."[32]

Along with using negation strategically to specify the policy's identity, Reagan also invokes the principle of consistency for strategic purposes of placement. One chief purpose is to locate the act of proposing such a policy within the tradition of past presidents' taking the lead in urban policy. Since the 1930s the public, local officials, and Congress have expected the president to lead in this area.[33] Staying consistent with this tradition makes Reagan seem responsive to the "severe problems" of inner cities, which in turn insulates him from criticism that he has neglected their plight. Hence, Reagan honors the tradition of federal urban leadership, while proposing a "bold, new concept" that is consistent with his own political values. The substance of enterprise zone policy paradoxically appears to be both traditional and "path-breaking," and this ambiguity of substance marks the rest of Reagan's use of consistency.

Given the immediacy of urban problems, Reagan declares that some "innovative" and timely response is required that departs from the "old approach." The primary flaw of the "old approach" was that the federal government, in Reagan's words, "dictated" policies to states and localities without regard for their particular "conditions and preferences." Reagan's alternative is a "partnership" among federal, state, and local governments that minimizes federal authority and maximizes the creative contributions of state and local governments.[34] With this scheme, he predicts, zones can be designed that will be consistent with the specific needs of local environments.

The principle of consistency as a "way of placement" appears finally in Reagan's description of the enterprise zone as "consistent with the Administration's policy of Federalism," which returns power to state and local governments.[35] The cornerstone of Reagan's "New Federalism" is the assumption that "state and local governments have demonstrated that, properly unfettered, they will make better decisions than the federal government acting for them."[36] With respect to urban policy, cities must stop depending on federal and state grants and instead develop "strategies" for coping with changing regional economies.[37] The enterprise zone proposal is consistent with the New Federalism because it requires state and local participation and is not a grant program.

Evident here is another paradox whereby the substance of the enterprise zone policy is not located exclusively at the federal, state, or local level. Distributing the policy's substance across three levels of government effectively transforms it into an intergovernmental, rather than a federal, initiative. Note as well the strategic ambiguity involved in spreading policy responsibility across three levels of government. If an individual zone fails, then the responsibility lies as much with state and local governments as with the federal government and, by extension, Reagan. If a program succeeds, then the responsibility lies at least partly with the federal government that originally proposed the policy. The ambiguity, then, produces a "no-lose" situation for the Reagan administration.

The policy story implicit in Reagan's use of the geometric strategy of naming is one of decline and reversal, and the underlying policy metaphor effecting that reversal is one of novelty. We have shown that Reagan portrays inner city decline in graphic terms, speaking of "decaying areas" that demand immediate attention. He also concentrates on the decline of urban policies, using the Model Cities program synecdochically to stand for an array of failed federal programs. The two strands of the story of decline intersect in Reagan's intimation that flawed federal policy exacerbated what was already a terrible problem.

Stone's account of one kind of policy story corresponds well with the interrelated strands of Reagan's story. According to Stone, the story goes: "In the beginning, things were pretty good. But they got worse. In fact, right now, they are nearly intolerable. Something must be done."[38] Rewriting Reagan's policy story in Stone's terms yields the following narrative: Cities once were thriving marketplaces, but they began to deteriorate. Accelerating their deterioration were the misguided policies of the 1960s that usurped local power and ignored local needs. Now cities are in an "intolerable" state of economic disrepair and "something must be done." That "something," if it is to reverse urban decline, must be a novel solution that differs from past approaches.

Indeed, the metaphor of novelty pervades Reagan's story of how enterprise zone policy will reverse urban economic decline. He calls the policy "fresh," "creative," "innovative," "path-breaking," "experimental," and "new." Reagan compares the novelty and "potential" of enterprise zones with the staleness and failure of earlier urban programs. Stone states that metaphors typically prescribe a course of action, and Reagan's novelty metaphor is no exception. It directly prescribes enterprise zones as a cure for urban ills by the optimistic interpretation it gives of the policy's "fresh approach."

Focusing on the policy's novelty is also strategic because, in contrast with earlier policies' record of failure, enterprise zones have as yet an untarnished record and thus seem to hold great promise. In many respects, then, the novelty metaphor at the heart of Reagan's policy stories makes for a dramatic conflict between the old and the new, with the former symbolizing the mistakes of the past and the latter the "innovative" prospects of the future.

The Familial Strategy of Naming

The familial strategy of naming, according to Durham, "stresses descent from a 'genetic' source."[39] It names a policy by identifying the family or source from which the policy's substance is derived. Burke explains the logic of this strategy by referring to Plato's notion that "the members of a class derive their generic nature from the 'idea' of the class in which they are placed."[40] The Model Cities program, for example, derives its meaning in Reagan's discourse from its membership in the class of Great Society programs, the central and erroneous idea of which is that "government expenditures and subsidies" solve urban problems. In contrast, the enterprise zone policy derives its meaning, or substance, from its membership in the class of "market-oriented" programs, the central idea of which is the moral necessity of embracing entrepreneurial capitalism.

Entrepreneurial capitalism is the family concept from which enterprise zone policy derives its substance, and in Reagan's "civil religion" its meaning transcends its status as an economic theory.[41] In familial strategies, Burke writes, the family concept is often "spiritualized," and entrepreneurial capitalism clearly holds a spiritual meaning for Reagan. It represents the antithesis of the "demonic power of big government" and the nexus of transcendent values such as freedom, work, and neighborhood.[42] Throughout Reagan's policy discourse is evidence of the religious zeal with which he condemns the evil of "big government" and proclaims his faith in the inherent goodness of the market. Reagan believes "big government" is a destructive force that restricts competition and opportunity, constrains progress and prosperity, and ultimately erodes the public's work ethic.

These lamentable consequences are evident throughout society, but nowhere are they more pronounced than in the "economically depressed areas" of America's inner cities. As noted, Reagan's policy stories imply that the root cause of urban decline is the government's imposition of "taxes, regulations, and other burdens on economic activity."[43] Restraining economic activity suppresses individual freedom and limits work opportunities, both of which contribute to the decline of urban neighborhoods. Also contributing to urban decline and the erosion of the work ethic is the federal government's promise "to solve all our problems for us without demanding individual initiative in return."[44]

Reagan's position is that government's promise to solve our economic problems has diminished the incentive to compete and undercut the belief that rewards accrue to individuals who work hard. Risk taking, personal industry and ambition, and innovation—all foundational values in entrepreneurial capitalism—seem pointless to pursue if individual needs will be met without individual effort. "Big government," in Reagan's view, has dissipated our "national will and purpose," leaving in its wake a dispirited citizenry and a weak economy.[45] Reagan's solution to these problems, as Robert Dallek reports, "was the same one he had been voicing since the 1950's: shrink the power and control of government and increase the freedom of individuals and private enterprise."[46]

Entrepreneurial capitalism was thus a "spiritualized" family concept for Reagan, embodying three of the values on which he had campaigned for twenty-five years: freedom, work, and neighborhood.[47] A free market stimulates competition and fosters innovation, which are the hallmarks of individual freedom. It invariably generates economic growth and thereby increases work opportunities. Furthermore, the prosperity resulting from competition, innovation, and growth strengthens neighborhood economies, making them more stable and secure. Since the enterprise zone policy is the direct descendant of entrepreneurial capitalism by virtue of sharing the same family substance, it too embodies these values.

The importance of the enterprise zone, then, lies in its embodiment of these values. Anything else would be intrusive and run counter to the values of freedom, work, and neighborhood. The enterprise zone thus eliminates the need for additional urban programs, and the administration could reduce or eliminate them without "hurting" cities. Explaining how these values function rhetorically in Reagan's policy discourse requires returning briefly to Burke's discussion of derivation.

Burke maintains that familial strategies work rhetorically by deriving "actualities from potentialities."[48] Similarly, Reagan used derivation rhetorically by establishing the family consubstantiality between entrepreneurial capitalism and the enterprise zone policy, detailing first the value "potentialities" of the former, including the aforementioned freedom, competition, growth, innovation, work opportunities, and better neighborhoods. Each one of these values, as shown, figured prominently in his enterprise zone policy discourse.

Reagan then derived his justification of the "actualities" of enterprise zones from these value "potentialities," implying that those values would be manifest in the operation of individual zones. Creating enterprise zones, in short, would magically transform potential values into actual economic practice. He described inner cities as places where potential economic power had become dormant because of "government barriers to entrepreneurs who can create jobs and economic growth."[49] This latent economic power was what Reagan hoped to unleash with enterprise zones, and leading the way would be the urban entrepreneur Reagan celebrated.

All family naming strategies are synecdochic in that they posit a part-whole relationship between a phenomenon and the family to which it supposedly belongs. The phenomenon, or part, derives its substance or meaning from its imputed membership in the family, or larger whole. Two synecdoches were critical to the rhetorical success of Reagan's enterprise zone policy discourse. By asserting that the zone policy belongs to the philosophical family of entrepreneurial capitalism, Reagan could claim that the values of entrepreneurial capitalism defined the zone policy's substance. Strategically, this value transfer did not need to be proven so much as merely asserted on the basis of common family membership.

A second synecdoche took the form of a microcosm-macrocosm relationship. Each enterprise zone represented a microcosm of the wider family, or macrocosm, of economic change that Reagan envisioned sweeping America. Just as the national Economic Recovery Program would restore entrepreneurial capitalism to the market as a whole, so would enterprise zones restore it to the urban marketplace. Reagan closed his congressional zone message by proclaiming that his "mission" was to "free enterprise so that together we can save America."[50] Each zone represented a testament to the virtues of entrepreneurial capitalism and to the wisdom of economic change, both of which were part of Reagan's larger "mission" to wrest control of the economy from the federal government's grip.

Reagan regularly employed metaphors such as "mission" to symbolize his quasi-religious quest to restore entrepreneurial capitalism to the marketplace. He used the similar metaphor of "crusade" for the same purpose in the 1980 presidential debates, where he announced: "I would like to have a crusade today, and I would like to lead that crusade with your help. And it would be one to take government off the backs of the great people of this country, and turn you loose again to do those things that I know you can do so well, because you did them and made this country great."[51] The story of this "crusade" overlapped with the policy story Reagan told through family naming strategies in his zone discourse. By replacing government with "opportunity," as Reagan declared in his congressional message, enterprise zones would "spark the latent talents and abilities already in existence in our Nation's most depressed areas."[52]

Those overlapping policy stories are both variations on what Stone labels "stories of control." The message of these stories, as Stone indicates, is that returning choice to individuals will rebuild their hope in a more promising future. Stories of control hinge on situations once thought to be "out of our control," which then become "controllable." Stone notes: "Stories about control are always gripping because they speak to the fundamental problem of liberty—to what extent do we control our own life conditions and destinies. Stories that purport to tell us of less control are always threatening, and ones that promise more are always heartening."[53] Reagan relied on just such a policy story to elicit support for enterprise zones.

Reagan's message, too, was liberty, and specifically how economic liberty in the inner city had been lost to "monopolized government services."[54] The loss of economic liberty meant that inner-city residents have sacrificed control of their "destinies" to government programs and planners. As bad as these circumstances may be, however, they are not intractable. If city residents were just turned loose, then they could regain control of their "life situation" by using their latent entrepreneurial abilities to take advantage of the opportunities created by enterprise zones. As individuals succeeded, moreover, they would begin to improve their neighborhoods by contributing to the overall economic growth of the city. Reagan's story was heartening in the end because its central metaphor was opportunity, especially in the sense of creating opportunities for individuals bypassed by the marketplace.

The opportunity metaphor is the rhetorical meeting ground of key symbols in Reagan's enterprise zone policy discourse. It is synonymous with economic liberty in his discourse, where it means "freeing individuals to create, produce, and earn their own wages and profits."[55] It is indispensable to the success of entrepreneurs, who cannot pursue their innovative solutions to urban economic problems unless they are presented an opportunity. Similarly, it promises a better future for neighborhoods by returning to individuals the responsibility for community improvement. Furthermore, the "opportunity" metaphor is integral to the latent-manifest symbolic theme running through Reagan's policy discourse. Only if individuals are supplied with opportunity can they make manifest their latent skills and creativity.

Reagan strategically offered no guarantee that all these changes would occur, leading to the reestablishment of choice and the rebirth of hope. What he did offer was his faith that individuals who once "made this country great" can do it again if given the opportunity created by enterprise zones. He also strategically underestimated the extent to which the substance of enterprise zone policy is a mixture of the values of entrepreneurial capitalism and traditional government assistance. His discourse's focus on those attractive values is not matched by a comparable focus on how much of the success of zones depends on "sustained government involvement and the expenditures of public resources."[56] This paradox, whereby the values imputed to enterprise zone policy conflict with what the policy requires to be practicable and effective, underlies all of Reagan's discourse. That paradox is made possible by the ambiguity of the policy's substance, which represents a strategic resource that Reagan exploited for rhetorical ends.

Conclusion

Stone points out that stories of decline and stories of control often are woven together in public discourse, and clearly this occurred in Reagan's enterprise-zone policy discourse. His story of urban decline set the stage for his story of economic salvation through enterprise zones. Reagan promoted belief in these stories and thereby won political support not only for his policy but for the marketplace approach to public problem solving.

Reagan's endorsement of enterprise zones represented more than a shift in urban policy. Indeed, enterprise zones were part of a larger shift in political values promoted by Reagan, whereby individual efforts would prevail, opportunities would abound, and the influence of government would diminish. These fundamental political values of Reagan are evident throughout his public discourse, and they help to explain his rhetorical success as president. Celebrating the heroic individual and describing America's limitless opportunities while derogating government and calling for its diminution have been essential elements of Reagan's discourse for more than twenty-five years. The intrinsic appeal of these values, along with Reagan's inimitable ability to express them clearly and forcefully, had much to do with Reagan's earning the title "the Great Communicator."

12

Civil Religion and Public Argument

Reagan as Public Priest of the Antiabortion Movement

Catherine Helen Palczewski

Although scholars now recognize the institution of the presidency as embodying the civil religion of America, they do not deem overtly religious arguments appropriate tools of the president. Robert Bellah argues that "the god of the civil religion is not only 'unitarian,' he is also on the austere side, much more related to order, law, and right than to salvation and love. . . . He is actively interested and involved in history, with a special concern for America."[1] Despite God's involvement, with the president acting as his priest,[2] the president needs to retain the appearance of being secular because the separation of church and state requires the segregation of the religious from the political sphere.[3]

Roderick Hart, in *The Political Pulpit*, argues that the intersection and segregation of civic and sacred arguments is primarily rhetorical. He writes: "Civic piety, in America at least, emerges not so much from blind, momentary passion, but from a knowing, practiced, thoroughly pragmatic understanding of the suasory arabesques demanded when God and country kick up their heels rhetorically."[4] The problem faced by rhetors is how to maintain the balance between the civic and the religious. Hart describes the process as a "rhetorical juggling act when dealing with politically tinged religious issues or with affairs of state bordering on the theological," with the rhetorical ideal being "to maintain the hyphen in civil-religious matters."[5] While Hart does much to elucidate the interaction of sacred religion and civic government on the grand scale, he does not delve into how the civil religion, or civic piety, affects the rhetorical presentation of particular political issues that address theological questions. This essay examines how President Ronald Reagan, by drawing on the analogical resources offered by the civil religion, was able to present a case against abortion that appealed to his religious constituency while not violating the separation of church and state. Reagan, through the skillful use of rhetoric, maintained the hyphen between *civil* and *religion.*

How could Reagan explicitly align himself with a religious issue while maintaining the presidency's secular veneer? This apparent dilemma may be resolved by analyzing how Reagan uses the civic-sacred analogue as a rhetorical resource in his abortion discourse. Reagan's ability to address an issue infused with religious undertones was facilitated by the way civic and sacred religions function as formal symbolic analogues. Reagan strategically structured his arguments so that he could maintain his status as priest of the civil religion while appealing to a religious constituency; avoiding the explicitly religious, Reagan's strategy made the civil religion, as codified in the Constitution and Declaration of Independence, the foundation for his arguments against abortion. Reagan, by using the framework of the American civil religion, could collapse the meaning of *civic* and *sacred,* using the suasory power of both values while not violating the separation of church and state.

In examining Reagan's use of the civil religion as a resource for argument, I will focus on his four "March for Life" speeches and his contribution to the book *Abortion and the Conscience of the Nation.* These texts exclusively address the issue of abortion, while most of Reagan's other statements on abortion are contained in speeches that cover a broad range of issues and hence only superficially discuss abortion. The "March for Life" speeches were delivered to the participants of the March for Life rally, held every year to protest the *Roe* v. *Wade* decision.

Religion's Moral Rule Against Abortion: Where the Religious Meets the Civic

President Ronald Reagan became involved in what many perceive to be an issue of the religious right: abortion. Being religious does not necessarily mean that one is antiabortion. However, being antiabortion does tend to imply that one is zealously religious. In the public mind, abortion and religion are unavoidably intertwined. For example, Mario Cuomo in a September 1984 speech at Notre Dame urged people to tolerate abortion on the grounds that such toleration would ensure religious pluralism. Another reason for the perceived connection between religion and an antiabortion stance is outlined by Kristin Luker in *Abortion and the Politics of Motherhood*. She explains: "The pro-life world view, notwithstanding the occasional atheist or agnostic attracted to it, is at the core one that centers around God: pro-life activists are on the whole deeply committed to their religious faith and deeply involved with it."[6]

The way in which this religious grounding permeates antiabortion rhetoric is analyzed in Randall A. Lake's essay "Order and Disorder in Anti-Abortion Rhetoric: A Logological View." He argues that the rhetoric of the antiabortion groups (primarily composed of religious fundamentalists, Catholics, Moral Majoritarians, and the New Right) may be understood by examining the "'moral landscape' painted by anti-abortion rhetoric."[7] Lake argues that antiabortion rhetoric is guided by an ascent-descent pattern akin to an archetypal metaphor that tends to infuse religious discourse.[8]

The connection between religious and civic antiabortion discourse is also noted by antiabortion activists. J. Robert Nelson, in an article entitled "Religion and Abortion," writes that abortion "is a political issue of first magnitude. For most people it is a moral question, provoking emotional disputes, and its moral character is determined for many by religious convictions."[9] Despite his recognition of the connection, Nelson does not believe that this excludes religious issues from discussion in the public realm. He writes, "The separation of church and state does not mean separation of religion and politics."[10] He argues that religious beliefs may transcend a particular denomination and hence function as universals; simply because the belief originated in one denomination does not mean that it does not have political application. Such a universal issue is life and death ethics, which, Nelson believes, cannot be "wholly divorced from religious teachings and insights."[11] Accordingly, Nelson sees a deontological approach as inherent to the abortion controversy.

Locating the Moral Rule Against Abortion in the Civil Religion: Translating the Religious into the Civic

Reagan's antiabortion rhetoric avoided any explicit linkage with religious objections by locating the moral rule against abortion in the Constitution and the Declaration of Independence. Although Reagan argued elsewhere for a constitutional amendment guaranteeing the fetus's right to life, he ultimately believed that the right already existed, even if only implicitly, in the Constitution and

Declaration of Independence. In his 30 July 1987 briefing of right to life activists, he urged the adoption of a human life amendment, while also appealing to a constitutionally recognized right to life: "No, ours is a nation founded upon a shared and basic law, the Constitution. And because it is the Constitution that must reflect our most fundamental values—freedom, equality before the law, and, yes, the dignity of human life—because of this, the duty of everyone here today is clear. We must not rest—and I pledge to you that I will not rest—until a human life amendment becomes part of our Constitution."[12] Although this statement appears to find a basis for the evilness of abortion in a higher moral law, the higher moral law is one that the Constitution already reflects: freedom, equality, and dignity. Ultimately, Reagan was able to rely on the Constitution; the Constitution either reflects or was intended to reflect the higher moral law.

Lake detailed the implications and form of moral rules in his study of the metaethical framework of antiabortion rhetoric. Lake argues that the ethical framework of antiabortion arguments begins with assumptions about the nature of morality. Antiabortionists rely primarily on deontological arguments. This entails examining the rightness or wrongness of an action rather than its goodness or badness.[13] Lake describes deontological ethics as an "ethics of duty" consisting of a theory of duty and moral obligation. Deontological ethics recognizes a threshold between right and wrong, whereas teleological ethics acknowledges gradations of right and wrong. Under deontological ethics, an action can be wrong even though it produces good effects. The rightness or wrongness of an action is inherent in the action, the goodness or badness is determined by the effects of the act. Under deontology, more importance is given to the rightness or wrongness than to the good or ill effects. The conflict between deontology and teleology is evident in the abortion controversy, the two sides respectively being categorized as "sanctity of life" or "quality of life."

Deontology is also an ethics of duty in which one judges an act by how it fulfills commitments or duties to others rather than by whether the act is in one's self-interest.[14] The duties are codified as moral rules that typically translate into positive law. In some instances this translation may fail, as antiabortionists believe it did with *Roe* v. *Wade*.

Rules exist in two forms: first-order rules, which prescribe or proscribe specific actions (for example, the Ten Commandments); and second-order rules, which specify the criteria by which an action may be judged right or wrong (for example, Kant's categorical imperative). The antiabortionists' claim of the fetus's right to life, a second-order rule, merely obscures their primary objection to abortion, which is abortion's violation of the first-order rule that prohibits "murder," "killing," or "the taking of an innocent human life."[15]

While typical antiabortion rhetors appeal to religiously based moral rules against murder, Reagan's interpretation of morality was based on the Constitution and Declaration of Independence; these documents codified moral rules that differentiated right from wrong. Adopting the persona of the priest of the civil religion, Reagan appealed to a "set of beliefs, symbols, and rituals," shaped by the Founding Fathers, which Bellah calls "the American civil religion."[16] These symbols in turn functioned as the basis for Reagan's moral rules. Reagan defined the belief central to the civil religion as the sacredness of all human life[17] and defined the evil of abortion in terms of its violation of the fetus's "rights to life, liberty, and pursuit of happiness," which the nation's founders proclaimed inviolable in the Declaration of Independence.[18] Reagan's constant reference to the Constitution and Declaration of Independence served to highlight the objective and legal basis of his moral claims; although he argued for the subjective belief in the sanctity of life, he grounded those beliefs in the civil religion and in documents that had objective existence.

In 1985, in his first public address to the March for Life, Reagan opened his statement by defining the actions of the marchers as a demonstration of the "overwhelming support for the right to life of the unborn."[19] Reagan located the rule against abortion in the Constitution, noting that no constitutional provision allowed it. He observed "that abortion is taking the life of a human being; that the right to abortion is not secured by the Constitution; and the state has a compelling interest in protecting the life of each person before birth."[20] In 1986 Reagan equated the antiabortion position with the defense of basic American principles; he proclaimed: "When you insist upon legal protection for all human life, you're simply being true to our most basic principles and convictions as Americans."[21] He also recalled the country's founding documents, explaining: "By your presence today, you reaffirm the self-evident truths set forth in our Declaration of Independence."[22] Reagan cast the right to life as not only the corollary to the *moral* rule against murder, but also as constituting "our most basic *civil* right" (emphasis added).[23] He repeated this philosophy in 1987; Reagan located the right to life in the civil religion by arguing that the right to life is "granted by our Creator,"[24] echoing the words of the Declaration of Independence and, as Bellah describes it, imparting to "the political realm a religious dimension" that highlighted "certain common elements of religious orientation that the great majority of Americans share."[25]

If one believes in the Declaration of Independence then it seems that one must believe that abortion is wrong. In fact, Reagan described abortion's violation of the civil religion as national civic injury and argued that, "with God's help," the people may "heal our wounded nation."[26]

Reagan's most overt appeal to the civil religion appeared in his 1988 speech. He argued:

> And yet our opponents tell us not to interfere with abortion. They tell us not to impose our morality on those who wish to allow or participate in the taking of the life of infants before birth. Yet no one calls it imposing morality to prohibit the taking of life after a child is born. We're told about a woman's right to control her own body. But doesn't the unborn child have a higher right, and that is to life, liberty, and the pursuit of happiness? Or would our critics say that to defend life, liberty, and the pursuit of happiness is to impose morality? Are we to forget the entire moral mission of our nation through its history? Well, my answer, and I know it's yours, is no. America was founded on a moral proposition that human life—all human life—is sacred. And this proposition is the bedrock of our national life, the foundation of our laws. It's the wellspring of our Constitution. Courts may ignore it, and they have. They cannot—and I should add—have not denied it. When reverence for life can have no boundaries, when we begin to take some life casually, we threaten all life.[27]

According to Reagan, the activities of the antiabortionists do not impose morality but rather preserve the belief upon which the nation was founded, the proposition that "human life—all human life—is sacred." This preservation of rights, however, is in itself a moral mission. Reagan exploited the rhetorical power of the civil religion; he wanted to appear objective in the sense that he was not imposing morality but was holding fast to the tenets of the founders, yet he also wanted to have the absolute suasory power of religiously based morality backing him. Reagan also characterized the right to life that inheres in the Constitution as one that cannot be denied but only ignored; it is a natural right. According to Reagan, the founders also believed in the sacredness of all human life despite their inability to recognize the humanity of either women or slaves.

The above quotation not only reveals the basis of Reagan's anti-abortion morality but also discloses some of the tensions in Reagan's position. Despite the deontological assumptions present in his appeals to the civil religion, Reagan argues from teleological assumptions when he employs the "slippery slope" argument that "when we begin to take some life casually, we threaten all life." Reagan appealed not only to the rightness or wrongness of the act of abortion but also argued that allowing abortion will have bad effects.

This moral ambiguity is intensified by his treatment of the issue of "compelling state interest."

Weighing Competing Moral Rules Within the Civil Religion: Woman's Place

Reagan argued that the compelling state interest of protecting the "life of each person before birth"[28] could not be outweighed by the rights of women. The "woman's right to control her own body" was distinguished from a right to life and liberty. Apparently, physical integrity did not coincide with those individual rights that Reagan found valuable in a civic/sacred sense. Reagan's ability to define the right to life as greater than "woman's right to control her own body" was enhanced by his definition of "woman." Reagan defined women as childbearers, and hence, all who chose not to carry fetuses to term had no weight in a moral calculus.

Typically, antiabortion forces treated the fetus as an absolute. However, the Supreme Court has held that a balancing of interests is necessary. Accordingly, antiabortion advocates have responded by weighting the scales in the fetus's favor by magnifying the rights of the fetus and claiming to "overweigh" the rights of women; Celeste Michelle Condit, in *Decoding Abortion Rhetoric*, describes this process:

> This stark difference between the women's weighing of the fetus and that of the public discourse arose from the different meanings possible for the public term, *mother's life* and the private experience of *my life*. Because the public term *mother's life* had to cover so many women, it could only be reduced to the common denominator of physical survival. However, the women's version of the term *my life* meant identity—the "everything I was." Although for the women, this was necessarily a higher and more differentiated concern, from a male or institutional perspective, the woman's life was just an abstraction to be weighed against the equal abstraction of the fetus. To the extent that women were categorized *essentially* as childbearers, the weighting was necessarily tilted in favor of the fetus.[29]

The importance the undervaluing of women plays in this process highlights the need for women to expand and develop a discourse that translates issues of personal identity into public values. In order to do this, women must understand the subtlety of the process of role imposition. Reagan's discourse provides an example of the definitional process that delimits women's identity to that of mother, and hence excludes them from consideration within the civil religion.

Reagan has entrenched the process of overweighing, not only by magnifying the rights of the fetus but also by minimizing the rights of women. Reagan minimizes women's rights via a skillful manipulation of the meaning of *woman*. Throughout his discourse on abortion, women did not appear unless they played the role of mother. In his 1985 March for Life speech, no mention of women appeared unless "womb" is considered synonymous to woman. In 1986 Reagan spoke of Mother Teresa and "women who choose life" by seeking alternatives to abortion.[30] "Women who choose life" are also the only women mentioned in 1987; he said, "Each woman who chooses life for her child affirms our reverence for human life and enobles our society."[31]

The casting of women who choose to carry a fetus to term as "women who choose life" creates an interesting dichotomy between types of women. It appears that if a woman who has a child is a woman who chooses life, then the woman who chooses not to have a child chooses not-life. This places not only women who choose abortions, but all childless women into the realm of women who choose not-life. Another interesting aspect of the choose life/choose not-life distinction is its double meaning; not only may it refer to a woman's choice concerning a fetus, but it may also refer to her choice of lifestyle. If a woman chooses not to bear children under this calculus, then she chooses the oxymoronic life-of-death.

Expanding the options for women beyond choosers of life or choosers of death, Reagan provided a third alternative: victim. In his 1988 "March for Life" speech, he mentioned women more than in any other, but all the references were to women who had been victimized by the not-life choice. Before addressing those who had been victimized, however, Reagan dismissed all the women who knowingly chose to terminate their pregnancies by remarking: "We're told about a woman's right to control her own body. But doesn't the unborn child have a higher right, and that is to life, liberty and the pursuit of happiness?"[32] While the "higher" right of the unborn child may imply that some comparatively lesser right exists for women, that right is minimized both by Reagan's definition of *woman* and by his referring to the right as one about which the public is "told." Reagan implies that women's rights have only been claimed (in that we are told about that right) and have not been recognized; therefore, no weight needs to be attached to that right. Reagan also fails to specify who has "told" the public about the right. Without a name, and hence without credibility "for without a name it is difficult to accept the existence of an object,"[33] the agents who attach rights to women "tell" (as in a lie); they do not speak (as in the truth). In Reagan's calculus a woman has no actual right to control her body.

Women are victims, according to Reagan, because they have been led to believe that abortion "is their best, if not their only option."[34] According to Reagan, conflict-of-interest cases attributed to Title X funding, which supports family-planning and birth-control clinics, have caused unnecessary abortions. Some counseling programs that receive Title X funding have financial ties to programs that perform abortions and hence artificially increase the demand for abortion.[35] Thus, Reagan allowed that women who have had abortions are not *all* death choosers but merely may have had the choice of life taken away from them.

Reagan also limited the role of women in his book *Abortion and the Conscience of the Nation* by denying women any authoritative role in decision making concerning abortion. This essay is much longer than any of the rally speeches, yet it developed a role for women that is no more elaborate. All authorities quoted except three are men. At the beginning of the essay, Reagan referred to the position of Mother Teresa and Margaret Heckler, both saying that abortion is the number one problem facing the country; toward the end he quoted a young pregnant woman named Victoria who decided against abortion even though "everyone wanted me to throw away my baby."[36] Throughout the remainder of the essay, men were the only authorities; they included Jesse Helms, Henry Hyde, Roger Jepsen, John A. Bingham, William Brennan (not the justice), Abraham Lincoln, Jeremiah Denton, John Powell, Malcolm Muggeridge, and William Wilberforce. In Reagan's discourse women were not authorities on the sanctity of human life. While I believe that all human beings are capable of speaking about the value of life, I find it ironic that Reagan, who defined women as the choosers of life, relied almost exclusively on men to define what life means; men are the authorities on life and women are only the choosers of what has been defined as life.

Also noteworthy is Reagan's inconsistent use of gender-neutral language. In Reagan's discourse the child aborted is "he or she," yet the people who choose to abort are "some men." The most notable shift occurred in Reagan's conclusion: "Likewise, we cannot survive as a free nation when some *men* decide that others are not fit to live and should be abandoned to abortion or infanticide" (emphasis added).[37] To Reagan, a woman is not a decision-making agent unless she "chooses" life; yet that is not really a choice because "real" women always choose life.

Reagan further limited women's role by defining a pregnant

woman as a mother and only a mother. Once a woman becomes pregnant, she loses her identity as an individual woman and is exclusively a mother. She is named in terms of her relationship to a child, regardless of what else she might be. In Reagan's words, "Why else do we call a pregnant woman a mother?"[38]

This limited view of women is most extensively developed in (of all places) Reagan's 1987 Woman's Equality Day speech. In the opening paragraph Reagan praised the "women of strength and determination" who have been "indispensable to this Nation's progress."[39] Why were these women strong and determined? According to Reagan, because their faith in God, their trust in the promise of the New World, and their love for their families steeled them against the rigors of daily living in a harsh and untamed land."[40] Why were these women indispensable? Because they were faithful, trusting and loving—not the typical qualities of a person who ventures forth into uncharted territories to live a virtually isolated existence facing unknown natural forces and an uncertain future.

Reagan also rewrote the suffrage battle in his Women's Equality Day speech. He turned a long and hard-fought battle led and orchestrated by women into a simple reward readily handed over by the men in power. Reagan said, "In recognition of these immeasurable contributions and to redress the injustice of denying American women the right to vote, the Nineteenth Amendment was adopted in 1920 to guarantee political equality, the very bedrock of all rights and liberties, to American women."[41] Apparently, two bedrocks of national life exist: the sanctity of human life (as Reagan proclaimed in his 1988 March for Life speech) and political equality. What one does when the earth quakes because the rocks collide is unclear. (One hopes that public discourse is one mechanism for riding out the shocks.)

The use of the passive voice in the above quotation enabled Reagan to obscure the responsibility for moral action concerning the Nineteenth Amendment. The Nineteenth Amendment "was adopted in 1920" by an actor or actors who remain unnamed. This description of the granting of suffrage makes it sound as though the politicians in Congress simply decided that women deserved a reward. It ignores the long state battles and the fierce opposition that the suffragists fought to overcome.

Also in the Women's Equality Day speech, Reagan listed the things that equality serves: "healthy families, good neighborhoods, productive work, true peace, and genuine freedom."[42] The interplay between these goals rests in the move from concrete to abstract. To the extent that people are better able to identify with the concrete, one goes away from this speech feeling that women's equality serves the family and the neighborhood, not quite understanding how it affects work and peace or how equality is a means to freedom. Reagan defined women's equality in terms of how it better enables women to be traditional women—mothers, serving the family and the neighborhood.

The final definition of womanhood appears at the conclusion of the speech, where women are praised for "all they have done, as pioneers, patriots, parents and partners."[43] Remembering that women contributed as pioneers by being faithful, trusting, and loving, the picture left of women is that they have done so much because they are patriots for their fatherland, parents to their sons, and partners to their husbands. A woman by any other name is still defined by her closest male relative.

Reagan's overarching definition of womanhood, which infused all of his speeches that addressed issues affecting or concerning women, named women in terms of their relationship to men and children. Insofar as women were defined as childbearers, Reagan more easily could argue that women's rights were overweighed by the fetus's. If a woman is one who has children, then a woman's rights are those that allow her to fulfill her definition; a woman's right is to have children. Once one accepts this interpretation, abortion, instead of being a privacy or equality right *for* women, is a *violation of* women's right to bear children. This move enabled Reagan to remove women from consideration under the Constitution and to place the full force of the Constitution behind the claims for the fetus.

Reagan's moral ambiguity and his reliance on arguments based on the Constitution and Declaration of Independence were not accidental. Reagan could not venture into the land of the overtly religious lest he foresake the civil religion; however, he could turn the Constitution and Declaration of Independence into a religion, or at least into documents sanctioned by religion. In his role as public priest, Reagan could preach the gospel of rights and not violate the separation of church and state. The reliance on rights language also emphasized Reagan's role as chief enforcer of the laws. Reagan was not speaking out against the "law" of *Roe* v. *Wade* but rather was defending the Constitution and Declaration of Independence as the higher law governing the situation. He then interpreted the dictates of the Constitution and Declaration of Independence as prohibiting abortion. In other words, Reagan substituted his interpretation of the Constitution for the Supreme Court's interpretation.

Insofar as one's moral character, under a deontological theory, is determined by adherence to rules, one is moral (under the civil religion) as long as one follows the rules set forth by the Constitution and Declaration of Independence. Reagan estabished the immorality of abortion by weighting the rules; he employs the first-order rule that all life is sacred, while assuming a second-order rule, that "one should err on the side of life," to be unquestionable. Reagan was able to present the second-order rule as unquestionable because no rules conflict existed between the fetus and the woman.

Although the use of the Constitution and Declaration of Independence as the foundation for his antiabortion arguments did not violate, and in fact may have strengthened, his role as public priest and secular enforcer of the laws, it was incomplete as an argument. In order to prove that the power of the sacred texts of the civil religion had to be exercised, Reagan needed to establish a violation of them; he had to prove that abortion violates the prohibition on killing based on the right to life, liberty, and the pursuit of happiness.

Establishing Abortion's Civil Sacrilege: The Use of Definitional Argument

If antiabortionists are to convince the confused, uncommitted, and prochoice that abortion is wrong, they must first prove the initial claim that the fetus is a human being. For antiabortionists, proper moral reasoning is definitional.[44] Antiabortionists know that the fetus is a human being by way of a universal. How does one discover this universal? Antiabortionists typically refer to positive law, which reflects morality. However, in this case, the positive law of *Roe* v. *Wade* does not reflect morality.[45] The antiabortionists then refer to intuition; for example, they appeal to conscience by referring to the guilt women felt after abortion as evidence of the power of intuitive conscience.[46]

Reagan approached the problem of moral reasoning by searching for proof of the humanity of the fetus in positive law as represented by the laws of the land. Hence, he appeals to the court's conscience by arguing that the court can ignore the right to life but cannot deny it. The Constitution and Declaration of Independence are more authoritative, and they establish the right to life. Reagan became not only the chief enforcer of the laws but also their primary interpreter.

Reagan also attempted to prove the humanity of the fetus without relying on religiously based arguments. In both 1985 and 1988 he used medical advances to establish the personhood of the fetus. In 1985 he said that "advances in medical technology have changed the debate" in the sense that we now speak of the "patient in the womb."[47] He also used anecdotal evidence that women are urged to play Mozart to soothe the unborn.[48] In 1988 he concluded: "Isn't there enough evidence for even skeptics to admit that those who assert the personhood of the fetus may be right? And if we are to err, shouldn't it be on the side of life? I believe it's time the law caught up with science."[49] The reliance on scientific data was a strategic choice. Law did not need to catch up with morality; rather, law needed to reflect the secular truths of science.

Reagan also established a criterion for balancing interests when making public policy. He argued that if doubt exists concerning the humanity of the fetus, then people should err "on the side of life," which he equated with the fetus. For example, in 1985 he argued, "We must not forget that in reality, if there is any justice in the abortionist position, it would require that they establish beyond a doubt that there is not life in the unborn—and they can't do that."[50] Reagan did not see a need to recognize the corollary responsibility that he must prove, beyond a doubt, that banning abortion will not harm women. Again, the rights of the fetus were given more weight than the rights of the woman. Reagan also defined the prochoice position as untenable; in this same passage, he established a standard of proof necessary for argument and then proclaimed that the standard of proof could not be met by the "abortionists."

Reagan also reinforced the notion of fetus-as-person by *not* referring to it as a fetus but rather by describing it in terms that grant it humanity, for example, "the personhood of the fetus." Insofar as humanity is a social construct, Reagan made the fetus human by referring to it as human. For example, in 1985 he described the fetus with phrases like the "unborn," "human life," "unborn child," "living human being," "life of each person before birth"; in 1986 with "human life," "a very young, very small, dependent and very vulnerable live member of the human family," "child about to be born," "child"; in 1987 with "America's unborn children," "young lives," "human life," "unborn"; in 1988 with "life of infants before birth," "unborn child," "human life," "unborn," and "personhood." Also, women "who choose life" choose to carry a fetus to term. The fetus is equated with life; the fetus is life.

Reagan also defined the fetus as an entity separate from the mother. Present in Reagan's description of the fetus was the notion of the public fetus; the fetus was one of "America's unborn children"[51] and a "member of the human family."[52] The fetus was not of the mother; the fetus was of the society. The public fetus is prevalent in antiabortion discourse. For example, when pictures of fetuses are shown, all one sees is the fetus; the fact that the fetus exists inside a woman is excised from the picture. The context of the fetus is the world; the context is not a woman.

The notion of the "public fetus" was developed by Barbara Duden in her paper "The Pregnant Woman and the Public Fetus." She argues that "signs of pregnancy [have] become symptoms for fetal life," an occurrence that is historically unique.[53] The fetus is separated from the body of the woman and granted a "life" of its own. Things that scientists considered tissue now become the upper and lower parts of a baby when magnified and viewed on a television screen, but it is not advanced scientific knowledge that calls for an antiabortionist to name a genome a life but is instead the dissociation from the scientific sphere and the entrance into the public realm that enables the tissue to become a child. Duden's is not an argument for ignorance; instead, she demonstrates how the scientific and quasi-scientific meaning of physical objects is socially constructed and changed.[54]

The rhetorical implications of this construction of the fetus are manifold. If the fetus is self-contained, then arguments concerning the competing interests of women become irrelevant. There is no other right against which the rights of the fetus need to be weighed. Additionally, the fetus is contextualized not as the child of a woman, but rather as a child of the world. The public is its caretaker; it is as though all of the potentially aborted fetuses have already been adopted, the preferred alternative to abortion. Therefore, the public is granted the right to decide the fetus's future, and no public would want to kill a helpless baby. The decision to disallow abortion becomes easier. One is asked if one wants to harm the public "baby," and one can more easily answer no if one need not consider the woman carrying the baby. The decision on abortion belongs to the public because the baby is of the public. Any decision-making role for women whose menses stops because of pregnancy is disregarded. The public is given the right to choose because the baby in question belongs to them; women are given no right to choose because the baby is no longer theirs.

In order to place the moral power of the civil religion's sacred documents behind the fetus, Reagan also equivocally used the phrase "right to life." In Reagan's discourse, "right to life" meant not only the right to keep the life one has but also the chance to have a life; it referred both to actual and to potential life. While antiabortionists may take exception to this idea of a right to live, I think that one hears the right to a chance at life in Reagan's rhetoric. Although only implicitly present in his March for Life speeches, the idea that all should have a chance to live is explicit in other antiabortion statements. In his remarks to the Annual Conservative Political Action Conference Dinner on 2 March 1984, he described abortion as a "national tragedy" in which "15 million children have been lost—15 million children who will never laugh, never sing, never know the joy of human love, will never strive to heal the sick or feed the poor or make peace among nations. They've been denied the most basic of human rights, and we're all the poorer for their loss."[55] The fetus is not able to experience those things that human beings experience as part of their lives; it appears that the fetus is not alive but is denied the right *to* live. In the same statement, he described life as an "opportunity" denied to the fetus, again intimating that the fetus is not denied a right already bestowed but is denied a chance to have the right. On 17 October 1988, in a speech to the students and faculty of Archbishop Carroll and All Saints High School, Reagan asked the audience to "imagine what so many, denied the right to life, have missed. Imagine what we've all missed for their absence."[56] Here, the right to life is not something the fetus already has, which is what the right to life means to the antiabortionists. Instead, life is something the fetus "misses." Reagan does not describe the fetus as alive and deserving to continue living; rather, the right to life is the right to be born.

This argument provides another means by which Reagan may tilt the scales. Even if the audience does not believe that a fetus is alive, Reagan's "right to life" equates the value of the right-to-continue-living with the right-to-have-a-life. The reasons given for why each person has a right-to-continue-living are the same reasons Reagan offers for why a fetus should have a right to live. Reagan de-emphasizes the ambiguities involved in determining the fetus's rights and highlights the ambiguities involved in determining the rights of women to control their bodies.

The definitional method of argument as used by Reagan to establish the humanity of the fetus is not neutral. Lake argues that its use by antiabortion rhetors eliminates the possibility of neutrality and dispute; it is impossible to resolve a definitional argument based on conscience.[57] Reagan, however, *appeared* to have held out the possibility of argument. Throughout his speeches, Reagan presented "proof" of the humanity of the fetus by arguing that it is a patient, feels pain, and likes music. Yet in 1986 Reagan argued, "Each year remarkable advances in prenatal medicine bring even more dramatic confirmation of what common sense has told us all along: that the child in the womb is [a] . . . live member of the human family."[58] However, Reagan actually foreclosed argument by defining his "proof" as merely reinforcing what people already *knew* via common sense, common sense being a necessary and sufficient proof of humanity. If his scientific proofs should be disproven or discounted, he could always revert to common-sense arguments.

Reagan's standards for evidence presented prochoice factions with a difficult rhetorical problem. As discussed earlier, Reagan established the argumentative standard for prochoice groups that they "establish beyond a doubt that there is not life in the unborn" in order to grant justice to their position.[59] Now it appears that if prochoice groups met his standard of proof, Reagan could still argue that they are wrong because common sense does not support their claims. Reagan's form of definitional argument relied on two forms of proof: scientific and common sense. While the support of either one of these proofs was sufficient to establish his position, opponents must garner the support of both merely to be considered defending a *just* position.

Even though one could assert that Reagan's proof of humanity also proved that dogs are human (they are considered patients, they feel pain, and they are soothed by music), this assertion is irrelevant to Reagan because he could revert to arguments based on common sense. This subversion of the debate process was demonstrated by Reagan's ingenuous appeal for understanding. He stated: "A year ago, in my State of the Union Address, I called on everyone in our country to rise above bitterness and reproach and seek a greater understanding of this issue. I believe that spirit of understanding begins with the recognition of the reality of life before birth and the reality of death by abortion."[60] Echoing Nixon's first inaugural promise of understanding protestors as long as protesters lowered their voices and hence adopted the standards of the establishment, Reagan promised understanding only if it was based on the opposition's surrender. Prochoice advocates had to admit the "truth" of the antiabortionists' underlying premise and accept the moral condemnation attached to it: the fetus is alive and abortion is murder. Reagan effectively precluded debate and compromise with the definitional constraints he placed on deliberation.

Locating the Citizen in Reagan's Civil Religion: The Constitution of the Congregation

A religion not only has moral rules, but it also has a congregation bound by those rules. Reagan's civil religion is no exception. But, while Reagan possessed an optimistic outlook for America, typical antiabortion discourse paints a very bleak picture of the moral status of human beings. Lake explains: "Anti-abortionists view human beings as at best weak, selfish, and callous, and at worst maliciously immoral."[61] This lends credence to the antiabortionists belief that if some "murder" is allowed, human beings will begin to commit murder of other types (infanticide and euthanasia) until

Nazi death camps reappear. Reagan accepted this premise to a certain extent; he did accept the argument that if abortion was allowed, and "we begin to take some life casually, we threaten all life."[62] However, the agents that threatened life were not weak, selfish, callous and immoral human beings, but rather were the evil institutions and programs that enabled abortions. Reagan removed human agency from the decision to abort. Reagan altered the agent involved in abortion in order to maintain the consistency of his worldview.

Reagan's optimistic outlook for America did not allow him to rhetorically accept the antiabortionists' negative view of humanity. According to Craig Allan Smith, Reagan created a "neighborhood" made up of "extraordinary ordinary Americans" who live out a proud heritage created by their moral purpose and who believe in heroism, faith and "can-do Americanism"; the United States possesses a self-evident morality.[63]

Reagan's abortion rhetoric was consistent with his neighborhood. In Reagan's world, a few "bad" entities exist, but only in an institutional form; whenever Reagan spoke of human agency as producing evil, that agency was contained in some institutionalized form, and despite these institutions' activities, the "noble marches" of the antiabortionists will prevail.[64] The noble marches, in many ways, offered the civic salvation of a nation wounded by abortion. Accordingly, Reagan could maintain his optimistic outlook, even in light of institutional evil, by recognizing the possibility for salvation contained in the marches and other activities of antiabortionists.

One example of civic salvation was provided in 1986, when Reagan praised the activities of women who care for other women who choose to carry a fetus to term. He argued that these women were "testimony to our reverence of human life. Each child about to be born is a unique, unrepeatable gift. Each child who escapes the tragedy of abortion is an immeasureable victory."[65] In 1988 Reagan attacked rather than praised, focusing on some disembodied other who used "coercion" to motivate a young pregnant woman to abort.[66] Reagan argued that women are not given a real choice by the system because abortion is presented as the best option by Title X programs. He explained: "Too often, the same Title X funded programs that give referrals have financial ties to programs that perform abortions. In practice, young women using their services have sometimes been led to believe that abortion is their best, if not their only option."[67] He then quoted a young woman who retrospectively saw her decision to abort as "no decision at all. It was a coercion."[68] It is unclear whether Reagan would ever see any decision to abort as the best solution; therefore the percentage of women who were not given a "real" choice is unclear.

In his description of coercion, Reagan did not cast an obvious human agent as the coercer. Rather, the enemy was the institutional form that received government funds and had ties to abortion clinics. This was a potentially powerful argument for two reasons.

First, it recognized the prochoice advocates' emphasis on choice and transformed it into an argument for the antiabortion side. Because abortion services are sometimes tied to counseling centers, the abortion decision is forced on women. Removing funding, thus removing the tie, resulted in "real" choice. The failure of the clinic may also result, but that is irrelevant to the choice issue. However, this attempt to co-opt the choice argument, in the final analysis, was not rhetorically successful because Reagan defined the only real choice as the choice of life. Yet in order to respond to the prochoice argument, he did not need to argue that absolute choice should be given to women or that his position offered choice. He may have sought only to limit the suasory power of the prochoice position by delimiting the degree of choice possible in a prochoice world.

Second, casting the enemy as the institutional form that mixed government funds with abortion clinics made the culprit in abortion the institutional structure, not a human being. Accordingly, Reagan could maintain the goodness of Americans while condemning abortion. Additionally, the choice of some Americans was denied not by other Americans but by institutions that grew beyond human control. Locating the evil in institutions also appealed to the democratic myth that government should reflect the will of the people. A chance at salvation remained.

Reagan also exonerated his Americans by blaming their inaction on ignorance rather than on evilness. In the book *Abortion and the Conscience of the Nation*, he argued that people are not motivated to fight against abortion because they do not completely understand the issue. He wrote: "But the great majority of the American people have not yet made their voices heard, and we cannot expect them to—any more than the public voice arose against slavery—*until* the issue is clearly framed and presented."[69]

Even though the public has allowed abortion to occur, its members are not accomplices. Reagan continued, "As a nation today, we have *not* rejected the sanctity of human life. The American people have not had an opportunity to express their view on the sanctity of human life in the unborn."[70] One should note that "view" is singular, indicating that a unified and coherent perspective on abortion is held by the American public. One also should note that Reagan knows what that view is; it is his. Again echoes of Nixon reverberate in Reagan's words; Reagan knows the mind of America, even though America's majority is silent.

Conclusion

By examining Reagan's rhetoric in light of the rhetoric of the antiabortion movement, one comes to understand how presidents employ civic-religious arguments to reconcile their office and their worldview with political positions, such that the president may rally support while not violating the dictates of the office. Reagan was the secular, public priest with the Constitution and Declaration of Independence as his Bible. Reagan's arguments had the persuasive power of religious appeals because of reliance upon notions of civic piety; but by not directly referring to sacred religion, and instead by relying on the civil religion, Reagan could garner the power of universal moral discourse while not appearing partisan, a tactic also used by antiabortion advocates.[71] Reagan's ability to claim the persuasive power of secular universals, which tended to mirror closely religious universals, was enabled by his use of the Constitution and the Declaration of Independence as sacred texts.

Notes

Introduction: Rhetorical Analysis

1. Giambattista Vico, *The New Science of Giambattista Vico,* trans. T. G. Bergin and Max H. Fisch (Ithaca, N.Y.: Cornell University Press, 1984); and Rom Harre, "A Metaphysic for Conversation: A Newtonian Model of Speech-Acts in People-Space," *Research on Language and Interaction* 22 (1988–89): 1–22.

2. Robert Dallek, *Ronald Reagan: The Politics of Symbolism* (Cambridge, Mass.: Harvard University Press, 1984); and John L. Palmer and Isabel V. Sawhill, eds., *The Reagan Record* (Cambridge, Mass.: Ballinger, 1984).

3. Peggy Noonan, *What I Saw at the Revolution: A Political Life in the Reagan Era* (New York: Random House, 1990), 263–71.

4. Nathaniel Nash, "Saving Crisis Politics," *New York Times,* 21 June 1990, D1 and D19.

5. Francis Fox Piven and Richard A. Cloward, *The New Class War* (New York: Pantheon, 1985).

6. Paul Kennedy, *The Rise and Fall of the Great Powers* (New York: Random House, 1987), 531–33.

7. Emil Arca and Gregory Pamel, eds., *The Triumph of the American Spirit: The Presidential Speeches of Ronald Reagan* (Detroit: National Reproductions, 1984).

8. Mary McGrory, "Gracious, Not Great," *Boston Globe,* 17 July 1990, 15.

9. Dennis Gilbert, *Sandinistas* (New York: Basil Blackwell, 1988), 164–68.

10. Theodore Draper, "Revelations of the North Trial," *New York Review of Books*, 17 August 1989, 54–59.

11. Paul Hernnson, "Assessing the Reagan Presidency," *Polity* 21 (Summer 1989): 809–19; and Barry Sussman, *What Americans Really Think* (New York: Basic Books, 1988).

12. David Johnston, "Reagan Is Said to Have Testified He Didn't Authorize Lawbreaking," *New York Times*, 18 February 1990, 28.

13. Noonan, *What I Saw*, 149–85.

14. Roderick P. Hart, *The Sound of Leadership: Presidential Communication in the Modern Age* (Chicago: University of Chicago Press, 1987).

15. Karlyn Kohrs Campbell and Kathleen Hall Jamieson, *Deeds Done in Words: Presidential Rhetoric and the Genres of Governance* (Chicago: University of Chicago Press, 1990).

16. Kathleen Hall Jamieson, *Eloquence in an Electronic Age: The Transformation of Political Speechmaking* (New York: Oxford University Press, 1988).

17. Michael Paul Rogin, "Ronald Reagan: 'The Movie,'" in *Ronald Reagan: The Movie, and Other Episodes in Political Demonology* (Berkeley and Los Angeles: University of California Press, 1987), 3–44.

18. Hernnson, "Assessing the Reagan Presidency," 809.

19. Arthur M. Schlesinger, Jr., *Cycles of American History* (Boston: Houghton Mifflin, 1986).

20. Tom Wicker, "The Other Guy Did It," *New York Times*, 16 July 1990, A15.

Chapter 1. Ceremonial Discourse

1. We use the terms *inventional* and *noninventional* in the classical sense. *Invention* describes the element of rhetoric comprising explicit appeals and their evidentiary support. These include Aristotle's three modes of proof; that is, invention can cover appeals to reason (logos), to the emotions (pathos), and to the personal credibility of the speaker (ethos). See *The Rhetoric and The Poetics of Aristotle*, trans. W. Rhys Roberts (New York: Modern Library, 1954). Noninventional rhetorical elements can include what is often referred to as style; that is, diction, the use of tropes and figures, rhythm, and the like, as well as the delivery skills of the speaker. Our term *ceremonial* again comes from Aristotle, this time from his classification of speaking situations. Ceremonial or epideictic situations were those in which a speaker praised or blamed someone or something, usually in ways that appealed to and reinforced an audience's preexisting ideological inclinations. Artistotle distinguished epideictic from deliberation (the determination of public policy in the legislature) and forensic (the determination of guilt or innocence in the law courts).

2. The phrase is Kathleen Jamieson's. See *Eloquence in an Electronic Age: The Transformation of Political Speechmaking* (New York: Oxford University Press, 1988), 76.

3. Bill Moyers, *The Public Mind: Leading Questions* (New York: Journal Graphics, 1989), 15.

4. Murray Edelman, *Political Language: Words That Succeed and Policies That Fail* (New York: Academic Press, 1977).

5. *Ronald Reagan: 1986*, vol. 2, Public Papers of the Presidents of the United States (Washington D.C.: U.S. Government Printing Office, 1989), 1575.

6. Ibid.

7. Michael Oreskes, "Study Finds Astonishing Indifference to Elections," *New York Times*, 6 May 1990, 32.

8. Goren Therborn, *The Ideology of Power and the Power of Ideology* (London: Verso, 1980), 94–98.

9. See Jurgen Habermas, "The Public Sphere: An Encyclopedia Article," in *Critical Theory and Society*, ed. Stephen Eric Bronner and Douglas McKay Kellner (New York: Routledge, 1989), 136–42; and Richard Sennett, *The Fall of Public Man* (New York: Vintage Books, 1978).

10. Michel Foucault, "The Order of Discourse," in *Language and Politics*, ed. Michael Shapiro (New York: New York University Press, 1984).

11. Ibid., 112–13.

12. Antonio Gramsci, *Selections from the Prison Notebooks*, ed. Quintin Hoare and Geoffrey Nowell Smith (New York: International, 1971), 450.

13. J. G. A. Pocock, "Verbalizing a Political Act: Towards a Politics of Speech," *Political Theory* 1 (1973): 27–43.

14. Jurgen Habermas, "The Scientization of Politics," in *Toward a Rational Society* (Boston: Beacon Press, 1971), 63–75.

15. Murray Edelman, "The Political Language of the Helping Professions," *Politics and Society* 4, no. 3 (1974): 296–97.

16. See J. L. Austin, *How to Do Things with Words* (Cambridge, Mass.: Harvard University Press, 1975).

17. Emil Arca and Gregory Pamel, eds., *The Triumph of the American Spirit: Presidential Speeches of Ronald Reagan* (Detroit: National Reproductions, 1984), 5.

18. Ibid., 4.

19. Ibid., 6.

20. "'Checkers' Speech," in Arthur M. Schlesinger, Jr., *History of American Presidential Elections: 1940–1968* (New York: McGraw-Hill, 1971).

21. *Ronald Reagan: 1984*, vol. 1, Public Papers of the Presidents of the United States (Washington, D.C.: U.S. Government Printing Office, 1986), 172–73.

22. *Ronald Reagan: 1981*, Public Papers of the Presidents of the United States (Washington, D.C.: U.S. Government Printing Office, 1982), 202.

23. *Ronald Reagan: 1982*, vol. 1, Public Papers of the Presidents of the United States (Washington, D.C.: U.S. Government Printing Office, 1983), 151.

24. Fred Block et al., *The Mean Season* (New York: Pantheon, 1987), xi–xii.

25. See Frances Fox Piven and Richard A. Cloward, *The New Class War* (New York: Pantheon, 1985), 159–63.

26. For an insightful analysis of the use of narrative in Reagan's rhetoric,

see William F. Lewis, "Telling America's Story: Narrative Form and the Reagan Presidency," *Quarterly Journal of Speech* 73 (1987): 280–302.

27. See *Ronald Reagan: 1982,* Public Papers of the Presidents, 1:144.

28. Ibid., 258.

29. *Ronald Reagan: 1981,* Public Papers of the Presidents, 557.

30. *Ronald Reagan: 1984,* vol. 2, Public Papers of the Presidents of the United States (Washington, D.C.: U.S. Government Printing Office, 1987), 1794.

31. Ibid.

32. Ibid., 1793.

33. *Ronald Reagan: 1982,* Public Papers of the Presidents, 1:147.

34. See Michael Calvin McGee and Martha Anne Martin, "Public Knowledge and Ideological Argumentation," *Communication Monographs* 50 (1983): 47–65.

35. Nicholas Jay Demerath and Rhys Williams, "Civil Religion in an Uncivil Society," *Annals of the American Academy* 480: 154–66.

36. For a useful discussion of the distinction between "America" as an idea at best only partially captured in the term *United States,* see Lester D. Langley, *America and the Americas: The United States in the Western Hemisphere* (Athens: University of Georgia Press, 1989).

37. This heritage raises particularly interesting questions for religion and politics in the twenty-first century. The long-term trend in U.S. history is clear: disestablishment. First, the de facto establishment of Protestantism was undone by including Catholics; then Judaism's inclusion disestablished Christianity. The third disestablishment, currently in progress, reflects what some scholars have described as a "turn East," and others as "desecularization." Either way, the hegemony of the Judeo-Christian tradition may well be replaced by institutionalizing a far more pluralistic religious environment. See Wade Clark Roof and William McKinny. *American Mainline Religion* (New Brunswick, N.J.: Rutgers University Press, 1987). The effects of the third disestablishment are hard to foresee, but Ronald Reagan's frequent use of Judeo-Christian formulations may be seen as a reaction against it.

38. *Ronald Reagan: 1982,* Public Papers of the Presidents, 1:147.

39. Arca and Pamel, *Triumph,* 8.

40. *Ronald Reagan: 1984,* Public Papers of the Presidents, 2:1796–97.

41. *Ronald Reagan: 1982,* Public Papers of the Presidents, 1:467.

42. *Ronald Reagan: 1987,* vol. 1, Public Papers of the Presidents of the United States (Washington, D.C.: U.S. Government Printing Office, 1989), 209.

43. For a provocative discussion of the sociopolitical implications of this emphasis on the inner rather than the outer self, see Christopher Lasch, *The Culture of Narcissism* (New York: Warner Books, 1979).

44. *Ronald Reagan: 1982,* Public Papers of the Presidents, 1:467.

45. Arca and Pamel, *Triumph,* 120.

46. *Ronald Reagan: 1984,* Public Papers of the Presidents, 2:1797.

47. *Ronald Reagan: 1983,* vol. 2, Public Papers of the Presidents of the

United States (Washington, D.C.: U.S. Government Printing Office, 1985), 1522.

48. See William Connolly, *The Terms of Political Discourse* (Princeton, N.J.: Princeton University Press, 1983), 9–44.

49. See Waltraud Morales, "The War on Drugs: A New U.S. National Security Doctrine?" *Third World Quarterly* 11 (1989): 147–69.

50. Michael Shapiro, *The Politics of Representation* (Madison: University of Wisconsin Press, 1988), 17.

51. Ibid.

52. *Ronald Reagan: 1986,* vol. 1, Public Papers of the Presidents of the United States (Washington, D.C.: U.S. Government Printing Office, 1988), 272.

53. *Ronald Reagan: 1986,* Public Papers of the Presidents, 2:1181.

54. David Landau, *The Dangerous Doctrine: National Security and U.S. Foreign Policy* (Boulder, Colo.: Westview Press, 1988), xii and 2–4.

55. *Ronald Reagan: 1983,* Public Papers of the Presidents, 2:1521.

56. "Central America and U.S. Security," *U.S. Department of State Bulletin,* May 1986, 28.

57. See especially Theodore Draper's analysis of Oliver North's rationale for his actions in the Iran-Contra affair. "Revelations of the North Trial," *New York Review of Books,* 17 August 1989, 54–59.

58. Morales, "The War on Drugs," 151.

59. Kenneth Burke, "The Rhetoric of Hitler's Battle," in *The Philosophy of Literary Form* (Berkeley and Los Angeles: University of California Press, 1973), 191.

60. *Ronald Reagan: 1981,* Public Papers of the Presidents, 193.

61. *Ronald Reagan: 1984,* Public Papers of the Presidents, 2:174.

62. *Ronald Reagan: 1986,* Public Papers of the Presidents, 2:1181.

63. *Ronald Reagan: 1981,* Public Papers of the Presidents, 559.

64. Ibid., 560.

65. See Michael Calvin McGee, "The Ideograph: A Link Between Ideology and Rhetoric," *Quarterly Journal of Speech* 66 (1980): 1–16.

66. Quoted in Moyers, *The Public Mind: Leading Questions,* 13.

67. Jamieson, *Eloquence in an Electronic Age,* 172.

68. Ibid., 84.

69. Ibid., 122.

70. Ibid., 76.

71. Ibid., 89.

72. Ibid., 81.

73. Quoted in Bill Moyers, *The Public Mind: Consuming Images* (New York: Journal Graphics, 1989), 2.

74. Quoted in Bill Moyers, *The Public Mind: Illusions of News* (New York: Journal Graphics, 1989), 5.

75. Ibid.

76. Quoted in Moyers, *The Public Mind: Leading Questions,* 5–7.

77. Ibid., 12.

78. Though President Bush and his party no longer attack domestic

spending initiatives with Reagan's fervor, we should remember that the budget crisis induced by Reaganomics has effectively "defunded" such initiatives for the foreseeable future. Thus, Reagan's policy legacy is substantial even if no longer embraced. See Daniel Patrick Moynihan, "Another War—the One on Poverty—Is Over Too," *New York Times*, 16 July 1990, A15.

79. Quoted in Moyers, *The Public Mind: Leading Questions*, 9–10.

80. Richard Gephardt, "How Dare I Criticize The President?" *Daily Hampshire Gazette*, 24 March 1990, 6.

Chapter 2. Reagan at the Moscow Summit

1. Phil Williams, "Carter's Defense Policy," in *The Carter Years: The President and Policy Making*, ed. Glen Abernathy, Dilys M. Hill, and Phil Williams (New York: St. Martin's Press, 1984), 84–105; Earl C. Ravenal, "Reagan's Failed Restoration: Superpower Relations in the 1980s," in *Assessing the Reagan Years*, ed. David Boaz (Washington, D.C.: Cato Institute, 1988), 125–42; Gordon R. Weimiller, *U.S.-Soviet Summits: An Account of East-West Diplomacy at the Top, 1955–1985* (New York: University Press of America, 1986), 109–10; Strobe Talbot, *Endgame: The Inside Story of SALT II* (New York: Harper and Row, 1979), 288–94.

2. "The President's News Conference, January 29, 1981," in *Ronald Reagan: 1981*, Public Papers of the Presidents of the United States (Washington, D.C.: U.S. Government Printing Office, 1982), 57. Reagan had defined his anti-Soviet themes in his 1976 and 1980 bids for the presidency. See Peter Hannaford, *The Reagans: A Political Portrait* (New York: Coward-McCann, 1983), 94–96, 287–88. Anti-Soviet themes were translated into a large defense budget modernizing the United States nuclear arsenal.

3. *Ronald Reagan: 1981*, Public Papers of the Presidents, 57. On these issues see, for example: Michael R. Gordon, "Weinberger's War Readiness Claims Spark Controversy Within the Pentagon," *National Journal* 16, no. 2 (1984), 1120–23; Daniel O. Graham, "National Defense: The Strategic Framework," in *The Future Under President Reagan*, ed. Wayne Valis (Westport, Conn.: Arlington House, 1981), 137–48; Richard Halloran, "Pentagon Draws Up First Strategy for Fighting a Long Nuclear War," *New York Times*, 30 May 1982; Jeff McMahan, *Reagan and the World: Imperial Policy in the New Cold War* (London: Pluto Press, 1984). Walter Isaacson says that Reagan "disparaged détente. He criticized arms control. He assaulted the three-decade-old doctrine of nuclear deterrence" (Plus Ça Change . . ." *Time*, 6 June 1988, 12).

4. Richard Smoke, *National Security and the Nuclear Dilemma: An Introduction to the American Experience* (New York: Random House, 1984), 241.

5. Ravenal, "Reagan's Failed Restoration," 134, 138. On the doctrine of massive retaliation see Lawrence Freedman, *The Evolution of Nuclear Strategy* (New York: St. Martin's Press, 1981), 76–90.

6. "Remarks at the Annual Convention of the National Association of Evangelicals in Orlando, Florida, March 3, 1983, *Ronald Reagan: 1983,* vol. 1, Public Papers of the Presidents of the United States (Washington, D.C.: U.S. Government Printing Office, 1984), 359–64. Though this address states Reagan's anti-Soviet views most boldly, it is not out of character. See Robert Dallek, *Ronald Reagan: The Politics of Symbolism* (Cambridge, Mass.: Harvard University Press, 1984), 130–31; G. Thomas Goodnight, "Ronald Reagan's Reformulation of the Rhetoric of War: Analysis of the 'Zero Option,' 'Evil Empire,' and 'Star Wars' Addresses," *Quarterly Journal of Speech* 72 (1986): 390–414.

7. Rowland Evans and Robert Novak, *The Reagan Revolution* (New York: E. P. Dutton, 1981), 157–61; Robert Tucker, "Reagan's Foreign Policy," in *The Reagan Foreign Policy,* ed. William G. Hyland (New York: New American Library, 1987), 1–27; David Mervin, *Ronald Reagan and the American Presidency* (New York: Longman, 1990), 166.

8. See, for example, Adam M. Garfinkle, *The Politics of the Nuclear Freeze* (Philadelphia: Foreign Policy Research Institute, 1984); Edward M. Kennedy and Mark D. Hatfield, *Freeze: How You Can Help Prevent Nuclear War* (New York: Bantam Books, 1982).

9. Bob Schieffer and Gary Paul Gates, *The Acting President* (New York: E. P. Dutton, 1989), 328; Mervin, *Reagan and the Presidency,* 168; I. M. Destler, "The Evolution of Reagan Foreign Policy," in *The Reagan Presidency: An Early Assessment,* ed. Fred I. Greenstein (Baltimore: Johns Hopkins University Press, 1983), 117–68; Strobe Talbot, *Deadly Gambits: The Reagan Administration and the Stalemate in Nuclear Arms Control* (New York: Knopf, 1984).

10. Mervin, *Reagan and the Presidency,* 165. See also *Reagan: The Next Four Years* (Washington, D.C.: Congressional Quarterly, 1985), 43; Weimiller, *U.S.-Soviet Summits,* 111–12.

11. Ronald Reagan, "Address to the Nation on Defense and National Security, March 23, 1983," in *Ronald Reagan: 1983,* Public Papers of the Presidents, 437–43; David Holloway, "The Strategic Defense Initiative and the Soviet Union," *Daedalus,* special issue, *Weapons in Space,* vol. 2, *Implications for Security* 114 (1985): 265–68.

12. Lawrence T. Caldwell, "Washington and Moscow: A Tale of Two Summits," *Current History* 87 (October 1988): 306.

13. Paul Marantz, "Soviet 'New Thinking' and East-West Relations," *Current History* 87 (October 1988): 309. On suspicions see Gerhard Wetting, "'New Thinking' on Security and East-West Relations," *Problems of Communism* 37 (March–April 1988): 1–14; Jean Quatras [Pseud.], "New Soviet Thinking Is Not Good News," *Washington Quarterly* 11 (Summer 1988): 171–83; Robert Legvold and the Task Force on Soviet New Thinking, "Gorbachev's 'New Thinking,' in *Gorbachev's Foreign Policy: How Should the United States Respond?,* Headline Series, no. 284 (New York: Foreign Policy Association 1988), 7–30. Gorbachev not only was facing suspicions in the United States, but also was working toward reform in the Soviet Union. Part of the immediate exigence of the trip was a meeting of the Soviet Congress scheduled for June.

14. Sidney Blumenthal, *Our Long National Daydream: A Political Pageant of the Reagan Era* (New York: Harper and Row, 1988), 307–27.

15. Russell Watson, Robert B. Cullen, and Thomas M. DeDrank, "Reagan in Moscow," *Newsweek*, 6 June 1988, 16–19; Alex Beam, "A Superpower Script of Pomp over Substance," *Boston Globe*, 2 June 1988, 1, 11.

16. Kenneth Burke, *A Rhetoric of Motives* (Berkeley and Los Angeles: University of California Press, 1969), 208. John L. Gaddis writes of the "rules" that "establish limits of acceptable behavior on the part of nations who acknowledge only themselves as the arbiters of behavior."

> These "rules" are, of course, implicit rather than explicit: they grow out of a mixture of custom, precedent, and mutual interest that take shape quite apart from the realm of public rhetoric, diplomacy, or international law. They require the passage of time to become effective; they depend, for that effectiveness, upon the extent to which successive generations of national leadership on each side find them useful. They certainly do not reflect any agreed-upon standard of international morality: indeed they often violate principles of "justice" adhered to by one side or the other. But these "rules" have played an important role in maintaining the international system that has been in place these past four decades: without them the correlation one would normally anticipate between hostility and instability would have become more exact than it has in fact been since 1945. [*Long Peace: Inquiries into the History of the Cold War* (New York: Oxford University Press, 1987), 242]

Gaddis lists the implicit rules: (1) respect spheres of influence; (2) avoid direct military confrontation; (3) use nuclear weapons only as an ultimate resort; (4) prefer predictable anomaly over unpredictable rationality; (5) do not seek to undermine the other side's leadership.

Rhetoric both tests and creates implicit rules through symbols and gestures of adherence and criticism. During the Reagan administration, the president challenged the limits of each of these rules with speeches and the disposition of military forces. Courtship involves a mixed-motive situation where the "rules" are simultaneously upheld and undermined so that one side cannot take the behavior of the other for granted.

17. Walter R. Fisher, "Romantic Democracy, Ronald Reagan, and Presidential Heroes," *Western Journal of Speech Communication* 46 (1982): 299–310.

18. For problems with *glasnost* and *perestroika* see David E. Powell, "Soviet Glasnost: Definitions and Dimensions," *Current History* 87 (October 1988): 321–24; Marshall I. Goldman, "Perestroika in the Soviet Union," *Current History* 87 (October 1988): 313.

19. Gordon A. Craig and Alexander George say: "Cold War is a descriptive term that was generally adopted in the late forties to characterize the hostile relationship that developed between the West and the Soviet Union. While loosely employed, the term had an exceedingly important connotation: it called attention to the fact that, however acute their rivalry and conflict, the two sides were pursuing it by means short of another war and

that, it was hoped, they would continue to do so" (*Force and Statecraft: Diplomatic Problems of Our Time* [New York: Oxford University Press, 1983], 118). Within the scope of cold war discourse, rhetoric is an instrument of competition, inducing cooperation as a means of gaining advantage in struggle. The question of the Reagan administration in general, and the Moscow summit in particular was this: Could *perestroika* and *glasnost* enable, and Reagan certify, a new "dialogue" that would outstrip the language of the cold war which dissolves all gestures of goodwill into power gambits?

20. For descriptions of cold war rhetoric, see Wayne Brockriede and Robert L. Scott, *Moments in the Rhetoric of the Cold War* (New York: Random House, 1970); Farrel Corcoran, "The Bear in the Back Yard: Myth, Ideology, and Victimage Rituals in Soviet Funerals," *Communication Monographs* 50 (1983): 305–20; John F. Cragan, "The Origins and Nature of the Cold War Rhetorical Vision, 1946–1972: A Partial History," in *Applied Communication Research: A Dramatistic Approach*, ed. John F. Cragan and Donald C. Shields (Prospect Heights, Ill.: Waveland Press, 1981); Philip Wander, "The Rhetoric of American Foreign Policy," *Quarterly Journal of Speech* 70 (1984): 339–61; Robert T. Oliver, "The Varied Rhetoric of International Relations," *Western Speech* 25 (1961): 213–21. Robert L. Scott lists ten characteristics of cold war rhetoric and invites a pluralistic critical response. The characteristics can be reformulated into an "argument formation" that sets out the parameters of controversy. Cold war rhetoric thus may lead to armed conflict or away from it; generate gestures of hostile or peaceful intentions; define foreign policy events or ripple through domestic agendas; lead toward the enhancement of certain national institutions or provide objectives of criticism of institutions; absorb "science" and the "arts" into military or diplomatic postures while inviting "pluralistic criticism"; bipolarize the world while giving rise to the emergence of the "Third World" ("Cold War and Rhetoric: Conceptually and Critically," in *Cold War Rhetoric: Strategy, Metaphor, and Ideology*, ed. Martin J. Medhurst et al. [New York: Greenwood Press, 1990], 11–13). Notably cold war rhetoric is not held to invite reciprocal self-criticism on the part of the United States and Soviet institutions.

21. Jurgen Habermas, *Communication and the Evolution of Society*, trans. Thomas McCarthy (Boston: Beacon Press, 1976), 178. Italics in original.

22. Ibid., 183. See also Jurgen Habermas, *The Theory of Communicative Action*, vol. 2, *Lifeworld and System: A Critique of Functionalist Reason*, trans. Thomas McCarthy (Boston: Beacon Press, 1981), 178–80.

23. "From a metaphorical perspective, the arguments in support of America's absolute superiority over its Cold War adversary have depended upon a fundamental distinction between civilization and savagery—an image advanced by Reagan and his predecessors through various categories of decivilizing vehicles, from darkness to demons" (Robert Ivie, "The Prospects of Cold War Criticism," in *Cold War Rhetoric*, ed. Medhurst et al., 205). Reagan's dramatistic task was twofold: first, to vary enough from the traditional script to be wisely induced into accepting U.S.-Soviet tensions as a

matter of accident that can be redressed, rather than tragedy that must be played out; second, to make the performance convincing by enacting a discourse that would spring him away from his own demonization of the Soviet Union and toward a sound relationship with Gorbachev and the Russian people.

24. Many sources that evaluate the Reagan presidency do not include a detailed examination of his legacy in foreign policy. See Charles O. Jones, ed., *The Reagan Legacy: Promise and Performance* (Chatham, N.J.: Chatham House, 1988); John Kenneth White, *The New Politics of Old Values* (Hanover, N.H.: University Press of New England, 1988); Dilys M. Hill and Phil Williams, "The Reagan Legacy," in *The Reagan Presidency: An Incomplete Revolution?*, ed. Dilys M. Hill, Raymond A. Moore, and Phil Williams (Hampshire, Eng.: University of South Hampton and Macmillan, 1990), 233–41; Robert Lekachman, *Visions and Nightmares: America After Reagan* (New York: Macmillan, 1987). Others examine the foreign policy legacy but do not explain how it worked rhetorically to encompass different attitudes and policies toward the Soviet Union. See Raymond A. Moore, "The Reagan Presidency and Foreign Policy," in *The Reagan Presidency*, ed. Hill, Moore, and Williams, 187.

25. Jim Hoagland, "An Interview with Mikhail Gorbachev," *Washington Post*, 22 May 1988, A32. Throughout the cold war the United States has been in a difficult position internationally in justifying its military policies, which from time to time feature the prospective use of nuclear weapons without finding a rationale for the public's constructive role in a nuclear exchange. Gorbachev made the most of this legitimation problem in the presummit interview. "I am convinced that strategic military parity can be maintained at a low level and without nuclear weapons. We have clearly formulated our choice: to stop, then reverse the arms race" (ibid).

26. Charles Krauthammer, "Too Hungry for START," *Washington Post*, 27 May 1988, A19; Henry Kissinger and Cyrus Vance, "An Agenda for 1989," *Newsweek*, 6 June 1988, 31–32; Constantine C. Menges, "That Old Summit Magic," *National Review*, 24 June 1988, 37–40. Gorbachev's main dramatic responsibilities included pressing the suit for a START agreement that would make progress toward the joint goal of a nuclear-free world and defending his policies of openness to Reagan and against those who would criticize Reagan's visit; thus he had to be both near to and distant from the American leader.

27. Charlotte Saikowski, "A New 'Détente' for the 1980s," *Christian Science Monitor*, 24 May 1988, 1, 7; and "Menu of Bilateral, Regional Issues Awaits Summiteers," *Christian Science Monitor*, 25 May 1988, 3.

28. "Remarks to the Paasikivi Society and the League of Finnish American Societies in Helsinki, Finland, May 27, 1988," in *Ronald Reagan: 1988*, vol. 1, Public Papers of the Presidents of the United States (Washington, D.C.: U.S. Government Printing Office, 1990), 656.

29. Ibid., 658.

30. "Arrival Remarks, the Kremlin, Moscow, May 29, 1988," *U.S. Department of State Bulletin* 88 (August 1988): 8.

31. Ibid., 9.

32. Robert G. Kaiser, "Leaders Team Up Against Cold War Dragon," *Washington Post*, 3 June 1988, 1, A26.

33. "Arrival Remarks," 9.

34. Ibid.

35. Ibid.

36. Steven V. Roberts, "Reagan and Gorbachev Begin Summit, Cordiality Fades," *New York Times*, 30 May 1988, 1.

37. Bill Keller, "Presidential Stroll: Chaos and Applause," *New York Times*, 30 May 1988, 1, 8.

38. "President's Remarks, Danilov Monastery, Moscow, May 30, 1988," *U.S. Department of State Bulletin* 88 (August 1988): 10.

39. Ibid., 10–11.

40. Lou Cannon and Don Oberdorfer, "The Superpowers' Struggle over 'Peaceful Coexistence,'" *Washington Post*, 3 June 1988, A26.

41. Paul D. Erickson, *Reagan Speaks: The Making of an American Myth* (New York: New York University Press, 1985), 2–3. The traditional language to justify arms increases was featured as "bargaining from strength." Objections to arms increases were featured as "unilateral disarmament." The onus put on liberals was to show why the Soviet Union should reduce its weapons if America refused to spend more. The "asymmetrical" reductions in numbers of weapons occasioned by the INF treaty blunted this argument somewhat because America was getting a better deal as a result of its deployment, putatively proving that "peace through strength" worked. However, the right viewed any deal with the Soviets with suspicion; therefore, it was important for Reagan to purify his intentions, distancing himself from Soviet ideology even while ringing in the new era.

42. Reagan was said to want to meet real Soviet citizens and made adaptations in his speeches to honor Soviet literature and the arts. These symbolic gestures softened the "human rights" onslaught. Suzanne Massie, his chief adviser on Russian culture, found that Reagan's mere presence in the Soviet Union was a key to the performance. "Let's face it," she said. "Reagan is going as an icon of the United States" (Beam, "Superpower Script," 6). If he did not win complete popularity, he at least erased much of his negative image. Alex Beam, "President Draws Mixed Reviews in Moscow Speeches," *Boston Globe*, 1 June 1988, 1, 12.

43. "President's Remarks to Selected Citizens, Spaso House, Moscow, May 30, 1988," *U.S. Department of State Bulletin* 88 (August 1988): 12.

44. Thom Shanker, "Soviet Media Slap at Activists Embraced by Reagan," *Chicago Tribune*, 1 June 1988, 14.

45. Ibid.

46. Felicity Barringer, "First Ladies' Traveling Road Show: We're Fine, Thank You Very Much," *New York Times*, 30 May 1988, 8. See Donald T. Regan, *For the Record: From Wall Street to Washington* (New York: Harcourt Brace Jovanovich, 1988), 314.

47. "Dinner Toasts, the Kremlin, Moscow, May 30, 1988," *U.S. Department of State Bulletin* 88 (August 1988): 12.

48. Ibid.

49. Ibid., 13.

50. Ibid.

51. Joseph G. Whelan, *The Moscow Summit, 1988: Reagan and Gorbachev in Negotiation* (Boulder, Colo.: Westview Press, 1990), 38.

52. In public Gorbachev expressed irritation with emphasis on human rights rhetoric at the expense of progress in arms control; in talks, he was quite concerned with slow progress on economic matters.

53. Whelan, *Moscow Summit*, 40.

54. Fred Kaplan, "'Evil Empire' Era Has Ended, Reagan Declares in Kremlin," *Boston Globe*, 1 June 1988, 1, 13.

55. Stanley Meisler, "Reagan Recants 'Evil Empire' Description," *Los Angeles Times*, 1 June 1988, 1:1, 13.

56. Schieffer and Gates, *The Acting President*, 335.

57. Tom Shales, "Media Glasnost, Soviet Savvy," *Washington Post*, 1 June 1988, B1, B10.

58. Ibid.

59. "Remarks at a Luncheon Hosted by Artists and Cultural Leaders in Moscow, May 31, 1988," in *Ronald Reagan: 1988*, Public Papers of the Presidents, 1:681.

60. Ibid.

61. Ibid., 682.

62. Ibid.

63. Ibid.

64. Garry Wills, *Reagan's America: Innocents at Home* (Garden City, N.Y.: Doubleday, 1987).

65. David Remnick, "Reagan Cites Roll Call of Cultural Icons," *Washington Post*, 1 June 1988, A30.

66. Whelan, *Moscow Summit*, 33–34.

67. Ideologically the speech is significant because the president abjures the right-wing position, which lets the Communist state sink until it dies. Instead, the speech emphasizes the forces of positive change rather than "Russian bashing" (Editorial, *Dallas Morning News*, 1 June 1988).

68. "President's Remarks, Moscow State University, Moscow, May 31, 1988," *U.S. Department of State Bulletin* 88 (August 1988): 17 (cited hereafter as "Moscow State Address"). Schieffer and Gates call the speech Reagan's "most impressive" of the conference (*The Acting President*, 334). The speech read against the press conference, however, contradicts this judgment. At the press conference the president demonstrated that he did not have full command of the very position he was espousing.

69. "Moscow State Address," 17.

70. Ibid., 16.

71. Hugh Duncan, *Communication and Social Order* (New York: Bedminster Press, 1962), 395–405.

72. To Reagan, bureaucrats were no joke. He had developed a dislike for government bureaucracy as far back as his days in the army. Laurence I. Barrett, *Gambling with History: Ronald Reagan in the White House* (Garden City, N.Y.: Doubleday, 1983), 55.

73. The press reported politely favorable responses on the part of the

students; many sorted out the president as a person, who was acceptable, from his ideas of reform, which were not. Gary Lee, "Students Find Reagan a Pleasant Surprise," *Washington Post*, 1 June 1988, 1, A26.

74. "Remarks at a Question and Answer Session with the Students and Faculty at Moscow State University, May 31, 1988," in *Ronald Reagan: 1988*, Public Papers of the Presidents, 1:690–91.

75. "Reagan Remark Angers Indians," *Chicago Tribune*, 1 June 1988, 1:15.

76. "In Leningrad Nancy Reagan Ruffles Feathers," *Boston Globe*, 1 June 1988, 12.

77. "Dinner Toasts, Spaso House, Moscow, May 31, 1988," *U.S. Department of State Bulletin* 88 (August 1988): 22.

78. Ibid.

79. Ibid., 23.

80. "Exchange of INF Treaty Documents, The Kremlin, Moscow, June 1, 1988," *U.S. Department of State Bulletin* 88 (August 1988): 23–24.

81. Ibid., 25.

82. Ibid., 25.

83. To a large extent, the debate of the 1980s replayed the issues of the 1950s and early 1960s. In different eras the same arguments were forwarded about the unreliability of verification procedures and the incautiousness of relying upon sheer trust. David Mervin attributes the INF agreement to Gorbachev's maneuvering (*Reagan and the Presidency*, 171); however, Reagan played a necessary part by brandishing his conservative credentials and supporting the treaty.

84. R. Jeffrey Smith, "Process of Inspecting, Destroying 2,400 Missiles Begins." *Washington Post*, 1 June 1988, A29. Charles Kauffman claims that "the realm of nuclear conflict is largely symbolic; aside from Hiroshima and Nagasaki, nuclear war has been fought with words rather than weapons" ("Names and Weapons," *Communication Monographs* 56 [1989]: 283). If this observation is correct, reciprocal control of nuclear weapons is a threshold movement away from the symbolic warfare and toward a limiting rationality structure. Of course, countries have been linked by emergency communication technology and "bean counting" to meet reciprocal numerical limitations on nuclear weapons for some time; however, the Moscow accords take steps toward routinizing relationships between nuclear establishments.

85. "Gorbachev 'Peaceful Coexistence' Draft: Final Text," *Washington Post*, 2 June 1988, A24.

86. Cannon and Oberdorfer, "Superpowers' Struggle," A26.

87. "Secretary's Interview, MacNeil/Lehrer Newshour, Moscow, May 31, 1988," *U.S. Department of State Bulletin* 88 (August 1988): 19–21.

88. "Gorbachev's Press Conference," *Current Digest of the Soviet Press* 40 (1988): 3. See Thom Shanker, "News Conference a Personal Best for Gorbachev," *Chicago Tribune*, 2 June 1988, 1:19.

89. "Gorbachev's Press Conference," 4.

90. Ibid.

91. Ibid., 5.

92. "President's News Conference, Spaso House, Moscow, June 1, 1988," *U.S. Department of State Bulletin* 88 (August 1988): 706.

93. Ibid., 707.

94. Ibid., 712.

95. Editorial, "Down from the Summit," *Washington Post*, 2 June 1988, A20; George de Lama, "Reagan Eases Up on Rights Crusade," *Chicago Tribune*, 2 June 1988, 1:19; Robert Gillette, "Reagan Blames Emigre Snags on Bureaucrats," *Los Angeles Times*, 2 June 1988, 1:8; Jack Nelson and Michael Parks, "Leaders Vow to 'Bang Fists' to Complete Strategic Pact," *Los Angeles Times*, 1 June 1988, 1:1. Fist banging is not without its comic element, and Reagan received applause when he used the phrase elsewhere. Here he could not distinguish between the Soviet government's responsibility and the bureaucracy.

96. Lou Cannon, "Blaming the 'Bureaucracy': Reagan Shuns Criticizing Gorbachev on Human Rights," *Washington Post*, 2 June 1988, 1.

97. For Reagan's attitudes see Dilys M. Hill, "Domestic Policy in an Era of 'Negative Government,'" in Hill, Moore, and Williams, *The Reagan Presidency*, 161–79.

98. Edward Walsh, "Conservatives Praise President for Advocacy of Rights, Democracy," *Washington Post*, 2 June 1988, A24.

99. The weakness of the Reagan position is that it asks for self-criticism on the part of the Soviet Union while accepting little or none in return, but it is difficult to see how Reagan could have taken up the burden of a thoroughgoing critique of American human-rights policy without undercutting his own posture. In another sense, the very enactment of Reagan's criticisms in the Soviet Union is a way of proving the legitimacy of Soviet change through *glasnost* and *perestroika*. While cold war discourse has thawed from time to time, there was no comparable invitation for self-criticism from the Soviet Union before Gorbachev.

100. "President's News Conference," 710. The president tried out the family metaphor at Moscow State without apparent result there. "Remarks and a Question and Answer Session with the Students at Moscow State University, May 31, 1988," 691.

101. "First Ladies in 'Mexican Standoff,'" *Los Angeles Times*, 1 June 1988, 1:1; "Mrs. Reagan, Mrs. Gorbachev Drop All Pretense of Friendship," *Boston Globe*, 2 June 1988, 14.

102. "Summit Ends with Smiles, Hugs and a Signed Treaty," *Los Angeles Times*, 1 June 1988, 1:1–2. Schieffer and Gates (*The Acting President*, 333–34) conclude: "In terms of style, the Moscow visit was a triumph for the President. Having been upstaged by Gorbachev on his home turf, the old actor now returned the compliment."

103. Walther Stützle, "Introduction: More Questions than Answers—How to Manage Change," *SIPRI Yearbook 1990: World Armaments and Disarmament* (Oxford: Oxford University Press, 1990), xxi–xxxi.

104. William Hyland, *The Cold War Is Over* (New York: Random House, 1990), 189.

105. Robert L. Ivie, "Metaphor and the Rhetorical Invention of Cold War 'Idealists,'" *Communication Monographs* 54 (1987): 179. Ivie argues that "some kind of a SYMBIOSIS metaphor must be identified and elaborated in order to move beyond the peril of pre-nuclear thinking in the nuclear age" (181). The Moscow summit eschewed a metaphor that would merge national views in favor of finding similarity-in-difference and an agenda of distinct areas for pursuing mutual interests, but not similar differences. Such a discourse is inherently incremental, but it might prove more durable in times of cooperation, and less enveloping in times of tensions, than one based on organic expectations.

106. Robert L. Ivie, "Prospects of Criticism," 204. I concur that cold war rhetoric is a symbolic resource for controversy that is still available for rhetoricians, should they choose to use it; however, the Moscow summit is not reducible to cold war discourse insofar as it put in place a reciprocal political commitment to new legitimation structures for nuclear discourse. Moreover, Gorbachev's rhetorical commitments to *glasnost,* and Reagan's tentative approval, interject a different sense of Soviet-American "dialogue" into international relations. Whether these new, rhetorical resources expand or contract depends upon which paths are followed in the postsummit world.

107. Cori E. Dauber, "Negotiating from Strength: Arms Control and the Rhetoric of Denial," *Political Communication and Persuasion* 7 (1990): 97–114.

108. Ronnie Dugger, *On Reagan: The Man and His Presidency* (New York: McGraw-Hill, 1983), 399–416.

109. Steven Ivan Griffiths, "The Implementation of the INF Treaty," in *SIPRI Yearbook 1990,* 443–58.

110. The rationalization of nuclear weapons systems and policy between the United States and the Soviet Union may or may not be a good idea. On the one hand, a common system of control could lead to minimal levels of survivable nuclear weapons, thereby reducing the chance of accidental war. On the other, rationalization assures that nuclear weapons development will continue to be an institutionally sanctioned activity for the foreseeable future. At some point, the reciprocal nuclear establishments may have more in common with one another than with their respective ideologies, nation-states, or protected publics.

111. Erickson, *Reagan Speaks,* 51.

112. Gary Lee and Robert G. Kaiser, "Public Impressed; President 'Stands Up Well to Everyone,'" *Washington Post,* 1 June 1988, 1, A31; Alex Beam, "President Draws Mixed Reviews in Moscow Speeches," *Boston Globe,* 1 June 1988, 12.

113. Kathleen Hall Jamieson, *Eloquence in an Electronic Age: The Transformation of Political Speechmaking* (New York: Oxford University Press, 1988), 246.

114. Schieffer and Gates, *The Acting President,* 335–36.

115. Jamieson, *Eloquence in an Electronic Age,* 246.

Chapter 3. Reagan at the London Guildhall

1. My approach is based on the kind of techniques used or recommended by various American scholars in works such as Lester Thonnsen, A. Craig Baird, and Waldo W. Braden, *Speech Criticism* (New York: Ronald Press, 1948); Bernard L. Brock and Robert L. Scott, *Methods of Rhetorical Criticism: A Twentieth Century Perspective* (Detroit: Wayne State University Press, 1989); Karlyn Kohrs Campbell, *Critiques of Contemporary Rhetoric* (Belmont, Calif.: Wadsworth, 1972); and Halford Ross Ryan, *American Rhetoric from Roosevelt to Reagan* (Prospect Heights, Ill.: Waveland, 1983).

2. David Dimbleby and David Reynolds, *An Ocean Apart* (London: British Broadcasting Corporation, 1988), 302.

3. Reagan's words conjure up "olde-worlde" images of friends and neighbors gathering in the town hall to be addressed by their leaders. Although a cinematic stock cliché, the historical reality has been rather different, the Guildhall being used as a recruiting center or for protest meetings rather than for "national" gatherings. I am indebted to the Guildhall's library and archive staff for clarification of this point.

4. Dimbleby and Reynolds, *An Ocean Apart*, 332.

5. My main source of material is the collection *President Ronald Reagan: The Quest for Peace, the Cause of Freedom* (Washington: U.S. Information Agency, 1988).

6. See my paper "Liberty: Elusive Concept as Rhetorical Device," in *Proceedings* of the 1989 Conference of the British Association for Applied Linguistics (forthcoming). The paper discusses Mrs. Thatcher's use of the term *liberty*.

7. See for example, "Televised Address, October 27, 1964," transcribed in *The Quest for Peace*, 14–24.

8. For the text of Reagan's 1982 speech to Parliament, see *The Quest for Peace*, 198–207. Charles Kennedy, M.P., analyzes this speech in "The Father of the New Right in the Mother of Parliaments," *World Communication* 14, no. 1 (1985): 43–52.

9. Cornelius Ryan, *A Bridge Too Far* (New York: Simon and Schuster, 1974), 539.

10. Dimbleby and Reynolds, *An Ocean Apart*, 161.

11. Peter Jenkins, *Mrs. Thatcher's Revolution* (Cambridge, Mass.: Harvard University Press, 1988), 210.

12. The findings of a 1990 national opinion poll are relevant to the question of British self-perception. When asked, "Which of these descriptions best fits Britain today?," 48 percent of all respondents chose "a middle-ranking European power," compared with 10 percent choosing "a major world power," 22 percent choosing "a major European but not world power," and 14 percent choosing "a minor European power." Conservative Party supporters were slightly more likely to choose the strongest option, with 16 percent choosing "a major world power." See *The Independent*, 20 January 1990.

13. I would like to thank the U.S. Information Service at the U.S. Embassy in London and the Political Office at 10 Downing Street, London, for providing transcripts of these speeches. Additionally, I am pleased to acknowledge the help of four students in my logic and rhetoric course at Portsmouth Polytechnic in 1988–89: Isabelle Augello, Caroline Boisserie, Ina Roepcke, and Andreas Urscheler. All provided valuable insights into Reagan's techniques as well as into impressions his rhetoric made on people from different cultures. I am also grateful to John Clements for reading a preliminary draft of this paper and offering many helpful suggestions. Any misjudgments in the foregoing analysis, however, are my own.

Chapter 4. Acting like a President

1. Ferdinand Mount, *The Theatre of Politics* (New York: Schocken Books, 1972), vii.

2. Richard E. Neustadt, *Presidential Power: The Politics of Leadership, with Reflections on Johnson and Nixon* (New York: Wiley and Sons, 1976).

3. Sidney Blumenthal, *The Permanent Campaign*, rev. ed. (New York: Simon and Schuster, 1982); Kathleen Hall Jamieson, *Eloquence in an Electronic Age: The Transformation of Political Speechmaking* (New York: Oxford University Press, 1988).

4. David S. Broder, "The Waning Days of Washington," *Washington Post National Weekly Edition*, 26 February–1 March 1990, 4, 9.

5. Quoted by Broder in "Waning Days."

6. David S. Broder, "Bush Speech Taken from Gipper's Playbook," *Sunday Herald-Times* (Bloomington, Ind.), 3 February 1991, A8.

7. John F. Stacks, *The Campaign for the Presidency, 1980* (New York: Times Books, 1981), 256.

8. Ronald Reagan, *An American Life* (New York: Simon and Schuster, 1990), 145.

9. *Sacramento Bee*, 3 August 1965.

10. Mark Green and Gail MacCool, *There He Goes Again: Ronald Reagan's Reign of Error* (New York: Pantheon, 1983), 8.

11. Charles D. Hobbs, *Ronald Reagan's Call to Action* (New York: McGraw-Hill, 1968), 143.

12. Paul Duke, ed., *Beyond Reagan: The Politics of Upheaval* (New York: Warner Books, 1986), 133.

13. Personal observation and inquiry in Shanghai. Also see *New York Times* report of Reagan visit, 30 May 1984.

14. Is this the source of George Bush's often-expressed concern with finding "the vision thing"?

15. A similar analysis is in Reagan, *An American Life*, 42.

16. Elizabeth Drew, "Washington Letter," *New Yorker*, 4 July 1988, 80.

17. On the PBS "Washington Week in Review" telecast, 3 June 1988.

18. Myles Martel, "Debate Preparations in the Reagan Camp: An In-

sider's View," *Speaker and Gavel* 18 (1981): 44; Lou Cannon, *Reagan* (New York: Putnam, 1982), 297; Laurence Leamer, *Make-Believe: The Story of Nancy and Ronald Reagan* (New York: Harper and Row, 1983), 280.

19. Elizabeth Drew, *Portrait of an Election: The 1980 Presidential Campaign* (New York: Simon and Schuster, 1981), 324.

20. Jack W. Germond and Jules Witcover, *Wake Us Up When It's Over: Presidential Politics of 1984* (New York: Macmillan, 1985), 496, 500; Elizabeth Drew, *Campaign Journal: The Political Events of 1983–1984* (New York: Macmillan, 1985), 690.

21. Joe McGinnis, *The Selling of the President, 1968* (New York: Trident Press, 1969), 30.

22. Jeff Greenfield, *Playing to Win: An Insider's Guide to Politics* (New York: Simon and Schuster, 1980), 109.

23. Myron G. Phillips, "William Jennings Bryan," in *A History and Criticism of American Public Address*, ed. William Norwood Brigance (New York: McGraw-Hill, 1943), 2:891.

24. Edward A. Weinstein, *Woodrow Wilson: A Medical and Psychological Biography* (Princeton, N.J.: Princeton University Press, 1981), 132.

25. Earnest Brandenburg and Waldo W. Braden, "Franklin Delano Roosevelt," in *A History and Criticism of American Public Address*, ed. Marie Kathryn Hochmuth (New York: Longmans, Green, 1955), 3:527.

26. Ronald Reagan and Richard C. Hubler, *Where's the Rest of Me?* (New York: Dell, 1965). See also Reagan, *An American Life*, 41–43.

27. Reagan and Hubler, *Where's the Rest of Me?*, 36–37.

28. Ibid., 48.

29. David Reisman, *The Lonely Crowd* (New Haven, Conn.: Yale University Press, 1953); Daniel J. Boorstin, *The Image: A Guide to Pseudo-Events in America* (New York: Harper Colophon, 1961).

30. J. Jeffery Auer, *The Rhetoric of Our Times* (New York: Appleton-Century-Crofts, 1969), 3–5.

31. Theodore Sorenson, *A Different Kind of Presidency* (New York: Harper and Row, 1984), 46–47.

32. Blumenthal, *The Permanent Campaign*; Hedrick Smith, *The Power Game: How Washington Works* (New York: Random House, 1988), 700.

33. Paul Taylor, *See How They Run: Electing the President in an Age of Mediaocracy* (New York: Knopf, 1990), 218.

34. Michael K. Deaver with Mickey Herskowitz, *Behind the Scenes: In Which the Author Talks About Ronald and Nancy Reagan . . . and Himself* (New York: William Morrow, 1987), 44.

35. Murray Edelman, *Constructing the Political Spectacle* (Chicago: University of Chicago Press, 1988). Creating the spectacle became known in the Reagan White House as "the art of the backdrop," staging even mundane events in dramatic, familiar, or novel settings. Robert Schmuhl, *Statecraft and Stagecraft: American Political Life in the Age of Personality* (Notre Dame, Ind.: University of Notre Dame Press, 1990), 32–34.

36. Donald T. Regan, *For the Record: From Wall Street to Washington* (New York: Harcourt Brace Jovanovich, 1988), 12, 20–21.

37. Ibid., 304–5.

38. Larry Speakes with Robert Pack, *Speaking Out: The Reagan Presidency from Inside the White House* (New York: Scribner's, 1988); Martin Anderson, *Revolution* (New York: Harcourt Brace Jovanovich, 1988); David A. Stockman, *The Triumph of Politics: How the Reagan Revolution Failed* (New York: Harper and Row, 1986).

39. Anthony R. Dolan, Letter to the editor, *New York Times Book Review,* 13 January 1991, 25. See Dolan, *Undoing the Evil Empire: How Reagan Won the Cold War* (Washington, D.C.: American Enterprise Institute Press, 1991). Mr. Dolan is widely regarded as the principal author of Reagan's "Evil Empire" speech.

40. Regan, *For the Record,* 376–77. For an early but discerning assessment of Reagan's role in history, see Haynes Johnson, *Sleepwalking Through History: America in the Reagan Years* (New York: Norton, 1991), 438–60.

41. In Jeremy Tunstall and David Walker, *Media Made in California: Hollywood, Politics, and the News* (New York: Oxford University Press, 1981), 160.

42. Deaver, *Behind the Scenes,* 77.

43. Schmuhl, *Statecraft and Stagecraft,* quotation from pp. 30–31.

44. Roderick P. Hart, *Verbal Style and the Presidency: A Computer-based Analysis* (Orlando, Fla.: Academic Press, 1984), 10.

45. *New York Times Magazine,* 26 January 1986, 8.

46. Smith, *The Power Game,* 381–82.

47. In *The Political Animal,* 29 March 1985, 1.

48. In Mark Hertsgaard, *On Bended Knee: The Press and the Reagan Presidency* (New York: Schocken Books, 1989), 5. This book is invaluable for reporting in detail how newspaper and television newspersons reacted to Reagan and his "handlers."

49. *New York Times,* 16 November 1986, A1.

50. Schmuhl, *Statecraft and Stagecraft,* 39.

51. *Washington Post,* 28 March 1984, 23.

52. Ernest Bormann, J. Jeffery Auer, and Franklyn S. Haiman, "Ghostwriting and the Cult of Leadership," *Communication Education* 33 (1984): 301–7.

53. AP dispatch, *Idaho State Journal* (Pocatello), 16 August 1988, B6.

54. For a detailed report on the process, see Peggy Noonan, *What I Saw at the Revolution: A Political Life in the Reagan Era* (New York: Random House, 1990), 298–317.

55. Reagan and Hubler, *Where's the Rest of Me?,* 299.

56. In L. William Troxler, ed., *Along Wit's Trail: The Humor and Wisdom of Ronald Reagan* (New York: Holt, Rinehart and Winston, 1984), 30.

57. Ronald Reagan, *Speaking My Mind* (New York: Simon and Schuster, 1989), 6.

58. Reagan, *An American Life,* 14.

59. Ibid., 7.

60. Ibid., 151, 195 and 203, 246. For Reagan's directions to his writers on speech composition (short sentences, one-syllable words whenever pos-

sible, examples, and humor), and delivery (audience adaptation, timing and cadence, and TelePrompTer management), and anecdotes about speaking, see 246–49.

61. Nancy Reagan with William Novak, *My Turn: The Memoirs of Nancy Reagan* (New York: Random House, 1989), xi, 150.

62. Maureen Reagan, *First Father, First Daughter* (Boston: Little, Brown, 1989), 150.

63. Nancy Reagan, *My Turn*, 118.

64. Ibid., 234.

65. This writer's tape-recorded interview with Khachigian in his San Clemente, California, office, 13 November 1981; *New York Times*, 20 January 1981, 14, and 21 January 1981, 1; *Newsweek*, 26 January 1981, 14. A diligent researcher can uncover the principal drafters of most other major Reagan speeches.

66. Nancy Reagan, *My Turn*, 41–42.

67. *New York Times*, 6 November 1981, 9.

68. William Safire, *Before the Fall: An Inside View of the Pre-Watergate White House* (Garden City, N.Y.: Doubleday, 1975).

69. Ibid., 13.

70. *Time*, 2 May 1988, 32.

71. Karlyn Kohrs Campbell and Kathleen Hall Jamieson, *Deeds Done in Words: Presidential Rhetoric and the Genres of Governance* (Chicago: University of Chicago Press, 1990), 11.

72. Anderson, *Revolution*, 255; J. Jeffery Auer, "Reagan's Speechwriting Staff" (Paper delivered at the Convention of the Speech Communication Association, Anaheim, Calif., 13 November 1981).

73. Noonan, *What I Saw*, 72–78.

74. Regan, *For the Record*, 77.

75. Ibid., 94.

76. Interview with Khachigian; and tape-recorded telephone interview from his office with Peter Hannaford, Los Angeles, Calif., 14 November 1981.

77. Quoted in Deborah Baldwin, "Funny Guys," *Common Cause Magazine*, July/August 1985, 30; J. Jeffery Auer, "Presidential Humor" (Manuscript).

78. William E. Leuchtenburg, *In the Shadow of FDR: From Harry Truman to Ronald Reagan* (Ithaca, N.Y.: Cornell University Press, 1983), 239. For reminding me of this volume I am indebted to an anonymous editorial reader.

79. *New York Times*, 20 July 1980, A1.

80. Duke, *Beyond Reagan*, 120.

81. AP dispatch, 1 March 1982.

82. Schmuhl, *Statecraft and Stagecraft*, 29.

83. Duke, *Beyond Reagan*, 131.

84. Michael Rogin, *Ronald Reagan: The Movie, and Other Episodes in Political Demonology* (Berkeley and Los Angeles: University of California Press, 1987), 11, and see 1–43. The reporter who knew him best recalled that "even before he became an actor, Reagan blurred the distinction between

reality and imagination when it suited his convenience," and that "the past was always present for Ronald Reagan, who took the fragments of his life that mattered most to him and fashioned them into powerful stories of personal experience" (Lou Cannon, *President Reagan: The Role of a Lifetime* (New York: Simon and Schuster, 1991), 223, 221.

85. Also see relevant statements in *The Tower Commission Report: The Full Text of the President's Special Review Board* (New York: Bantam Books and Times Books, 1987), 79–82, 489, 496, 498–99, 502–3, 504–7.

86. John G. Tower, *Consequences: A Personal and Political Memoir* (Boston: Little, Brown, 1991), 282–89. Reagan's account of these meetings is in *An American Life*, 540–43. A full and heavily documented account of the three Tower Commission meetings with the president is in Cannon, *President Reagan*, 708–29. Cannon also reveals the previously unreported fact that John Tower collaborated with Leonard Parvin and Stu Spencer in writing Reagan's nationwide televised speech of 4 March 1989, denying complicity in or knowledge of the trading of arms for hostages in the Iran-Contra affair (734–35).

87. *Wall Street Journal*, 1 January 1977, A11.

88. *Time*, 23 November 1981, 16.

89. Peter Jenkins, *Mrs. Thatcher's Revolution: The Ending of the Socialist Era* (London: Jonathan Cape, 1987), 210.

90. Said to Andrew King, and cited in Kenneth Hudson, *The Language of Politics* (London: Macmillan, 1978), 105.

91. Quoted in "What They Say About Bush" in *World Press Review*, September 1989, 13, 12.

92. *Sunday Herald-Times* (Bloomington, Ind.), 3 February 1991, A8.

93. See Reagan and Hubler, *Where's the Rest of Me?*, and Reagan, *An American Life*.

94. Eleanor Clift and Thomas M. DeFrank, "Ghosting for the Master," *Newsweek*, 8 April 1985, 24.

95. *U.S. News*, 11 August 1980, 8.

96. Anthony Lewis, "Abroad at Home: The Dignified Part," *New York Times*, 30 January 1986, Y27.

97. William Safire, "Essay: Mr. Bush Hires a Writer," *New York Times*, 11 February 1991, A15.

Chapter 5. Antithesis and Oxymoron

1. Ronald Reagan, "Remarks to Republican Local Officials During a White House Briefing on Federalism and Aid to the Nicaraguan Democratic Resistance (22 March 1988)," in *Weekly Compilation of Presidential Documents* (hereafter *WC*) 24:12, 375.

2. Robert L. Ivie, "Speaking 'Common Sense' About the Soviet Threat: Reagan's Rhetorical Stance," *Western Journal of Speech Communication* 48 (1984): 40. Examples of studies that represent the trends I describe include (in addition to the Ivie study): William F. Lewis, "Telling America's Story:

Narrative Form and the Reagan Presidency," *Quarterly Journal of Speech* 73 (1987): 280–302; Janice Hocker Rushing, "Ronald Reagan's 'Star Wars' Address: Mythic Containment of Technical Reasoning," *Quarterly Journal of Speech* 72 (1986): 415–33; Paul D. Erickson, *Reagan Speaks: The Making of an American Myth* (New York: New York University Press, 1985); Martin J. Medhurst, "Postponing the Social Agenda: Reagan's Strategy and Tactics," *Western Journal of Speech Communication* 48 (1984): 262–76; G. Thomas Goodnight, "Ronald Reagan's Re-formulation of the Rhetoric of War: Analysis of the 'Zero Option,' 'Evil Empire,' and 'Star Wars' Addresses," *Quarterly Journal of Speech* 72 (1986): 390–414; and Kathleen Hall Jamieson, *Eloquence in an Electronic Age: The Transformation of Political Speechmaking* (New York: Oxford University Press, 1988), 118–64.

3. In emphasizing Reagan's narrative strategies and structures, many commentators implicitly pass over the "tropical" aspects of Reagan's discourse. Paul Erickson, for example, writes: "Reagan is in at least one sense our most literary president. Others have used more poetically splendid tropes, and we will probably never see Reagan's speeches in anthologies of American literature. . . . Not a man of letters by the refined aesthetic standards of belles-lettres, Reagan is still supremely literary by virtue of the methods which he employs in his rhetoric. I do not refer to the technical devices of classical oratory—zeugma, apostrophe, and chiasmus, for example—but to the simple fact that Ronald Reagan tells stories to make his points" (*Reagan Speaks*, 5). In the way Erickson emphasizes Reagan's narrativity in this passage, he seems to imply that tropes do not play an important role in the force or coherence of Reagan's discourse.

4. Hans Kellner, "The Inflatable Trope as Narrative Theory: Structure or Allegory," *Diacritics* 11 (1981): 15. Examples of tropological analyses in this tradition are Kenneth Burke's discussion of the master tropes in *A Grammar of Motives* (Berkeley and Los Angeles: University of California Press, 1969), 503–17; Hayden White, *Tropics of Discourse: Essays in Cultural Criticism* (Baltimore: Johns Hopkins University Press, 1978); and James M. Mellard, *Doing Tropology: Analysis of Narrative Discourse* (Urbana: University of Illinois Press, 1987). For a less "inflated" approach to tropological analysis, see Richard A. Lanham, *Analyzing Prose* (New York: Scribner's, 1983).

5. In the literature of rhetorical criticism, this issue is developed by Lloyd F. Bitzer, "The Rhetorical Situation," *Philosophy and Rhetoric* 1 (1968): 1–14; and Richard Vatz, "The Myth of the Rhetorical Situation," *Philosophy and Rhetoric* 6 (1973): 154–61. Other useful studies that address the issue of the text/context relationship include: Kenneth Burke, "The Philosophy of Literary Form" in *The Philosophy of Literary Form* (New York: Vintage Books, 1957); J. Robert Cox, "Argument and the 'Definition of the Situation,'" *Central States Speech Journal* 32 (1981): 197–205; and Frederic Jameson, *The Political Unconscious: Narrative as a Socially Symbolic Act* (Ithaca, N.Y.: Cornell University Press, 1981), esp. 81–82.

6. Ibid., 82.

7. As Hayden White notes, tropes function as different strategies for

constituting 'reality' in thought [and practice] so as to deal with it in different ways" (*Topics of Discourse*, 22).

8. See Richard A. Lanham, *A Handlist of Rhetorical Terms* (Berkeley and Los Angeles: University of California Press, 1969), 12.

9. Cf. Chaim Perelman and L. Olbrechts-Tyteca, *The New Rhetoric: A Treatise on Argumentation*, trans. John Wilkinson and Purcell Weaver (Notre Dame, Ind.: University of Notre Dame Press, 1969) on "arguments by division," 234–41.

10. The studies by Ivie, Erickson, and Goodnight cited in n. 2 attest to Reagan's frequent use of antithetical opposition.

11. Ronald Reagan, "Acceptance Address (17 July 1980)," in *Vital Speeches of the Day* 46:21, 643–44; Ronald Reagan, "Remarks to the World Affairs Council (21 April 1988)," In *WC* 24:16, 503–7. Reagan's remark in the 21 April address is reminiscent of the "Ivan and Anya" story he told at the end of his 16 January 1984 address.

12. Erickson, *Reagan Speaks*, 61–62.

13. Ronald Reagan, "Remarks at the Dedication Ceremony for the Knute Rockne Commemorative Stamp (9 March 1988)," in *WC* 24:10, 313.

14. Chaismus (a term derived from the Greek letter χ) describes a syntactical inversion (ABBA) very useful for establishing contrasts. See Lanham, *Handlist*, 22.

15. The quote is from Reagan's 27 October 1984 radio address (in *Ronald Reagan's Weekly Radio Addresses: The President Speaks to America*, ed. Fred L. Israel [Wilmington, Del.: Scholarly Resources, 1987], 256); this opposition was a major theme of Reagan's 1984 campaign (see, for example, the radio addresses of 13 October and 3 November 1984). A variation on this antithetical chiasmus can be found in Reagan's 15 September 1984 radio address: "We believe in knowing when opportunity knocks, they [Democrats] knock opportunity" (245).

16. Ronald Reagan, "Remarks at the NRA Annual Members Banquet (6 May 1983)," in *WC* 19:19, 672–73.

17. The presence of polyptoton (the play on the root *end*) in this passage is interesting because it seems to recognize an underlying similarity between the two approaches to policy—ending *versus* continuing a dependence relationship—that are set in opposition. Reagan's figurative depiction of the matter seems to work against his overt rhetorical purpose (unlike the use of chiasmus noted earlier, which reinforced the established opposition).

18. Ronald Reagan, "Remarks at the 105th Meeting of the American Bar Association (1 August 1983)," in *WC* 19:31, 1082.

19. Ibid., 1080.

20. Ronald Reagan, "Remarks at the Annual Convention of the National Association of Evangelicals (8 March 1983)," in *WC* 19:10, 366; Ronald Reagan, "Radio Address," 16 June 1984 in *Weekly Radio Addresses*, 219.

21. Ronald Reagan, "Address to the Members of Parliament (8 June 1982)," in *WC* 18:23, 764–65.

22. Reagan does not elaborate on the connection between different forms of totalitarianism (Communist and others). Reagan does, however, arrange

his speeches in a way that leads audiences to perceive connections that are never discursively developed. For example, Reagan employs antithetical contrasts between the Middle East and Central America that effectively obscure the differences between the regions. In remarks to civic leaders on 20 January 1982, Reagan asserted that "the kind of turmoil that exists in the Persian Gulf cannot be allowed to exist in the Gulf of Mexico" (*WC* 24:3, 64). The impression Reagan conveys is that these various forms of totalitarianism are cut from the same cloth.

23. Reagan, "Address to Parliament," 768.

24. On *God* and *devil* terms, see Richard Weaver's essay "Ultimate Terms in Contemporary Rhetoric," in *The Ethics of Rhetoric* (Chicago: Henry Regnery, 1953), 211–32.

25. Hugh Heclo, "Reaganism and the Search for a Public Philosophy," in *Perspectives on the Reagan Years*, ed. John L. Palmer (Washington, D.C.: Urban Institute Press, 1986), 41. For a discussion of the power of oxymora, or "harmonizing discordancies," see David Green, *Shaping Political Consciousness: The Language of Politics in America from McKinley to Reagan* (Ithaca, N.Y.: Cornell University Press, 1987).

26. Goodnight, "Re-formulation"; Rushing, "'Star Wars' Address," 425.

27. Ronald Reagan, "Remarks to Students and Guests at Walt Disney's EPCOT Center (8 March 1983)," in *WC* 19:10, 363; Reagan, 9 March 1988, 315–16.

28. Ronald Reagan, "Remarks at the New York Partnership Luncheon (14 January 1982)," in *WC* 18:2, 30.

29. Reagan, 14 January 1982, 27–29; Heclo, "Reaganism," 44–46.

30. At the conclusion of the address, Reagan recounts the parable of the Good Samaritan. In Reagan's version, emphasis is placed on what we get out of charity as opposed to what we give to others.

31. Ronald Reagan, "Radio Address (4 August 1984)," in *Weekly Radio Addresses*, 233.

32. Michael Weiler, "The Reagan Attack on Welfare," in this volume. During the discussion at the Massachusetts symposium, Weiler described how "cruel to be kind" seems to be a "representative oxymoron" in Reagan's welfare rhetoric.

33. Thomas B. Farrell, "Knowledge, Consensus, and Rhetorical Theory," *Quarterly Journal of Speech* 62 (1976): 1–14.

34. The war in the Persian Gulf, just beginning as I complete the final version of this essay, helps illustrate the point that no amount of strength is sufficient to secure peace. History will show, I believe, that "peace through strength" is as problematic as fighting a war to end all wars.

35. Ronald Reagan, "Soviet-American Relations: Address to the Nation (16 January 1984)," in *WC* 20:3, 40–45; Reagan, "Remarks to World Affairs Council (21 April 1988)," 503–7.

36. Reagan, 16 January 1984, 44–45.

37. See Reagan, 8 June 1982, 764–70; Reagan, 16 January 1984, 40–45.

38. Ronald Reagan, "Remarks to the World Affairs Council (31 March 1983)," in *WC* 19:3, 484–89.

39. I discuss the idea of intimate distance as a manifestation of linguistic

privatism in "The 'Privatization' of Rhetorical Language" (Paper delivered at the Convention of the Speech Communication Association, New Orleans, November 1988).

40. Reagan, 17 July, 1980, 643; Reagan, 22 March 1988, 374.

41. Cf. Murray Edelman, *Constructing the Political Spectacle* (Chicago: University of Chicago Press, 1988), 37–65.

42. Robert G. Gunderson, "The Oxymoron Strain in American Rhetoric," *Central States Speech Journal* 28 (1977): 92–95.

43. Karlyn Kohrs Campbell, "The Rhetoric of Women's Liberation: An Oxymoron," *Quarterly Journal of Speech* 59 (1973): 84–85.

44. Ibid., 82.

45. On the limitations of pluralism, see Ellen Rooney, *Seductive Reasoning: Pluralism as the Problematic of Contemporary Literary Theory* (Ithaca, N.Y.: Cornell University Press, 1989).

46. Reagan, 22 March 1988, 374.

47. My point here is similar to Goodnight's argument ("Re-formulation") that Reagan's discourse inadvertently perpetuates the very grounds of controversy it had hoped to tame. See also Green, *Shaping Political Consciousness*, 258–59.

Chapter 6. Transformation of Actor to Scene

1. Kenneth Burke, *A Grammar of Motives* (Berkeley and Los Angeles: University of California Press, 1969), xv.

2. See, for example, David A. Ling, "A Pentadic Analysis of Senator Edward Kennedy's Address to the People of Massachusetts, July 25, 1969," *Central States Speech Journal* 21 (1970): 81–86; or David S. Birdsell, "Ronald Reagan on Lebanon and Grenada: Flexibility and Interpretation in the Application of Kenneth Burke's Pentad," *Quarterly Journal of Speech* 73 (1987): 267–79.

3. Burke, *A Grammar of Motives*, xvi.

4. Ibid.

5. Consider for example, Jane Blankenship et al., "One Year and Counting: A Transformational Analysis of the Pre-Primary and Early Primary Period of the 1988 Presidential Campaign" (in progress), which also argues for lateral transformation, particularly in issue analysis, where, for example, day care moves from a cluster of terms constituting the compassion issue to a cluster of terms constituting the competitiveness issue.

6. Included in a memo by Richard G. Darman, a senior aide to Ronald Reagan, written in June 1984 and reprinted in Peter Goldman and Tony Fuller, *The Quest for the Presidency* (New York: Bantam Books, 1985), 413.

7. Peggy Noonan, *What I Saw at the Revolution: A Political Life in the Reagan Era* (New York: Random House, 1990), 67.

8. Steven V. Roberts, "Reagan's Final Rating Is Best of Any President Since 40's," *New York Times*, 17 January 1989, 1, 14.

9. See for example "The News of the Week Was a Medical Miracle:

Reagan Suddenly Aged from Midlife Macho to Doddering Senility," in *Atlanta Constitution*, 13 October 1984, 2D. On 9 October the *Wall Street Journal* asked in a headline: "Is Oldest U.S. President Now Showing His Age?" (p. 1). They noted that in 1980 Reagan had "pledged to undergo regular tests for senility if he became president" (p. 64) and quoted him as saying in 1984 that "he would take the test only if there was 'some indication that I was drifting. . . . Nothing like that has happened" (p. 64).

10. A much longer treatment of this particular transformation is in Jane Blankenship, Marlene G. Fine, and Leslie Davis, "The 1980 Republican Primary Debates: The Transformation of Actor to Scene," *Quarterly Journal of Speech* 69 (1983): 25–36. Verbatim material reprinted by permission of the Speech Communication Association. Material quoted directly from the candidates was taken from videotapes of the debates.

11. The term *presence* has been recovered in Chaim Perelman and L. Olbrechts-Tyteca, *The New Rhetoric: A Treatise on Argumentation*, trans. John Wilkinson and Purcell Weaver (Notre Dame, Ind.: University of Notre Dame Press, 1969). See also Louise Karon, "Presence in *The New Rhetoric*," *Philosophy and Rhetoric* 9 (1976): 96–111. It is from Perelman that we borrow the term *foreground of consciousness*.

12. The smallest unit of meaning we have labeled "frame" whether that unit is exemplified by verbal or visual entities.

13. Burke, *A Grammar of Motives*, 77.

14. Kenneth Burke, *The Philosophy of Literary Form* (New York: Vintage Books, 1957), 23. We are, here, also talking about vessel not simply as "container" but as a symbolic vessel, "ritualistically loaded" and "charismatic" (p. 34) as in "vessel of grace" or, in Reagan's case, Vessel of the American Dream.

15. Richard Wirthlin memo, "Voters are Looking for a Leader," in Peter Goldman and Tony Fuller, *The Quest for the Presidency 1984* (New York: Bantam Books, 1985), 378.

16. Goldman and Fuller, *Quest*, 379.

17. Ibid., 380.

18. Ibid., 382.

19. The term *warrant* is taken from Stephen Toulmin, *The Uses of Argument* (Cambridge: Cambridge University Press, 1964), 97–107.

20. All quoted material is taken from a videotaped copy of the Reagan film at the 1980 GOP convention.

21. Roberts, "Reagan's Final Rating," 14.

22. Goldman and Fuller, *Quest*, 413.

23. See, e.g., Susan B. Mackey, "An Analysis of the Eighteen-Minute Film Preceding Ronald Reagan's Acceptance Speech at the 1984 National Convention" (Ph.D. diss., Pennsylvania State University, 1988).

24. For useful treatments of narrative see Walter R. Fisher, *Human Communication as Narrative: Toward a Philosophy of Reason, Value, and Action* (Columbia: Unversity of South Carolina Press, 1987); and William F. Lewis, "Telling America's Story: Narrative Form and the Reagan Presidency," *Quarterly Journal of Speech* 73 (1987): 280–302.

25. Roberts, "Reagan's Final Rating," 14.

26. Quoted material from this speech was taken from the text printed in the *New York Times,* 14 November 1986, A8. For a useful treatment of apologia see B. L. Ware and Wil A. Linkugel, "They Spoke in Defense of Themselves: On Generic Criticism of Apologia," *Quarterly Journal of Speech* 59 (1973): 278–83; and Ellen Reid Gold, "Political Apologia: The Ritual of Self Defense," *Communication Monographs* 45 (1978): 306–16.

27. Quoted material was taken from a transcript of the press conference printed in the *New York Times,* 20 November 1986.

28. Quoted material was taken from the text printed in the *Washington Post,* 12 January 1989, A8, A9.

29. Darman as quoted in Goldman and Fuller, *Quest,* 413.

Chapter 7. Rhetorical Ambush at Reykjavik

1. *U.S. Department of State Bulletin* 86, no. 2116 (November 1986): 60–64.

2. Ibid., 60.

3. Ibid., 86, no. 2117 (December 1986): 7.

4. At least more specific than the summaries of it given by the president and, in greater detail, by Secretary Shultz. Ibid., 2.

5. Mikhail Gorbachev, "The Geneva Meeting: Domestic and Foreign Policies" (Speech delivered at the session of the USSR Supreme Soviet, Moscow, 27 November 1985), published in *Vital Speeches of the Day* 52 (15 January 1986): 7, 196–99. In this speech, almost eleven months before Reykjavik, Gorbachev detailed all of the positions that apparently caught the American team by surprise. See particularly 196–202.

6. Mikhail Gorbachev, *Perestroika: New Thinking for Our Country and the World,* rev. ed. (New York: Harper and Row, 1988), 218–19, 221.

7. Ibid., 223.

8. Ibid. Gorbachev seemed surprised that Reagan agreed to the meeting.

9. Ibid., 223, 227. Gorbachev is referring to a position taken by the Twenty-seventh Congress of the Communist Party of the Soviet Union about international security. The four points refer to the military, political, economic, and humanitarian spheres and are summarized in *Perestroika,* 217–18.

10. The caption of the *New York Times* (14 October 1986), A11, photo of Reagan and Gorbachev leaving the Reykjavik talks on Sunday evening reads, "Nothing to Say."

11. These results are based on data published in the *Washington Post,* 3 November 1986, in which William F. Buckley, Jr., reported a classified survey conducted by an unnamed federal agency that found that more people in Great Britain and Germany blamed Reagan than Gorbachev for "not accomplishing much" in nuclear arms reductions. Further, over twice as many (46 percent to 20 percent in Great Britain and 42 percent to 18 percent in Germany) believed that the U.S.S.R. was "making a greater effort toward nuclear arms agreements." While Reagan was clearly perceived as

the more insistent proponent of human rights, he was not necessarily more "trustworthy." Twenty-nine percent of British respondents had more confidence in Reagan but 21 percent had more in Gorbachev. In Germany, the figures were reversed: 33 percent had more confidence in Gorbachev while 26 percent rated Reagan as more trustworthy.

12. *U.S. Department of State Bulletin* 86, no. 2117 (December 1986): 22.

13. Gorbachev, *Perestroika*, 227.

14. *U.S. Department of State Bulletin* 86, no. 2117 (December 1986): 17.

15. Sally A. Freeman, Stephen W. Littlejohn, and W. Barnett Pearce, "Communication and Moral Conflict" (Manuscript, 1990).

16. Vernon E. Cronen and W. Barnett Pearce, "Logical Force in Interpersonal Communication: A New Concept of the 'Necessity' in Social Behavior," *Communication* 6 (1981): 5–67.

17. Reagan consistently proclaimed a "mission for peace." On his departure for Reykjavik, 9 October 1986, he declared: "We go to Reykjavik for peace. We go to this meeting for freedom, and we go in hope" (*U.S. Department of State Bulletin* 86, no. 2117 [December 1986]: 8). Addressing the USSR Supreme Soviet, Gorbachev stated, "Our policy is clearly a policy of peace and cooperation" ("The Geneva Meeting," 27 November 1985, *Vital Speeches of the Day* 52, no. 7 [15 January 1986]: 202).

18. Gorbachev noted that "Today, there is no shortage of statements professing commitment to peace. What is really in short supply is concrete action to strengthen its foundations" ("Gorbachev's Arms Control Proposal," 15 January 1986, in *Historic Documents of 1986*, 20 January). See also Reagan's speeches published in *Vital Speeches of the Day* 52, no. 6 (15 November 1985): 67, and 53, no. 2 (1 November 1986): 35; and Gorbachev's speech published in *Vital Speeches of the Day* 52, no. 7 (15 January 1986): 198.

19. "In Geneva we failed," said Gorbachev. Moreover, he believed that the continuing disarmament talks, "where 50–100 various options are being bandied about," are "choking on dead issues." This was the reason Gorbachev cited for proposing the Reykjavik meeting. See "Remarks After Reykjavik Talks," in *New York Times*, 13 October 1986, A10; and Gorbachev, *Perestroika*, 213, 223. See also *U.S. Department of State Bulletin* 86, no. 2116 (November 1986): 60, and President's Address to the Nation, 13 October 1986, *Vital Speeches of the Day* 53, no. 2 (1 November 1986): 35.

20. *Vital Speeches of the Day* 52, no. 7 (15 January 1986): 196, and *U.S. Department of State Bulletin* 86, no. 2117 (December 1986): 1.

21. Gorbachev first introduced the Soviet unilateral moratorium in the summer of 1985 "to pave the way for Geneva," hoping to influence the United States to reciprocate. In January 1986 the moratorium was extended with the qualification that the USSR "could not be expected to indefinitely display unilateral restraint." The United States continued to conduct nuclear tests. On 11 April 1986 the Soviet Union announced that it would resume testing so as not to "forgo its own security and that of its allies." No tests were conducted, however. After the 26 April accident at Chernobyl, Gorbachev extended the test ban until August 1986, convinced of the devastating effects of nuclear power. On 18 August Gorbachev extended the

deadline a third time, until 1 January 1987 (*Historic Documents of 1986*, January, 11).

22. "It seemed that the US should have responded to our initiatives and moves since the Geneva summit by meeting us halfway and reacting to the aspirations of the people. But that was not the case. . . . The hopes that arose after the Geneva summit . . . soon gave way to disillusionment, because everything in US real politics remained as it had been" (Gorbachev, *Perestroika*, 215, 221).

23. Reagan regularly recited a litany of bitter disappointments concerning Soviet rejection of U.S. proposals for arms reductions. Reagan claimed, "We've gone the extra mile in arms control, but our offers have not always been welcome. In 1977 and again in 1982, the United States proposed to the Soviet Union deep reciprocal cuts in strategic forces. These offers were rejected, out-of-hand. In 1981, we proposed the complete elimination of a whole category of intermediate range nuclear forces. Three years later, we proposed a treaty for a global ban on chemical weapons. In 1983, the Soviet Union got up and walked out of the Geneva nuclear arms control negotiations altogether." Address to the American People, 14 November 1985, *Vital Speeches of the Day* 52, no. 4 (1 December 1985): 99.

24. Reagan announced the Strategic Defense Initiative (SDI) in March 1983. In his November 1986 report that followed the breakdown of talks at Reykjavik, Reagan reclaimed the purpose of SDI: "Believing that a policy of mutual destruction and slaughter of their citizens and ours was uncivilized, I asked our military a few years ago to study and see if there was a practical way to destroy nuclear missiles after their launch but before they can reach their targets rather than just destroy people. This is the goal for what we call SDI" (*Vital Speeches of the Day* 53, no. 2 [1 November 1986]: 34–35).

25. Both Reagan and Gorbachev frequently accused the other of not practicing what he preached. Reagan maintained that "When it comes to human rights and judging Soviet intentions, we are all from Missouri: you have got to show us" (*Vital Speeches of the Day* 53, no. 2. [1 November 1986]: 36). Gorbachev denounced the use of words to obscure political action: "All too often peaceful words conceal war preparations and power politics. Moreover, some statements made from high rostrums are in fact intended to eliminate any trace of that new 'spirit of Geneva' which is having a salutary effect on international relations today. It is not only a matter of statements . . . we reject such ways of acting and thinking" (*Historic Documents of 1986*, 20).

26. This is called "cultural theories of communication" in chap. 2 of W. Barnett Pearce and Vernon E. Cronen, *Communication, Action, and Meaning* (New York: Praeger, 1980).

27. Reagan insisted that progress is possible "if we continue to pursue a prudent, deliberate and realistic approach with the Soviets." He said that "what Mr. Gorbachev was demanding at Reykjavik was that the United States agree to a new version of a 14-year-old ABM treaty that the Soviet Union has already violated. I told him we don't make those kinds of deals in the United States" (*Vital Speeches of the Day* 52, no. 2 [1 November 1986]: 36).

Gorbachev "would not idealize each step in Soviet foreign policy over the past several decades." He admitted that "mistakes also occurred." But he reasoned that these "were the consequence of an improvident reaction to American actions, to a policy geared by its architects to "roll back communism" (Gorbachev, *Perestroika*, 204).

28. Vernon E. Cronen and W. Barnett Pearce, "Toward an Explanation of How the Milan Method Works: An Invitation to a Systemic Epistemology and the Evolution of Family Systems," in *Applications of Systemic Family Therapy: The Milan Approach*, ed. David Campbell and Rosalind Draper (London: Grune and Stratton, 1985), 69–84.

29. Gorbachev, *Perestroika*, 198.

30. "A number of years ago, I heard a young father, a very prominent young man in the entertainment world, addressing a tremendous gathering in California. It was during the time of the cold war, and communism and our way of life were very much on people's minds. And he was speaking to that subject. And suddenly, though, I heard him saying 'I love my little girls more than anything. . . .' And I said to myself, 'Oh, no, don't. You can't—don't say that.' But I had underestimated him. He went on: 'I would rather see my little girls die now, still believing in God, than have them grow up under communism and one day die no longer believing in God.'

"There were thousands of young people in that audience. They came to their feet with shouts of joy. They had instantly recognized the profound truth in what he had said, with regard to the physical and the soul and what was truly important" (*Weekly Compilation of Presidential Documents* 19, no. 10 [8 March 1983]: 368–69). We read this to imply the slogan Better Dead than Red.

31. President's Address to the Nation, 14 November 1985, in *U.S. Department of State Bulletin* 86, no. 2106 (1 January 1986): 3.

32. Gorbachev, *Perestroika*, 199–204.

33. Reagan claimed to have "no illusions about the Soviets or their ultimate intentions." He said there are "critical moral distinctions between totalitarianism and democracy" and "the principal objective of American foreign policy [is] not just the prevention of war but the extension of freedom . . . and democratic institutions around the world." On these bases, Reagan legitimized U.S. actions in "assist[ing] freedom fighters who are resisting the imposition of totalitarian rule in Afghanistan, Nicaragua, Angola, Cambodia, and elsewhere" (*Vital Speeches of the Day* 53, no. 2. [1 November 1986]: 35).

34. For the specific text of the "National Security doctrine" see the National Security Act of 1947. See also Waltraud Queiser Morales, "The War on Drugs: A New US National Security Doctrine?" *Third World Quarterly* 11, no. 3 (July 1989): 147–69; and "The Framework for Policy: The National Interest," in *Changing Course: Blueprint for Peace in Central America and the Caribbean* (Washington, D.C.: Institute for Policy Studies, 1984), 41–61.

35. The "Zero-Option" proposal was unveiled 18 November 1981 in a speech delivered at the National Press Club and transmitted to the European Broadcasting Union. As Mark Heertsgaard reports, this speech was the

first to be targeted directly to Europe: it was delivered at 10 A.M. in the United States, which was "drive-time" in Europe. See Mark Heertsgaard, *On Bended Knee: The Press and the Reagan Presidency* (New York: Farrar, Straus & Giroux, 1988), 273.

36. *Weekly Compilation of Presidential Documents*, 9 May 1982, 602.

37. Reagan said, "I support a zero option for all nuclear arms. As I've said before, my dream is to see the day when nuclear weapons will be banished from the face of the Earth" (Address to the Nation, Allies, and the Soviet Union, *Weekly Compilation of Presidential Documents*, 16 January 1984, 41).

38. Reagan demonized the Soviet Union to suggest the hopelessness of reaching an ideal state of total elimination: "They reserve unto themselves the right to commit any crime, to lie, to cheat" (quoted in "Where's the Rest of Him?" *New Republic*, 6 February 1984, 7).

39. "It would be wonderful if we could restore our balance with the Soviet Union without increasing our own military power. And ideally, it would be a long step in ensuring peace if we could have significant and verifiable reductions of arms on both sides. But let's not fool ourselves" (President's Radio Address, 26 April 1982, *Weekly Compilation of Presidential Documents* 18, no. 16 (2 March 1982): 504.

40. "Governments will always try to co-opt public sentiment when it reaches a certain intensity by using negotiations to propose cosmetic changes" (Pam Solo, *From Protest to Policy: Beyond the Freeze to Common Security* [Cambridge, Mass.: Ballinger, 1988], 2).

41. "Where's the Rest of Him?" 7–8.

42. Solo, *From Protest to Policy*, 135.

43. Editors, "Reagan and the World," *The Nation*, 2 November 1985, 425.

44. William F. Buckley, Jr., "The Meaning of Eureka," *National Review*, 11 June 1982, 719.

45. *Weekly Compilation of Presidential Documents*, 4 October 1982, 1260.

46. Robert J. Branham, "Roads Not Taken: Counterplans and Opportunity Costs," *Argumentation and Advocacy* (in press).

47. Mark Heertsgaard, *On Bended Knee*, 271.

48. W. Barnett Pearce, Stephen W. Littlejohn, and Alison Alexander, "The New Christian Right and the Humanists' Response: Reciprocated Diatribe," *Communication Quarterly* 35 (1987): 171–92.

49. William Connolly, *The Terms of Political Discourse* (Princeton, N.J.: Princeton University Press, 1983), 1.

50. Moshe Lewin, *The Gorbachev Phenomenon: A Historical Interpretation* (Berkeley and Los Angeles: University of California Press, 1988), 114.

51. That is, it stands between "fundamentalists" who would give an excessively literal reading to the classic texts, and the "liberals" who would read them poetically, if at all.

52. Gorbachev, *Perestroika*, 15.

53. Ibid., 11–14.

54. Ibid., 149, 204.

55. Ibid., xi.

56. Ibid., 215, 230.

57. Ibid., 223.

58. Ibid., 219–20.

59. Ibid., 223.

60. John Newhouse, *War and Peace in the Nuclear Age* (New York: Alfred A. Knopf, 1988), 395–97; see also Gorbachev, *Perestroika*, 223.

61. President's Remarks, 14 October 1986, *U.S. Department of State Bulletin* 86, no. 2117 (December 1986): 20–21.

62. Gorbachev, *Perestroika*, 230.

63. White House Statement, 15 October 1986, *Department of State Bulletin* 86, no. 2117 (December 1986): 21.

64. Gorbachev's Television Address, 14 October 1986, in *Soviet Life*, November 1987, 2.

65. Newhouse, *War and Peace*, 401.

66. Ibid., 397.

67. Ibid.

68. Ibid., 398.

69. Ibid., 396.

70. Ibid.

71. Michael Holquest, Introduction, to *Rabelais and His World*, by Mikhail Bakhtin (Bloomington: Indiana University Press, 1984).

72. Jean Bethke Elshtain, *Women and War* (New York: Basic Books, 1987).

Chapter 8. Paranoid Style in Foreign Policy

1. J. A. Nathan and J. K. Oliver, *Foreign Policy Making and the American Political System* (Boston: Little, Brown, 1987), 222.

2. See esp. E. B. Burns, *At War in Nicaragua: The Reagan Doctrine and the Politics of Nostalgia* (New York: Harper and Row, 1987), 41–43.

3. Ibid., 36.

4. See Richard H. Ullman, "At War with Nicaragua," *Foreign Affairs* 62 (Fall 1983): 39–58; and Joel Muravchik, "The Nicaragua Debate," *Foreign Affairs* 65 (Winter 1986/87): 366–82.

5. Walter LaFeber, *Inevitable Revolutions: The United States in Central America* (New York: W. W. Norton, 1984), 276.

6. Ibid., 277.

7. Ullman, "At War with Nicaragua," 54.

8. Tony Smith, *The Pattern of Imperialism: The United States, Great Britain, and the Late-Industrializing World Since 1815* (Cambridge: Cambridge University Press, 1981), 153–54.

9. LaFeber, *Inevitable Revolutions*, 277.

10. Peter Kornbluh, "The Covert War," in *Reagan Versus the Sandinistas: The Undeclared War on Nicaragua*, ed. Thomas Walker (Boulder, Colo.: Westview Press, 1987), 22.

11. Ullman, "At War with Nicaragua," 54.

12. Eldon Kenworthy, "Selling the Policy," in *Reagan Versus the Sandinistas*, 162.

13. James Rosenau, *Public Opinion and Foreign Policy* (New York: Random House, 1961), 40.
14. Nathan and Oliver, *Foreign Policy Making*, 230.
15. Kornbluh, "The Covert War," 23.
16. Quoted in Kornbluh, "The Covert War," 23.
17. Ibid.
18. Richard Hofstadter, *The Paranoid Style in American Politics and Other Essays* (New York: Alfred A. Knopf, 1966), 3.
19. Ibid., 4.
20. Ibid., 29–30.
21. The speeches employed in this analysis and their respective sources and pagination are as follows: "Aid to the Caribbean Basin," *Vital Speeches of the Day* 48 (1982): 322–25; "The Problems of Central America," *Vital Speeches of the Day* 49 (1983): 450–54; "Address to the Nation on United States Policy in Central America, May 9, 1984," *Ronald Reagan: 1984*, Public Papers of the Presidents of the United States (Washington, D.C.: U.S. Government Printing Office, 1985), 659–65; and "Nicaragua," *Vital Speeches of the Day* 52 (1986): 386–89. All subsequent references come from these texts.
22. Bruce E. Gronbeck, "Rhetorical Invention in the Regency Crisis Pamphlets," *Quarterly Journal of Speech* 58 (1972): 418–30.
23. Nathan and Oliver, *Foreign Policy Making*, 236.
24. Hofstadter, *Paranoid Style*, 32.
25. Burns, *At War in Nicaragua*, 58.
26. Hofstadter, *Paranoid Style*, 36.
27. Ibid., 36.
28. Ullman, "At War with Nicaragua," 43–47.
29. Ibid., 47.
30. Ibid., 50.
31. Muravchik, "The Nicaragua Debate," 375.
32. Ibid., 371.
33. Ibid.
34. See Jeff D. Bass, "The Mass Public and the Dominican Intervention of 1965" (Paper presented at the annual convention of the Speech Communication Association, Chicago, 1990).
35. Philip Wander, "The Rhetoric of American Foreign Policy," *Quarterly Journal of Speech* 70 (1984): 339–61.
36. Ibid., 342–47.
37. See Michael T. Klare and Peter Kornbluh, eds. *Low Intensity Warfare: Counterinsurgency, Proinsurgency, and Antiterrorism in the Eighties* (New York: Pantheon, 1988).

Chapter 9. The Shootdown of Iran Air 655

1. Marilyn J. Young and Michael K. Launer, "KAL 007 and the Superpowers: An International Argument," *Quarterly Journal of Speech* 74 (1988): 288. For an expanded discussion, see Marilyn J. Young and Michael

K. Launer, *Flights of Fancy, Flight of Doom: KAL 007 and Soviet-American Rhetoric* (Lanham, Md.: University Press of America, 1988).

2. All quotations are taken from U.S. Department of State, *KAL Flight #007: Compilation of Statements and Documents, 1–16 September 1983* (Washington, D.C.: Bureau of Public Affairs, 1983), reprinted from the *U.S. Department of State Bulletin*, October 1983.

3. Department of State, *KAL Flight 007*, 1.

4. Donald Smith, "KAL 007: Making Sense of the Senseless" (Paper presented at the annual convention of the Speech Communication Association, Denver, November 1985).

5. Lloyd F. Bitzer, "The Rhetorical Situation," *Philosophy and Rhetoric* 1 (1968): 1–14 (for discussion, see vol. 3 [Summer 1970]: 165–68). Reprinted in Walter R. Fisher, ed. *Rhetoric: A Tradition in Transition* (East Lansing: Michigan State University Press, 1974), 247–60; also, Lloyd F. Bitzer, "Rhetoric and Public Knowledge," in *Rhetoric, Philosophy, and Literature: An Exploration*, ed. Don M. Burks (West Lafayette, Ind.: Purdue University Press, 1978), 67–94.

6. See Young and Launer, "KAL 007 and the Superpowers," 271. It is important to realize that Shultz himself had been left out of the planning. Casey, Burt, and Eagleburger had worked through the night developing the initial U.S. response. According to Seymour Hersh, the first time Shultz saw the text of the announcement was in his briefing book on the morning of 1 September. See Seymour M. Hersh, *The Target Is Destroyed* (New York: Random House, 1986), 99.

7. Although the administration was successful in avoiding an air of crisis, haste was still present, with its attendant errors of fact and judgment. See Anthony Lewis, "Question of Honor," *New York Times*, 4 August 1988, A25. As we document in *Flights of Fancy*, such haste contributed to the erosion of the administration's position on KAL.

8. The pilot study, which is reported here, was done by Jim Kuypers (a graduate student in rhetoric and public address at Florida State University) for a content analysis class. It is our intention to expand this study as part of a long-term analysis of Soviet-American rhetoric.

9. This analysis includes all examples of official administration rhetoric, unlike the earlier analysis reported by Donald Smith (see n. 4), which dealt solely with the discourse of President Reagan. However, these results do not contradict Smith's findings; rather, they lend support to his primary thesis.

10. Of the approximately sixty texts printed in *Krasnaya zvezda, Izvestiya*, and *Pravda* in the weeks immediately following the Airbus shootdown, twenty mentioned the number of dead.

11. See Young and Launer, *Flights of Fancy.*

12. Examples include House Majority Leader Jim Wright's unexpected revelation of the presence of the RC-135 reconnaissance aircraft near the Kamchatka Peninsula and the timing of revisions to the transcript of the Soviet interceptor pilots talking with their ground control. For details and a complete discussion, see Young and Launer, "KAL 007 and the Superpowers."

13. *Weekly Compilation of Presidential Documents* 24 (11 July 1988): 896.

14. See *New York Times*, 5 July 1988. Reagan stated that the airliner was headed toward the *Vincennes* in an attack mode; in fairness, he may not—at that point—have been aware of the contradictory information.

15. *Weekly Compilation of Presidential Documents* 24 (18 July 1988): 911.

16. NBC News, "Network Report," 3 July 1988, 1:22 P.M., 7.

17. Press briefing, 3 July 1988.

18. CBS News, "CBS Evening News with Susan Spencer," Sunday, 3 July 1988, 6:30 P.M., 4.

19. CBS News, "CBS Sunday Night News with Bill Plante," 3 July 1988, 11:00 P.M., 2.

20. NBC News, "Network Report," 4 July 1988, 12:30 P.M., 2. See also *Washington Post*, 5 July 1988. The truly significant thing about this passage is Reagan's insistence that the Soviets had identified KAL as a civilian aircraft. Within days of the 1983 shootdown, the *New York Times* had reported that intelligence analysts believed the Soviets did not realize the airliner was a passenger plane; this ultimately became the official U.S. position and was confirmed by Hersh in *The Target Is Destroyed.* Final confirmation of the Defense Intelligence Agency assessment was released by Congressman Lee Hamilton in January 1988. President Reagan should have known better in 1983; he certainly should have been aware by July 1988. One can only conclude that it suited his purpose to so misstate the circumstances in this instance.

21. See *New York Times*, 4 July 1988.

22. This move was also motivated by a desire to avoid comparisons with the Carter administration's handling of the hostage crisis, when the president appeared to be "swept along by foreign events." See *Washington Post*, 4 July 1988. An interesting comparison can be drawn to 1983: after deciding initially to remain on vacation in Santa Barbara, Reagan returned to Washington under pressure from media criticism (criticism that did not occur in 1988). One wonders who had made the decision to promote the crisis atmosphere generated after the KAL tragedy: five years later (5 July 1988) the *New York Times* reported that it was Reagan himself who decided to stay at Camp David.

23. CBS News, "CBS Sunday Night News with Bill Plante," 3 July 1988, 11:00 P.M., 2.

24. *Washington Post*, 4 July 1988.

25. *New York Times*, 18 July 1988.

26. *Washington Post*, 4 July 1988.

27. Ibid.

28. See, for example, *Washington Post*, 5 July 1988.

29. Voice of America, CN-047, 4 July 1988, 12:43 P.M.

30. Voice of America, Background Report 5-04651, 3 July 1988, 10:00 P.M.

31. Voice of America, CLOSE-UP, 4-02923, 3 July 1988.

32. VOA and RFE/RL have very different goals: VOA is the official radio voice of the U.S. government; RL, on the other hand, seeks greater neutrality and provides information about conditions, attitudes, and trends

within the target countries as well as news about international developments "as they relate to the special interests of the listeners." *The Right to Know* (Washington, D.C.: U.S. Government Printing Office, 1973), 2.

33. Radio Liberty, "Daily Broadcast Analysis, Russian Service," Monday, 4 July 1988, 5.

34. Ibid., Tuesday, 5 July 1988, 1–2.

35. Ibid., Thursday, 7 July 1988, 6.

36. Raymie McKerrow, private communication.

37. See United Nations Organization, International Civil Aviation Organization, *Destruction of Korean Air Lines Boeing 747 over Sea of Japan, 31 August 1983, Report of the ICAO Fact-finding Investigation*, 30 December 1983. See also United Nations Organization, International Civil Aviation Organization, Air Navigation Commission, *1818th Report to Council by the President of the Air Navigation Commission*, Document C-WP/7809, 16 February 1984.

38. CBS News, "CBS Evening News with Dan Rather," Monday, 4 July 1988, 6:30 P.M., 11.

39. NBC News, "Network Report," 3 July 1988, 2:31 P.M., 2.

40. *New York Times*, 7 July 1988.

41. *New York Times*, 4 July 1988. This response raises an interesting rhetorical problem, however. One observer pointed out that Middle Easterners are convinced that U.S. equipment is so sophisticated it could not make such an error. The Soviets made a similar argument in 1983 in claiming that (a) the triply redundant inertial navigation system aboard the Boeing 747 was so sophisticated the plane could not fly off course by accident, and (b) that the U.S. had KAL 007 under radar surveillance throughout its flight. Such beliefs go a long way toward explaining the genesis and longevity of conspiracy theories when tragedies such as these occur.

42. For a refutation of this claim see Michael K. Launer, Marilyn J. Young, and Sugwon Kang, "The KAL Tapes: What They Really Say About the Tragedy," *Bulletin of Concerned Asian Scholars*, 18, no. 3 (1986): 67–71. See also Young and Launer, *Flights of Fancy*, 83–100.

43. Such claims were ultimately rejected by the U.S. government; the final report of the naval investigation of the tragedy concluded that the crew of the USS *Vincennes* had mistaken the radar signal of a distant Iranian fighter for that of the airbus. One needs considerably more technical expertise than I possess to understand how this can happen.

44. *New York Times*, 8 August 1988.

45. Press briefing, 3 July 1988.

46. Cited in Richard Rohmer, *Massacre 747: The Story of Korean Air Lines Flight 007* (Markham, Ont.: PaperJacks, 1984), 76.

47. *Pravda*, 28 September 1983; see also *New York Times*, 29 September 1983.

48. For a description of the Soviet statement of 3 September 1983 (the third day *after* the shootdown), see Young and Launer, *Flights of Fancy*, 140. The original of this statement appeared in *Pravda*, as a TASS release, 3 September 1983. The relevant portion reads, "TASS was authorized to express condolences regarding the casualties, but resolutely condemns

those who, whether consciously or through criminal negligence, had allowed lives to be lost." The Soviets, one must remember, did not officially admit shooting down KAL 007 until 6 September and did not begin their defensive campaign until three days after that—9 September—when Marshal Ogarkov held his press conference.

49. *Weekly Compilation of Presidential Documents* 24 (11 July 1988): 895–908.

50. The most obvious examples of this type of transformation are instances of "military aid" to "requesting governments" such as the Soviets in Hungary and Czechoslovakia and the United States in Grenada; a similar situation obtained with the Iran-Contra scandal.

51. The Soviets suffered a similar failure with the Chernobyl disaster: there was no one else to blame for the accident in the number 4 reactor and the radiation threat that resulted. The erosion of the strategy in Afghanistan may be largely responsible for the Soviet withdrawal there. The United States is no stranger to this approach either; consider U.S. military police actions in places such as Vietnam, Lebanon, the Dominican Republic, and, of course, Grenada. It is when the interpretation breaks down at home that the operation ultimately fails: as long as we perceived ourselves as helping the government of South Vietnam, or the Soviets saw themselves as assisting the government of Afghanistan, the interpretive fiction prevailed.

52. Grant Hugo, *Appearance and Reality in International Relations* (London: Chatto and Windus, 1970). See especially pp. 10–20. On p. 19 Hugo notes: "Cant, therefore, will be considered here as a mode of expression, or a cast of thought, of which the effect—irrespective of the motive—is to create a misleading discrepancy between the natural meaning of words and their practical significance, a discrepancy even more dangerous when, as often happens, the speaker is as much misled as his audience."

53. Hugo, *Appearance and Reality*, 10–11.

54. Also significant is the final report of the investigation, but its rhetorical importance is not relevant to the purpose of this paper.

55. This version was broadcast by Radio Liberty for several hours following the first reports of the incident.

56. U.S. Department of State, Daily Press Briefing (Phyllis Oakley, briefer), Wednesday, 6 July 1988 (Office of Public Affairs), 4.

57. Soviet rhetoric regarding KAL underwent a similar evolution: the shootdown began as a nonevent; became an event that might have been an accident but also might have been intentional; then became a planned spy mission; and finally became a spy mission with the secondary motive of embarrassing the Soviet Union. Ultimately, it was characterized as a provocation intentionally designed to trap the Soviet Union into destroying a passenger plane for the United States' own purposes.

58. United Nations Organization, Security Council, *Provisional Verbatim Record*, 20 July 1988 (S/PV.2821), 14–15.

59. For a discussion of public knowledge in this context, see Young and Launer, "KAL 007 and the Superpowers." See also Bitzer, "The Rhetorical Situation," and "Rhetoric and Public Knowledge." For a discussion of the future rhetorical functions served by such normalizing discourse, see

Thomas Kane, "Rhetorical Histories and Arms Negotiations," *Journal of the American Forensic Association* 24 (1988): 143–54.

60. The Soviet Union vetoed the Security Council resolution in 1983, but it was much stronger than the results of the 1988 debate.

61. Certainly, the role of the *Stark* in establishing the context for the public understanding of Airbus should not be overlooked: there were ten references to that tragic incident in the Reagan administration's Airbus rhetoric.

62. In fairness it must be noted that the Soviet Union assisted in this, by demurring rather than seizing the opportunity to berate the United States before a world audience. This is the most fascinating aspect of the Soviet side of this episode; undoubtedly its moderate international stance reflects the current rapproachement in U.S.-Soviet relations. Domestically, however, Soviet commentary was predictably quite biting. See *New York Times*, 4 July 1988.

63. Ibid., 14 July 1988.

Chapter 10. Reagan Attack on Welfare

1. Murray Edelman, *Political Language: Words That Succeed and Policies That Fail* (New York: Academic Press, 1977), 2.

2. In 1988, 13.1 percent or 31.9 million Americans fell below the poverty line. See Center on Budget and Policy Priorities, "Poverty Rate and Household Income Stagnant As Rich-Poor Gap Hits Post-War High" (Press release, Washington, D.C., 20 October 1989).

3. See Hugh Heclo, "The Political Foundations of Antipoverty Policy," in *Fighting Poverty*, ed. Sheldon H. Danziger and Daniel H. Weinberg (Cambridge, Mass.: Harvard University Press, 1986), 328–29. Heclo's most recent figures (from the Roper Poll) are from 1984 and show that 51 percent favor more spending and another 41 percent favor maintaining current spending levels. These figures do not differ substantially from results obtained in 1948 and 1961.

4. See David T. Ellwood, *Poor Support* (New York: Basic Books, 1988). Ellwood describes the book as his "attempt to understand the widespread disdain for welfare that exists in spite of the professed desire of most Americans to help the poor" (p. x). See also Heclo, "Political Foundations," 330.

5. Ellwood, *Poor Support*, 36–37.

6. Heclo, "Political Foundations," 330.

7. Ibid., 329.

8. See Mary Jo Bane, "Politics and Policies of the Feminization of Poverty," in *The Politics of Social Policy in the United States*, ed. Margaret Weir, Ann Shola Orloff, and Theda Skocpol (Princeton, N.J.: Princeton University Press, 1988), 381–96.

9. See John Schwartz, *America's Hidden Success: A Reassessment of Twenty Years of Public Policy* (New York: Norton, 1983). Schwartz argues that antipoverty policy since the 1960s reduced poverty significantly de-

spite its image of waste and failure. For the definitive study of how Lyndon Johnson's poverty rhetoric created unrealistic expectations of the War on Poverty, see David Zarefsky, *President Johnson's War on Poverty* (Tuscaloosa: University of Alabama Press, 1986).

10. Heclo, "Political Foundations," 329–32.

11. See David T. Ellwood and Lawrence H. Summers, "Poverty in America: Is Welfare the Answer or the Problem?" in *Fighting Poverty*, ed. Danziger and Weinberg, 86–87.

12. William Julius Wilson, *The Truly Disadvantaged* (Chicago: University of Chicago Press, 1987), 94–95.

13. Murray Edelman,*Constructing the Political Spectacle* (Chicago: University of Chicago Press, 1988), 60–61.

14. D. Lee Bawden and John L. Palmer, "Social Policy: Challenging the Welfare State," in *The Reagan Record*, ed. John L. Palmer and Isabel V. Sawhill (Cambridge, Mass.: Ballinger, 1984), 187.

15. Ibid., 194–201.

16. Ibid., 198.

17. See Nathan H. Schwartz, "Reagan's Housing Policies," in *The Attack on the Welfare State*, ed. Anthony Champagne and Edward J. Harpham (Prospect Heights, Ill.: Waveland, 1984), 157–63.

18. A reduction in benefit *levels* does not mean that total social welfare *spending* declines by a similar proportion or even at all. During recessions, for example, some unemployed workers join the ranks of the poor and demand benefits. When programs such as Food Stamps that benefit the near-poor as well as poor are cut, more people become poor and are therefore eligible for programs such as AFDC that benefit the poor only. During the 1970s the welfare rights movement caused a higher percentage of those actually eligible for benefits to apply for them. Finally, the skyrocketing costs of medical care have made the Medicaid program for the poor increasingly expensive per recipient.

19. Ellwood, *Poor Support*, 40.

20. See Theda Skocpol, "The Limits of the New Deal System and the Roots of Contemporary Welfare Dilemmas," in *Politics of Social Policy*, ed. Weir, Orloff, and Skocpol, 293–311.

21. See Ann Shola Orloff, "The Political Origins of America's Belated Welfare State," in *Politics of Social Policy*, ed. Weir, Orloff, and Skocpol, 37–80.

22. On 12 December 1945 Reagan addressed a meeting of the Hollywood Independent Citizens Committee of Arts, Sciences and Professions (HICCASP) on the topic of government waste in wartime contracts. See Anne Edwards, *Early Reagan* (New York: William Morrow, 1987), 293.

23. Ibid., 418. Reagan returned to the SAG presidency briefly in 1959–60; ibid., 465, 467.

24. Garry Wills, *Reagan's America: Innocents at Home* (Garden City, N.Y.: Doubleday, 1987), 248–49.

25. Edwards, *Early Reagan*, 453–57.

26. "A Time for Choosing," 27 October 1964, San Francisco, Calif., in Edwards, *Early Reagan*, 562.

27. Ibid., 565.

28. See Zarefsky's excellent discussion of the inception of the War on Poverty (*Johnson's War on Poverty*, 21–56).

29. See Skocpol, "Limits of the New Deal," 296.

30. Zarefsky, *Johnson's War on Poverty*, 48.

31. Ellwood, *Poor Support*, 32.

32. Since 1970 the largest increases in expenditures for "means-tested" programs aimed at the poor have occurred in Medicaid and Food Stamps. Yet public hostility has been greatest regarding the "welfare" program that has increased the least, AFDC. See Ellwood and Summers, "Poverty in America," 83–88.

33. Retold in Elizabeth Drew, *American Journal: The Events of 1976* (New York: Random House, 1977), 51–52.

34. See Thomas Byrne Edsall, *The New Politics of Inequality* (New York: Norton, 1984), 39–40.

35. *Political Language: Words That Succeed and Policies That Fail* (New York: Academic Press, 1977).

36. See Ernesto Laclau, *Politics and Ideology in Marxist Theory* (London: New Left Books, 1977), 157–58.

37. Marx noted that all ruling classes make this claim and must do so to rule legitimately. See Karl Marx, "The German Ideology, Part I," in *The Marx-Engels Reader*, ed. Robert C. Tucker (New York: Norton, 1978), 174.

38. I use these terms as Aristotle did in distinguishing between rhetoric that advocates or dissuades from action (deliberative) and rhetoric that praises and blames (epideictic). See Aristotle, *Rhetoric* 1.3.

39. See Michael Calvin McGee, "In Search of 'The People': A Rhetorical Alternative," *Quarterly Journal of Speech* 61 (1975): 235–38.

40. For a particularly insightful comparison of these two kinds of populism and their rhetorical forms, see Ronald Lee, "The New Populist Campaign for Economic Democracy: A Rhetorical Explanation," *Quarterly Journal of Speech* 72 (1986): 274–89.

41. In *The Triumph of the American Spirit: The Presidential Speeches of Ronald Reagan*, ed. Emil Arca and Gregory Pamel (Detroit: National Reproductions, 1984), 4.

42. Ibid., 72.

43. Ibid., 91.

44. For an explication of this doctrine by two of its most articulate supporters, see George Gilder, *Wealth and Poverty* (New York: Basic Books, 1981); and Michael Novak, *The Spirit of Democratic Capitalism* (New York: AEI/Simon and Schuster, 1982).

45. *Ronald Reagan's Weekly Radio Addresses* (Wilmington, Del.: Scholarly Resources: 1987), 103.

46. Michael Paul Rogin, "Ronald Reagan: The Movie," in *Ronald Reagan: "The Movie," and Other Episodes of Political Demonology* (Berkeley and Los Angeles: University of California Press, 1987), 3–44.

47. *Ronald Reagan: 1981*, Public Papers of the Presidents of the United States (Washington, D.C.: U.S. Government Printing Office, 1982), 883.

48. On the importance of volunteerism, see Friedrich A. von Hayek,

Law, Legislation, and Liberty (Chicago: University of Chicago Press, 1976), 2:150–51.

49. Bawden and Palmer, "Social Policy," 187.

50. See, for example, William F. Lewis, "Telling America's Story: Narrative Form and the Reagan Presidency," *Quarterly Journal of Speech* 73 (1987): 280–302.

51. Arca and Pamel, *Triumph*, 81–82.

52. Ibid., 85.

53. Edsall, *New Politics*, 39–41.

54. See Bernadyne Weatherford, "The Disability Insurance Program: An Administrative Attack on the Welfare State," in *Attack*, ed. Champagne and Harpham, 37–60.

55. Gregory B. Mills, "The Budget: A Failure of Discipline," in *The Reagan Record*, ed. Palmer and Sawhill, 131.

56. David Zarefsky, Carol Miller-Tutzauer, and Frank E. Tutzauer, "Reagan's Safety Net for the Truly Needy: The Rhetorical Uses of Definition," *Central States Speech Journal* 35 (1984): 113–19.

57. Ibid., 117.

58. Arca and Pamel, *Triumph*, 336.

59. *Ronald Reagan: 1981*, Public Papers of the Presidents, 882.

60. Ibid.

61. Zarefsky, Miller-Tutzauer, and Tutzauer, "Reagan's Safety Net," 118.

62. *Radio Addresses*, 15.

63. Arca and Pamel, *Triumph*, 81.

64. Edelman, *Constructing the Political Spectacle*, 17.

65. See Zarefsky, Miller-Tutzauer, and Tutzauer, "Reagan's Safety Net," 114–15.

66. *Ronald Reagan: 1981*, Public Papers of the Presidents, 882.

67. Bawden and Palmer, "Social Policy," 189n.

68. Zarefsky, Miller-Tutzauer, and Tutzauer, "Reagan's Safety Net," 116.

69. Edelman, *Constructing the Political Spectacle*, 16–17.

70. *Ronald Reagan: 1981*, Public Papers of the Presidents, 1077.

71. Arca and Pamel, *Triumph*, 81.

72. *Ronald Reagan: 1981*, Public Papers of the Presidents, 882.

73. Kevin Phillips, among others, suggests that David Stockman, Reagan's notorious director of the Office of Management and Budget, was well aware that the early Reagan budgets would produce large deficits, supply-side economics notwithstanding. See *The Politics of Rich and Poor* (New York: Random House, 1990), 89.

74. See Daniel Patrick Moynihan, "Another War—The One on Poverty—Is Over, Too," *New York Times*, 16 July 1990, A15.

75. *Edelman, Constructing the Political Spectacle*, 32.

Chapter 11. The City as Marketplace

1. Kenneth Burke, *A Grammar of Motives* (Berkeley and Los Angeles: University of California Press, 1969), 21.

2. Ibid., 23.
3. Ibid., xix.
4. Ibid., 51.
5. Ibid., 32.
6. Kenneth Burke, *A Rhetoric of Motives* (Berkeley and Los Angeles: University of California Press, 1969), 21.
7. Burke, *A Grammar of Motives*, 503.
8. Ibid., 517.
9. Ibid., 506.
10. Burke's definitions of the tropes we use in our analysis—metaphor, synecdoche, and metonymy—are fairly ambiguous. Terence Hawkes supplies clearer definitions. He defines metaphor as a "particular set of linguistic processes whereby aspects of one object are 'carried over' or transferred to another object, so that the second object is spoken of as if it were the first." Synecdoche too works through transference, but "here the transference takes the form of a part of something being 'carried over' to stand in place of the whole thing, or vice versa." In metonymy "the name of a thing is transferred to take the place of something else with which it is associated: 'The White House' for the President of the United States." Terence Hawkes, *Metaphor* (New York: Methuen, 1972), 1, 2.
11. Burke, *A Grammar of Motives*, xix.
12. Deborah Stone, *Policy Paradox and Political Reason* (Glenview, Ill.: Scott, Foresman, 1988), 4.
13. Ibid., 123.
14. Ibid.
15. Burke, *A Rhetoric of Motives*, 43.
16. Stone, *Policy Paradox*, 108.
17. Kenneth Burke, *Counter-Statement* (Berkeley and Los Angeles: University of California Press, 1968), 31.
18. Burke, *A Grammar of Motives*, 57.
19. Weldon Durham, "Kenneth Burke's Concept of Substance," *Quarterly Journal of Speech* 66 (1980): 355.
20. Myron Levine, "The Reagan Urban Policy," *Journal of Urban Affairs* 5 (1983): 17.
21. U.S. Department of Housing and Urban Development, *The President's National Urban Policy Report* ([Washington, D.C.]: U.S. Government Printing Office, 1982), 1.
22. Ronald Reagan, "Message on Enterprise Zones," *Congressional Quarterly* 40 (1983): 705.
23. Richard Mounts, "Urban Enterprise Zones: Will They Work?" *American Review of Public Administration* 15 (1981): 86–96.
24. Levine, "The Reagan Urban Policy," 21.
25. Michael Brintnall and Roy Green, "Comparing State Enterprise Zone Programs," *Economic Development Quarterly* 2 (1988): 51.
26. Ronald Reagan, "The State of the Union," in *American Intergovernmental Relations*, ed. Laurence O'Toole (Washington, D.C.: CQ Press, 1985), 247–48.
27. Reagan, "Message on Enterprise Zones," 705.

28. Stuart Butler, "Free Zones in the Inner City," in *Urban Economic Development*, ed. Richard Bingham and John Blair (Beverly Hills, Calif.: Sage, 1984), 141.
29. Reagan, "Message on Enterprise Zones," 705.
30. Jack Kemp, "Lighting Candles from Harlem to Watts," *Washington Post National Weekly*, 25 September–1 October 1989, 29.
31. Reagan, "Message on Enterprise Zones," 705.
32. Ibid.
33. Mark Gelfand, *A Nation of Cities* (New York: Oxford University Press, 1975).
34. Reagan, "Message," 705.
35. Ibid., 706.
36. Department of Housing and Urban Development, *Urban Policy Report*, 2.
37. E. S. Savas, "A Positive Urban Policy for the Future," *Urban Affairs Quarterly* 18 (1983): 450.
38. Stone, *Policy Paradox*, 109.
39. Durham, "Kenneth Burke's Concept," 356.
40. Burke, *A Grammar of Motives*, 28.
41. Robert Bellah, *Broken Covenant* (New York: Seabury Press, 1975).
42. Sidney Blumenthal, *The Rise of the Counter-Establishment* (New York: Times Books, 1986), 253.
43. Reagan, "Message on Enterprise Zones," 705.
44. Blumenthal, *Rise*, 255.
45. Ibid., 258.
46. Robert Dallek, *Ronald Reagan: The Politics of Symbolism* (Cambridge, Mass.: Harvard University Press, 1984), 63.
47. John Kenneth White, *The New Politics of Old Values* (Hanover, N.H.: University Press of New England, 1988), 4.
48. Burke, *A Grammar of Motives*, 30.
49. Reagan, "Message on Enterprise Zones," 706.
50. Ibid.
51. Elizabeth Drew, *Portrait of an Election: The 1980 Presidential Campaign* (New York: Simon and Schuster, 1981), 325.
52. Reagan, "Message on Enterprise Zones," 706.
53. Stone, *Policy Paradox*, 113.
54. Reagan, "Message on Enterprise Zones," 705.
55. Ibid.
56. Timothy Barnekov, Robin Boyle, and Daniel Rich, *Privatism and Urban Policy in Britain and the United States* (New York: Oxford University Press, 1989), 122.

Chapter 12. Civil Religion and Public Argument

1. Robert N. Bellah, "Civil Religion in America," *Daedalus* 96 (Winter 1967): 7.

2. James David Fairbanks, "The Priestly Functions of the Presidency," *Presidential Studies Quarterly* 11 (1981): 214–32.

3. Bellah, "Civil Religion in America," 3.

4. Roderick P. Hart, *The Political Pulpit* (West Lafayette, Ind.: Purdue University Press, 1977), 45.

5. Ibid., 61.

6. Kristin Luker, *Abortion and the Politics of Motherhood* (Berkeley and Los Angeles: University of California Press, 1984), 186.

7. Randall A. Lake, "Order and Disorder in Anti-Abortion Rhetoric: A Logological View," *Quarterly Journal of Speech* 70 (1984): 426.

8. Ibid.

9. Robert J. Nelson, "Religion and Abortion," *Center Magazine* 14 (July/August 1981): 51.

10. Ibid., 54.

11. Ibid., 55.

12. Ronald W. Reagan, *Weekly Compilation of Presidential Documents* (1987), 880.

13. Randall A. Lake, "The Metaethical Framework of the Anti-Abortion Debate," *Signs* 11 (Spring 1986): 481.

14. Ibid., 480–81.

15. Ibid., 482.

16. Bellah, "Civil Religion in America," 4–6.

17. *Weekly Compilation of Presidential Documents* (1988), 73.

18. *Weekly Compilation of Presidential Documents* (1986), 80; (1987), 42–43; (1988), 73.

19. *Weekly Compilation of Presidential Documents* (1985), 74.

20. Ibid.

21. *Weekly Compilation of Presidential Documents* (1986), 80.

22. Ibid.

23. Ibid., 81.

24. *Weekly Compilation of Presidential Documents* (1987), 83.

25. Bellah, "Civil Religion in America," 3.

26. *Weekly Compilation of Presidential Documents* (1987), 43.

27. *Weekly Compilation of Presidential Documents* (1988), 73–74.

28. *Weekly Compilation of Presidential Documents* (1985), 74.

29. Celeste Michelle Condit, *Decoding Abortion Rhetoric: Communicating Social Change* (Urbana: University of Illinois Press, 1990), 159–60, 184–85.

30. *Weekly Compilation of Presidential Documents* (1986), 81.

31. *Weekly Compilation of Presidential Documents* (1987), 43.

32. *Weekly Compilation of Presidential Documents* (1988), 73.

33. Dale Spender, *Man Made Language* (Boston: Routledge and Kegan Paul, 1980), 163.

34. *Weekly Compilation of Presidential Documents* (1988), 74.

35. Ibid.

36. Ronald W. Reagan, "Abortion and the Conscience of the Nation," reprinted in *Catholic Lawyer* 30 (Spring 1986): 105.

37. Ibid., 106.
38. Ibid., 100.
39. *Weekly Compilation of Presidential Documents* (1987), 901.
40. Ibid.
41. Ibid.
42. Ibid.
43. Ibid.
44. Lake, "Metaethical Framework," 490.
45. Ibid., 494.
46. Ibid.
47. *Weekly Compilation of Presidential Documents* (1985), 74.
48. Ibid.
49. *Weekly Compilation of Presidential Documents* (1988), 74.
50. *Weekly Compilation of Presidential Documents* (1985), 74.
51. *Weekly Compilation of Presidential Documents* (1987), 42.
52. *Weekly Compilation of Presidential Documents* (1986), 80.
53. Barbara Duden, "The Pregnant Woman and the Public Fetus" (Typescript, October 1988, STS-Program, Penn State University), 8.
54. Ibid., 29–30.
55. *Weekly Compilation of Presidential Documents* (1984), 293.
56. *Weekly Compilation of Presidential Documents* (1988), 1328.
57. Lake, "Metaethical Framework," 493.
58. *Weekly Compilation of Presidential Documents* (1986), 80.
59. *Weekly Compilation of Presidential Documents* (1985), 74.
60. In his 1984 State of the Union address, Reagan said, "During our first three years, we have joined bipartisan efforts to restore protection of the law to unborn children. Now, I know this issue is very controversial. But unless and until it can be proven that an unborn child is a living human being, can we justify assuming without proof that it isn't? No one has yet offered such proof; indeed, all the evidence is to the contrary. We should rise above bitterness and reproach, and if Americans could come together in a spirit of understanding and helping, then we could find positive solutions to the tragedy of abortion" (*Weekly Compilation of Presidential Documents* [1984], 91). Implicit in this statement is the precondition for understanding that those who are prochoice recognize that abortion is a tragedy. Insofar as prochoice advocates are unwilling to do this, they are cast as undermining understanding and compromise. Reagan shifted the blame away from the antiabortionists and placed it on prochoice advocates. The quotation is from *Weekly Compilation of Presidential Documents* (1985), 74.
61. *Weekly Compilation of Presidential Documents* (1986), 487.
62. *Weekly Compilation of Presidential Documents* (1988), 74.
63. Craig Allan Smith, "Mister Reagan's Neighborhood: Rhetoric and National Unity," in *Essays in Presidential Rhetoric*, ed. Theodore Windt and Beth Ingold (Dubuque, Iowa: Kendall-Hunt, 1987).
64. *Weekly Compilation of Presidential Documents* (1988), 73.
65. *Weekly Compilation of Presidential Documents* (1986), 81.
66. *Weekly Compilation of Presidential Documents* (1988), 74.

67. Ibid.
68. Ibid.
69. Reagan, "Abortion and the Conscience of the Nation," 100.
70. Ibid., 103.
71. Condit, *Decoding Abortion Rhetoric*, 162.

Bibliography

Austin, J. L. *How to Do Things with Words.* Cambridge, Mass.: Harvard University Press, 1975.

Bane, Mary Jo. "Politics and Policies of the Feminization of Poverty." In *The Politics of Social Policy in the United States,* edited by Margaret Weir, Ann Shola Orloff, and Theda Skocpol. Princeton, N.J.: Princeton University Press, 1988.

Barrett, Laurence I. *Gambling with History: Ronald Reagan in the White House.* New York: Doubleday, 1983.

Bellah, Robert. *Broken Covenant.* New York: Seabury Press, 1975.

———. "Civil Religion in America." *Daedalus* 96 (Winter 1967): 1–21.

Bitzer, Lloyd F. "The Rhetorical Situation." *Philosophy and Rhetoric* 1 (1968): 1–14.

———. "Rhetoric and Public Knowledge." In *Rhetoric, Philosophy, and Literature: An Exploration,* edited by Don M. Burks. West Lafayette, Ind.: Purdue University Press, 1978.

Blankenship, Jane, Marlene G. Fine, and Leslie Davis. "The 1980 Republican Primary Debates: The Transformation of Actor to Scene." *Quarterly Journal of Speech* 69 (1983): 25–36.

Blumenthal, Sidney. *Our Long National Daydream: A Political Pageant of the Reagan Era.* New York: Harper and Row, 1988.

———. *The Permanent Campaign.* Rev. ed. New York: Simon and Schuster, 1982.

———. *The Rise of the Counter-Establishment.* New York: Times Books, 1986.

Boorstin, Daniel J. *The Image: A Guide to Pseudo-Events in America.* New York: Harper Colophon, 1961.

Bormann, Ernest, J. Jeffery Auer, and Franklyn S. Haiman. "Ghostwriting and the Cult of Leadership." *Communication Education* 33 (1984): 301–7.
Burke, Kenneth. *Counter-Statement.* Berkeley and Los Angeles: University of California Press, 1968.
———. *A Grammar of Motives.* Berkeley and Los Angeles: University of California Press, 1969.
———. *The Philosophy of Literary Form.* Berkeley and Los Angeles: University of California Press, 1973.
———. *A Rhetoric of Movies.* Berkeley and Los Angeles: University of California Press, 1969.
Burns, E. B. *At War in Nicaragua: The Reagan Doctrine and the Politics of Nostalgia.* New York: Harper and Row, 1987.
Campbell, Karlyn Kohrs, and Kathleen Hall Jamieson. *Deeds Done in Words: Presidential Rhetoric and the Genres of Governance.* Chicago: University of Chicago Press, 1990.
Cannon, Lou. *President Reagan: The Role of a Lifetime.* New York: Simon and Schuster, 1991.
———. *Reagan.* New York: Putnam, 1982.
Condit, Celeste Michelle. *Decoding Abortion Rhetoric: Communicating Social Change.* Urbana: University of Illinois Press, 1990.
Connolly, William. *The Terms of Political Discourse.* Princeton, N.J.: Princeton University Press, 1983.
Cox, J. Robert. "Argument and the 'Definition of the Situation.'" *Central States Speech Journal* 32 (1981): 197–205.
Cragan, John F. "The Origins and Nature of the Cold War Rhetorical Vision, 1946–1972: A Partial History." In *Applied Communication Research: A Dramatistic Approach,* edited by John F. Cragan and Donald C. Shields. Prospect Heights, Ill.: Waveland Press, 1981.
Craig, Gordon A., and Alexander George. *Force and Statecraft: Diplomatic Problems of Our Time.* New York: Oxford University Press, 1983.
Dallek, Robert. *Ronald Reagan: The Politics of Symbolism.* Cambridge, Mass.: Harvard University Press, 1984.
Dauber, Cori E. "Negotiating from Strength: Arms Control and the Rhetoric of Denial." *Political Communication and Persuasion* 7 (1990): 97–114.
Deaver, Michael K., with Mickey Herskowitz. *Behind the Scenes: In Which the Author Talks About Ronald and Nancy Reagan . . . and Himself.* New York: William Morrow, 1987.
Demerath, Nicholas Jay, and Rhys Williams. "Civil Religion in an Uncivil Society." *Annals of the American Academy* 480 (July 1985): 154–66.
Drew, Elizabeth. *Portrait of an Election: The 1980 Presidential Campaign.* New York: Simon and Schuster, 1981.
Dugger, Ronnie. *On Reagan: The Man and His Presidency.* New York: McGraw-Hill, 1983.
Duke, Paul, ed. *Beyond Reagan: The Politics of Upheaval.* New York: Warner Books, 1986.
Edelman, Murray. *Constructing the Political Spectacle.* Chicago: University of Chicago Press, 1988.

———. *Political Language: Words That Succeed and Policies That Fail.* New York: Academic Press, 1977.
Edwards, Anne. *Early Reagan.* New York: Morrow, 1987.
Ellwood, David T. *Poor Support.* New York: Basic Books, 1988.
Erickson, Paul D. *Reagan Speaks: The Making of an American Myth.* New York: New York University Press, 1985.
Evans, Rowland, and Robert Novak. *The Reagan Revolution.* New York: Dutton, 1981.
Fairbanks, James David. "The Priestly Functions of the Presidency." *Presidential Studies Quarterly* 11 (1981): 214–32.
Fisher, Walter R. *Human Communication as Narration: Toward a Philosophy of Reason, Value, and Action.* Columbia: University of South Carolina Press, 1987.
———. "Romantic Democracy, Ronald Reagan, and Presidential Heroes." *Western Journal of Speech Communication* 46 (1982): 299–310.
Foucault, Michel. "The Order of Discourse." In *Language and Politics,* edited by Michael Shapiro. New York: New York University Press, 1984.
Gaddis, John L. *Long Peace: Inquiries into the History of the Cold War.* New York: Oxford University Press, 1987.
Gilbert, Dennis. *Sandinistas.* New York: Basil Blackwell, 1988.
Goodnight, G. Thomas. "Ronald Reagan's Re-formulation of the Rhetoric of War: Analysis of the 'Zero Option,' 'Evil Empire,' and 'Star Wars' Addresses." *Quarterly Journal of Speech* 72 (1986): 390–414.
Gorbachev, Mikhail. *Perestroika: New Thinking for Our Country and the World.* Rev. ed. New York: Harper and Row, 1988.
Graham, Daniel O. "National Defense: The Strategic Framework." In *The Future Under President Reagan,* edited by Wayne Valis. Westport, Conn.: Arlington House, 1981.
Green, David. *Shaping Political Consciousness: The Language of Politics in America from McKinley to Reagan.* Ithaca, N.Y.: Cornell University Press, 1987.
Green, Mark, and Gail MacCool. *There He Goes Again: Ronald Reagan's Reign of Error.* New York: Pantheon, 1983.
Gunderson, Robert G. "The Oxymoron Strain in American Rhetoric." *Central States Speech Journal* 28 (1977): 92–95.
Habermas, Jurgen. *Legitimation Crisis.* Translated by Thomas McCarthy. Boston: Beacon Press, 1973.
———. "The Public Sphere: An Encyclopedia Article." In *Critical Theory and Society,* edited by Stephen Eric Bronner and Douglas MacKay Kellner. New York: Routledge, 1989.
Hart, Roderick P. *The Political Pulpit.* West Lafayette, Ind.: Purdue University Press, 1977.
———. *The Sound of Leadership: Presidential Communication in the Modern Age.* Chicago: University of Chicago Press, 1987.
Heclo, Hugh. "Reaganism and the Search for a Public Philosophy." In *Perspectives on the Reagan Years,* edited by John L. Palmer. Washington, D.C.: Urban Institute Press, 1986.

Heertsgaard, Mark. *On Bended Knee: The Press and the Reagan Presidency.* New York: Schocken Books, 1989.

Hersh, Seymour M. *The Target Is Destroyed.* New York: Random House, 1986.

Hill, Dilys M., and Phil Williams. "The Reagan Legacy." In *The Reagan Presidency: An Incomplete Revolution?*, edited by Dilys M. Hill, Raymond A. Moore, and Phil Williams. Hampshire, Eng.: University of Southhampton and Macmillan, 1990.

Hobbs, Charles D. *Ronald Reagan's Call to Action.* New York: McGraw-Hill, 1968.

Hofstadter, Richard. *The Paranoid Style in American Politics and Other Essays.* New York: Knopf, 1966.

Hudson, Kenneth. *The Language of Politics.* London: Macmillan, 1978.

Hugo, Grant. *Appearance and Reality in International Politics.* London: Chatto and Windus, 1970.

Ivie, Robert. "Metaphor and the Rhetorical Invention of Cold War 'Idealists.'" *Communication Monographs* 54 (1987): 165–82.

———. "The Prospects of Cold War Criticism." In *Cold War Rhetoric: Strategy, Metaphor, and Ideology*, edited by Martin J. Medhurst et al. New York: Greenwood Press, 1990.

Jameson, Fredric. *The Political Unconscious: Narrative as a Socially Symbolic Act.* Ithaca, N.Y.: Cornell University Press, 1981.

Jamieson, Kathleen Hall. *Eloquence in an Electronic Age: The Transformation of Political Speechmaking.* New York: Oxford University Press, 1988.

Johnson, Haynes. *Sleepwalking Through History: America in the Reagan Years.* New York: Norton, 1991.

Kane, Thomas. "Rhetorical Histories and Arms Negotiations." *Journal of the American Forensic Association* 24 (1988): 143–54.

Kauffman, Charles. "Names and Weapons." *Communication Monographs* 56 (1989): 273–85.

Kellner, Hans. "The Inflatable Trope as Narrative Theory: Structure or Allegory." *Diacritics* 11 (1981): 14–28.

Kennedy, Paul. *The Rise and Fall of the Great Powers.* New York: Random House, 1987.

Kornbluh, Peter. "The Covert War." In *Reagan Versus the Sandinistas: The Undeclared War on Nicaragua*, edited by Thomas Walker. Boulder, Colo.: Westview Press, 1987.

Laclau, Ernesto. *Politics and Ideology in Marxist Theory.* London: New Left Books, 1977.

LaFeber, Walter. *Inevitable Revolutions: The United States in Central America.* New York: Norton, 1984.

Lake, Randall A. "The Metaethical Framework of the Anti-Abortion Debate." *Signs* 11 (Spring 1986): 478–99.

———. "Order and Disorder in Anti-Abortion Rhetoric: A Logological View." *Quarterly Journal of Speech* 70 (1984): 425–43.

Landau, David. *The Dangerous Doctrine: National Security and U.S. Foreign Policy.* Boulder, Colo.: Westview Press, 1988.

Launer, Michael K., Marilyn J. Young, and Sugwon Kang. "The KAL Tapes: What They Really Say About the Tragedy." *Bulletin of Concerned Asian Scholars* 18, no. 3 (July–September 1986): 67–71.

Leamer, Laurence. *Make-Believe: The Story of Nancy and Ronald Reagan.* New York: Harper and Row, 1983.

Lee, Ronald. "The New Populist Campaign for Economic Democracy." *Quarterly Journal of Speech* 72 (1986): 274–89.

Lekachman, Robert. *Visions and Nightmares: America After Reagan.* New York: Macmillan, 1987.

Leuchtenburg, William E. *In the Shadow of FDR: From Harry Truman to Ronald Reagan.* Ithaca, N.Y.: Cornell University Press, 1983.

Lewis, William F. "Telling America's Story: Narrative Form and the Reagan Presidency." *Quarterly Journal of Speech* 73 (1987): 280–302.

Luker, Kristin. *Abortion and the Politics of Motherhood.* Berkeley and Los Angeles: University of California Press, 1984.

McGee, Michael. "The Ideograph: A Link Between Rhetoric and Ideology." *Quarterly Journal of Speech* 66 (1980): 1–16.

McGee, Michael Calvin, and Martha Anne Martin. "Public Knowledge and Ideological Argumentation." *Communication Monographs* 50 (1983): 47–65.

McGinnis, Joe. *The Selling of the President, 1968.* New York: Trident Press, 1969.

McMahan, Jeff. *Reagan and the World: Imperial Policy in the New Cold War.* London: Pluto Press, 1984.

Martel, Myles. "Debate Preparations in the Reagan Camp: An Insider's View." *Speaker and Gavel* 18 (1981): 34–46.

Medhurst, Martin J. "Postponing the Social Agenda: Reagan's Strategy and Tactics." *Western Journal of Speech Communication* 48 (1984): 262–76.

Mellard, James M. *Doing Tropology: Analysis of Narrative Discourse.* Urbana: University of Illinois Press, 1987.

Mervin, David. *Ronald Reagan and the American Presidency.* New York: Longman, 1990.

Mount, Ferdinand. *The Theatre of Politics.* New York: Shocken Books, 1972.

Muravchik, Joel. "The Nicaragua Debate." *Foreign Affairs* 65 (Winter 1986/87): 366–82.

Nathan, J. A., and J. K. Oliver. *Foreign Policy Making and the American Political System.* Boston: Little, Brown, 1987.

Nelson, Robert J. "Religion and Abortion." *Center Magazine* 14 (July/August 1981): 51–55.

Newhouse, John. *War and Peace in the Nuclear Age.* New York: Knopf, 1988.

Noonan, Peggy. *What I Saw at the Revolution: A Political Life in the Reagan Era.* New York: Random House, 1990.

Oliver, Robert T. "The Varied Rhetoric of International Relations." *Western Speech* 25 (1961): 213–21.

Palmer, John L., and Isabel V. Sawhill, eds. *The Reagan Record.* Cambridge, Mass.: Ballinger, 1984.

Perelman, Chaim, and L. Olbrechts-Tyteca. *The New Rhetoric: A Treatise on Argumentation*. Translated by John Wilkinson and Purcell Weaver. Notre Dame, Ind.: University of Notre Dame Press, 1969.

Piven, Francis Fox, and Richard A. Cloward. *The New Class War.* New York: Pantheon, 1985.

Public Papers of the Presidents, 1981–88. Washington, D.C.: U.S. Government Printing Office, 1982–90.

Ravenal, Earl C. "Reagan's Failed Restoration: Superpower Relations in the 1980's." In *Assessing the Reagan Years*, edited by David Boaz. Washington D.C.: Cato Institute, 1988.

Reagan, Maureen. *First Father, First Daughter*. Boston: Little, Brown, 1989.

Reagan, Nancy, with William Novak. *My Turn: The Memoirs of Nancy Reagan*. New York: Random House, 1989.

Reagan, Ronald. *An American Life*. New York: Simon and Schuster, 1989.

———. *Speaking My Mind*. New York: Simon and Schuster, 1989.

———, and Richard C. Hubler. *Where's the Rest of Me?* New York: Dell, 1965.

Regan, Donald T. *For the Record: From Wall Street to Washington*. New York: Harcourt Brace Jovanovich, 1988.

Rogin, Michael. *Ronald Reagan: The Movie, and Other Episodes in Political Demonology*. Berkeley and Los Angeles: University of California Press, 1987.

Rohmer, Richard. *Massacre 747: The Story of Korean Air Lines Flight 007*. Markham, Ont.: PaperJacks, 1984.

Rushing, Janice Hocker. "Ronald Reagan's 'Star Wars' Address: Mythic Containment of Technical Reasoning." *Quarterly Journal of Speech* 72 (1986): 415–33.

Schieffer, Bob, and Gary Paul Gates. *The Acting President*. New York: Dutton, 1989.

Schmuhl, Robert. *Statecraft and Stagecraft: American Political Life in the Age of Personality*. Notre Dame, Ind.: University of Notre Dame Press, 1990.

Scott, Robert L. "Cold War and Rhetoric: Conceptually and Critically." In *Cold War Rhetoric: Strategy, Metaphor, and Ideology*, edited by Martin J. Medhurst et al. New York: Greenwood Press, 1990.

Smith, Craig Allen. "Mister Reagan's Neighborhood: Rhetoric and National Unity." In *Essays in Presidential Rhetoric*, edited by Theodore Windt and Beth Ingold. Dubuque, Iowa: Kendall-Hunt, 1987.

Smith, Hedrick. *The Power Game: How Washington Works*. New York: Random House, 1988.

Sorenson, Theodore. *A Different Kind of Presidency*. New York: Harper and Row, 1984.

Speakes, Larry, with Robert Pack. *Speaking Out: The Reagan Presidency from Inside the White House*. New York: Scribner's, 1988.

Stockman, David A. *The Triumph of Politics: How the Reagan Revolution Failed*. New York: Harper and Row, 1986.

Stone, Deborah. *Policy Paradox and Political Reason*. Glenview, Ill.: Scott, Foresman, 1988.

Talbot, Strobe. *Deadly Gambits: The Reagan Administration and the Stalemate in Nuclear Arms Control.* New York: Random House, 1985.

———. *Endgame: The Inside Story of SALT II.* New York: Harper and Row, 1980.

Toulmin, Stephen. *The Uses of Argument.* Cambridge: Cambridge University Press, 1964.

Troxler, L. William, ed. *Along Wit's Trail: The Humor and Wisdom of Ronald Reagan.* New York: Holt, Rinehart and Winston, 1984.

Tucker, Robert. "Reagan's Foreign Policy." In *The Reagan Foreign Policy,* edited by William G. Hyland. New York: New American Library, 1987.

Tulis, Jeffrey K. *The Rhetorical Presidency.* Princeton, N.J.: Princeton University Press, 1987.

Ullman, Richard H. "At War with Nicaragua." *Foreign Affairs* 62 (Fall 1983): 39–58.

Wander, Philip. "The Rhetoric of American Foreign Policy." *Quarterly Journal of Speech* 70 (1984): 339–61.

Weaver, Richard. "Ultimate Terms in Contemporary Rhetoric." In *The Ethics of Rhetoric.* Chicago: Regnery, 1953.

White, Hayden. *Tropics of Discourse: Essays in Cultural Criticism.* Baltimore: Johns Hopkins University Press, 1978.

White, John Kenneth. *The New Politics of Old Values.* Hanover, N.H.: University Press of New England, 1988.

Wiemiller, Gordon R. *U.S.-Soviet Summits: An Account of East-West Diplomacy at the Top, 1955–1985.* New York: University Press of America, 1986.

Wills, Garry. *Reagan's America: Innocents at Home.* Garden City, N.Y.: Doubleday, 1985.

Young, Marilyn J., and Michael K. Launer. *Flights of Fancy, Flight of Doom: KAL 007 and Soviet-American Rhetoric.* Lanham, Md.: University Press of America, 1988.

Zarefsky, David. *President Johnson's War on Poverty.* Tuscaloosa: University of Alabama Press, 1986.

Contributors

J. Jeffery Auer is Professor Emeritus of Speech Communication at Indiana University. He is the author of numerous articles on political communication topics, including Ronald Reagan, Margaret Thatcher, and political speechwriting. He is coeditor of *Antislavery and Disunion: 1858–1861,* and *The Rhetoric of Our Times.*

Jeff D. Bass is Associate Professor of Communication Studies at Baylor University. His research interests include the rhetoric of foreign policy and reactionary political discourse. His work has appeared in the *Quarterly Journal of Speech, Southern Speech Communication Journal,* and *Western Journal of Speech Communication.*

Jane Blankenship is Professor of Communication at the University of Massachusetts-Amherst. Her research covers a broad range of topics in contemporary political communication, and she is a leading scholar of the rhetorical theories of Kenneth Burke. She is author or coeditor of six books, and her essays have appeared in the *Quarterly Journal of Speech, Communication Monographs,* and *Philosophy and Rhetoric.*

Robert J. Branham is Professor of Rhetoric at Bates College. His critiques of political speeches and film documentaries have appeared in the *Quarterly Journal of Speech* and elsewhere. A longtime director of intercollegiate debate, he is the author of the recent *Debate and Critical Analysis: The Harmony of Conflict.*

DeLysa Burnier is Assistant Professor of Political Science at Ohio University. Her research interests include economic development and interpretive policy analysis. She is the author of "The Uncertain World of Economic Development" in *Policy Studies Review*, and of "State Economic Development Policy" in *Public Administration Review.*

Robin Carter was Principal Lecturer in the School of Languages and Area Studies at Portsmouth Polytechnic University in England. His research, in addition to emphasizing contemporary political rhetoric, has essayed the thought and literature of Renaissance Spain. His recent publications include "Liberty: Abstract Concept as Rhetorical Device," and "The Concept of Human Liberty in Golden Age Spain."

David Descutner is Associate Professor in the School of Interpersonal Communication at Ohio University. His research interests include political communication and popular culture. His work has appeared in various national and regional communication journals, and he has contributed chapters to several books.

G. Thomas Goodnight is Associate Professor of Communication Studies at Northwestern University. His work has emphasized the concept of "the public": its nature, functions, and forms. He has written extensively on nuclear weapons strategy as a public issue, and his essays have appeared in *Communication Monographs*, the *Quarterly Journal of Speech*, and the *Journal of the American Forensic Association.*

James Jasinski is Assistant Professor of Speech Communication at the University of Illinois. His work concentrates on the history and criticism of American public address, and he has published essays on American political ideology and other topics.

Deborah K. Johnson is a Ph.D. candidate at the University of Massachusetts-Amherst. Her research interests are rhetoric and gender, and conflict/negotiation.

Janette Kenner Muir is Basic Course Director at George Mason University. Her research interests include political rhetoric, citizen empowerment, and visual communication, particularly political cartoons. Her work has appeared in *Futurist, Communication Quarterly*, and *Journal of Communication*.

Catherine Helen Palczewski is Assistant Professor of Communications at St. John's University in Minnesota. Her research has emphasized gender issues as they arise in the context of political rhetoric. She has produced critical studies on abortion, pornography, and other issues.

W. Barnett Pearce is Professor and Chair of the Department of Communication at Loyola University of Chicago. His interdisciplinary research program has ranged across political rhetoric, communication theory, and critical cultural studies. He is author of the recent *Communication and the Human Condition* and coauthor of *Culture, Politics and Research Programs: An International Assessment of Practical Problems in Field Research*.

Michael Weiler is Assistant Professor of Communication Studies at Emerson College. His research interests include contemporary political ideology and the rhetoric of American religious groups. His work has appeared in the *Quarterly Journal of Speech* and the *Journal of the American Forensic Association*.

Marilyn J. Young is Professor of Communication at Florida State University. Her research interests include foreign policy rhetoric and political argumentation, especially in the context of U.S.-Soviet relations. Her essays have appeared in the *Quarterly Journal of Speech*, the *Journal of Communication*, and *Advocacy and Argument*. She is coauthor of the recent *Flights of Fancy, Flight of Doom: KAL 007 and Soviet-American Rhetoric*.

Index

About the Series

STUDIES IN RHETORIC AND COMMUNICATION
General Editors:
E. Culpepper Clark, Raymie E. McKerrow, and David Zarefsky

The University of Alabama Press has established this series to publish major new works in the general area of rhetoric and communication, including books treating the symbolic manifestations of political discourse, argument as social knowledge, the impact of machine technology on patterns of communication behavior, and other topics related to the nature of impact of symbolic communication. We actively solicit studies involving historical, critical, or theoretical analyses of human discourse.